THE VASCULAR PLANTS OF
SAN LUIS OBISPO COUNTY, CALIFORNIA

San Luis Obispo Lupine (*Lupinus ludovicianus*) official flower of San Luis Obispo County. (Photo by Mrs. Mary Ford.)

The Vascular Plants of San Luis Obispo County, California

Robert F. Hoover

University of California Press
Berkeley, Los Angeles, and London
1970

University of California Press
Berkeley and Los Angeles, California
University of California Press, Ltd.
London, England
Copyright © 1970, by
The Regents of the University of California
Standard Book Number 520-01663-7
Library of Congress Catalog Card Number: 71-104883
Printed in the United States of America

Contents

Introduction

For many years, the plants of San Luis Obispo County have attracted my particular interest. Although I seldom visited this part of California in my early years, and then only in transit, one of my well-remembered botanical experiences in the 1920's was growing the strange and beautiful localized species *Calochortus (Cyclobothra) obispoensis* at Modesto. The first time I gave serious attention to the county flora was during a spring trip in 1939. On this occasion the succession of plant communities from the coast to the eastern border made a notable impression. The differences in vegetation from the forested Santa Lucia Range through the upper Salinas River canyon (now submerged in the Santa Margarita Reservoir) and the chaparral-covered La Panza Range to the desert-like country of the interior were unmatched by anything in my previous experience. When, in 1946, I found permanent employment at San Luis Obispo, I had already developed a strong desire to make a detailed study of the local flora. This work, with various obstacles and interruptions, has continued over the past twenty years.

As compared with other parts of California, San Luis Obispo County has had the advantage that large areas were kept, until recently, in a condition closely resembling their primitive state. That is to say, plants which had been extensively, or even completely, destroyed in other parts of California were still to be found here. In recent years, however, San Luis Obispo County has increasingly shared in the general destruction of natural resources which is everywhere occurring. At least one species is definitely known to have disappeared from the county during the period of my field investigations here, and I suspect that the number may actually be larger. The botanist, therefore, here as elsewhere, has an important task in making as complete a record as possible

1

of the plants which at one time were present. This information has great historic interest, in addition to other values. It must be remembered that civilized man has been in contact with the native plants of California for only a short period. Very little is known about the possible economic importance of most of the species. Every effort should therefore be made to keep all these plants in existence, at least until adequate scientific research can be done on them.

This work is based primarily on study of plants in their natural habitats. However, since the ability of one individual to cover a large area within a given time is strictly limited, it has been necessary to supplement field observations with study of herbarium specimens. Also, the identity of a plant could often be determined only after extensive herbarium and library work. The authorities of California State Polytechnic College have made available a small room which has served to house an herbarium. During the past few years, considerable time has been devoted to herbarium work at the California Academy of Sciences. The herbaria of the University of California at Berkeley and of Stanford University have also been freely consulted.

The owners of the land where my field studies have been conducted have been, almost without exception, cordial and cooperative. It must be a little disturbing to have a stranger on one's property, busy with activities that one does not fully understand. Friendly landowners have therefore made an essential contribution toward the completion of this work.

So many people have helped in various ways that it would be impossible to list them all. It is probably advisable, in order to avoid slighting anyone not mentioned by name, to make only a general acknowledgement at this point. To all those who have furnished information, advice, or other forms of help I am sincerely thankful.

In a special way, I am indebted to my wife, Betty Louise Hoover. She has worked with me on numerous field trips and over a period of many months of herbarium studies. The work really represents a cooperative effort, and it is only because of my wife's modesty that she is not listed as a joint author.

San Luis Obispo, California

History of Botanical Exploration

As far as is definitely known, the first person to collect plants in what is now San Luis Obispo County was David Douglas. In 1831 he travelled by land between Monterey and Santa Barbara. No record has come down to us as to exactly where his specimens were collected, although in some instances it is possible to make a likely guess from the localized occurrence of a particular form. For example, strictly typical *Eriastrum densifolium* is virtually confined to Nipomo Mesa, which was on Douglas's route and so may be assumed to be the type locality. Many other species found by Douglas occur here, and the original collections could have been made in San Luis Obispo County as likely as elsewhere. A few examples in this category are *Calochortus albus, Quercus Douglasii, Lotus junceus, Astragalus Douglasii, A. macrodon, Gilia tenuiflora, Phacelia Douglasii,* and *Malacothrix californica.*

Dr. Thomas Coulter travelled southward from Monterey in 1832. No information is available as to how many specimens he may have collected in this area, but he did discover *Pinus muricata* near San Luis Obispo. *Pinus Coulteri,* another familiar species here, was first collected in the Santa Lucia Mountains, presumably in Monterey County.

Thomas Nuttall collected along the California coast in 1835, and in 1848 published under the title *Plantae Gambelianae* an account of some of his own collections as well as those of William Gambel. Frequent mention of the locality "St. Simeon" indicates that one or both of these collectors visited the coast of San Luis Obispo County.

W. H. Brewer collected for the Geological Survey of California from 1860 to 1864 and during this period passed through San Luis Obispo County at least once. His discoveries locally include *Chorizanthe Breweri* and *Senecio Breweri.*

Dr. Edward Palmer made extensive botanical explorations in or about 1874. Among his discoveries are *Chorizanthe Palmeri, C. rectispina, Psoralea californica* (first collected apparently in La Panza Range but not found here since) and *Monardella Palmeri* ("coast range north of San Luis Obispo").

J. G. Lemmon, an early California botanist who was active in various parts of the state, seems to have resided for a short time in the northeastern part of the county. The type collection of *Amsinckia Lemmonii* (generally regarded as a synonym of *A. Douglasiana*) is one of the specimens labelled as from "Lemmon's Ranch, Cholame." Lemmon also collected near San Luis Obispo.

During the 1880's, Mrs. Katharine Layne Curran (later Mrs. T. S. Brandegee) was collecting in the county, probably on more than one occasion. *Chorizanthe diffusa* var. *nivea* and *Lupinus Ludovicianus* were originally based on her collections. Also, near San Simeon she discovered some notable shrubs which were not named until much later: *Ceanothus maritimus, C. Hearstiorum,* and *Arctostaphylos Hearstiorum.*

Lorenzo Jared made several important collections from interior San Luis Obispo County some time prior to 1894. The highly localized distribution of his discovery *Lepidium Jaredii* gives a clue to the site of Jared's collecting locality "Goodwin," which would otherwise be an unsolved mystery.

Specimens at Stanford University indicate that W. R. Dudley collected in 1902 in the Santa Lucia Mountains above San Simeon.

The first really extensive collections in the county were made by Alice Eastwood. Beginning in 1902 (possibly earlier) she made important contributions to the knowledge of the flora, especially in the southern part of the county near Santa Maria. Later Miss Eastwood collected near Cambria and elsewhere. Most of her specimens are available at the California Academy of Sciences and have formed an important part of the basis for this work.

During the early years of the California Polytechnic School, I. J. Condit was an active collector in the vicinity of San Luis Obispo. Most of his specimens are now in the herbarium of the University of California at Berkeley.

More recently, a large number of California botanists have included San Luis Obispo County in their field studies. Those represented by considerable numbers of specimens include W. L. Jepson, I. L. Wiggins, John Thomas Howell, and many others. The list is far from complete.

In the contemporary period, the most diligent field botanists have been Mrs. Clare Hardham (mainly in the Santa Lucia Range) and Ernest Twisselmann (mainly in the Temblor Range and vicinity). This work has been aided immeasurably by the numerous records provided by these collectors. Significant contributions have also been made by Clifton Smith and Eben McMillan.

My own collections from San Luis Obispo County are estimated at about five thousand numbers. An attempt has been made to collect every species represented in the flora in at least one locality. Large areas have still not been visited at all by anyone prepared to collect plants or to make careful records as to what species are present. Even places which are visited frequently are found on each visit to harbor different plants. This work, then, though drawing on the findings of so many pioneers, must nevertheless be regarded as a preliminary sketch. Exciting discoveries are still being made and doubtless will continue to be made for a long time.

The Geographic Setting

The entire area of the county is characterized by a Mediterranean climate, which implies that almost all the rainfall comes during the cooler part of the year. Beyond this one generalization, diversity is the rule. In its greatest dimension, the diagonal from the northwest corner to the southeast corner, the county measures about 120 miles. The difference in environmental conditions is greater than this distance would suggest. It would be hard to find a more marked contrast than that between the cool damp north coast and the intensely hot and arid Cuyama Valley.

Among the factors which strongly influence local climates are the proximity of the Pacific Ocean on the southwest side and the arrangement of the mountain ranges, which stand approximately parallel to the coast. Dominating the western part is the Santa Lucia Range, extending almost the entire length of the county until it merges toward the south into a mass of rugged country, where separate ranges are not readily distinguishable, on the north side of Cuyama Canyon. Rising sharply from the shore at our northern boundary, the Santa Lucia Range trends gradually away from the coast toward the south. South of Morro Bay a region of high hills called the San Luis Range is situated next to the coastline.

In the east the boundary with Kern County runs back and forth across the Temblor Range. A northern extension of the main Temblor Range is known as the Cholame Hills. The extreme northeastern corner of the county is occupied by the southern end of the Diablo Range, which lies east of the Cholame Hills and next to the San Joaquin Valley but is replaced south of Polonio Pass by the Temblor Range.

Between the Santa Lucia Range and the Cholame Hills in the

northern part of the county is an extensive area of plains and low roll-
ing hills, which is here regarded as a part of the Salinas Valley. Southward the La Panza Range and the Caliente Range are roughly parallel
to the main mountain ridges. The Caliente Range bends toward the
east, following the course of the Cuyama River, and therefore parallels
the coast of southern Santa Barbara County more than that of San Luis
Obispo County. Carrizo Plain is a closed drainage basin bounded by
the Temblor Range on the northeast and partly by the Caliente Range
on the southwest.

The Santa Lucia Range intercepts a large portion of the rain-
bearing clouds moving in from the ocean and therefore has the heaviest
precipitation in the region. From such records as are available, it appears that the average annual rainfall varies around 30 inches in most
of this range. Much higher amounts are measured in some localities and
in exceptional years. The highest point in the San Luis Obispo County
portion of the Santa Lucia Range, Pine Mountain above San Simeon,
reaches an altitude of 3600 feet and is often covered with snow in winter. Climatic conditions are such that this northwestern corner of the
county is richest in plants which thrive mainly in the Sierra Nevada
and in mountains far to the north, despite the fact that the La Panza,
Temblor, and Caliente ranges reach elevations higher than Pine Mountain. Examples of such "montane" or "northern" plants, which in San
Luis Obispo County have been found only in the mountains of the
northwestern corner, include *Pinus ponderosa, Bromus Orcuttianus,
Melica Geyeri, M. Harfordii, Carex Brainerdii, Hypericum formosum*
var. *Scouleri, Viola purpurea,* and *Apocynum pumilum.*

The highest point in the county, the summit of Caliente Mountain
in the southeast (5104 feet), is notable for the complete absence of coniferous forest such as characterizes the Sierra Nevada at a comparable altitude. The flora of Caliente Mountain is very poor in "montane" species
but remarkably rich in "desert" species (e.g., *Arabis pulchra, Phacelia
affinis, Eriogonum Heermannii, Parishella californica, Anisocoma acaulis, Chaenactis Xantiana, Lomatium mohavense,* etc.).

The entire area east of the Santa Lucia Range can properly be
described as arid in varying degrees. Rainfall is probably lowest (in
some spots evidently averaging less than 5 inches a year, although official records are lacking) in the southern part of the Temblor Range
and in Cuyama Valley and the adjacent Caliente Range. Here the landscape is so barren as to recall the hills bordering Death Valley. Nevertheless, this driest region is the home of plants which are among the
most interesting, ecologically and geographically considered. In a year

of favorable rainfall, the otherwise barren hills are splashed with the colors of many flowers, and the number of species to be found is quite surprising.

A locally important ecological influence, particularly near the coast, is the frequency of fog in summer. Moisture in this form (as well as dew) is doubtless absorbed to some extent directly through the leaves of plants, but in addition it condenses on the leaves and branchlets of trees and drips to the ground. On some mornings during the long rainless summer, following a heavy fog, the ground is actually wet, and certain herbaceous plants are green and vigorous which otherwise would be dormant in summer. Coniferous trees, such as Monterey pine and Sargent cypress, seem to be particularly effective as fog-condensers, but probably all trees and shrubs show this phenomenon to different degrees. Very likely trees and shrubs with broad leaves absorb most of the condensed water directly, but those with needle-like or scale-like leaves let most of it drip to the ground. The abundant chaparral shrub *Adenostoma* has been observed to act in this way, the ground under these shrubs being damp after a heavy fog. The water thus supplied to plants must surely be the equivalent of several inches of rain a year. Summer fog is a significant influence on plant life even as far inland as the summits of the La Panza Range, although of course it is more frequent nearer the coast.

In the Salinas Valley, winter fog is in the same way a partial substitute for rain. Where the oaks have been allowed to remain, water is often seen dripping from them in foggy weather, and the herbaceous plants around the trees show a corresponding luxuriance of growth.

Life-Zones

The life-zone concept of C. H. Merriam, which is highly useful in some regions, such as the Sierra Nevada, has only limited application to San Luis Obispo County conditions. That is, "index species" of the three zones represented are often associated in such a manner that the zonal position of a particular locality can be stated only by an arbitrary decision.

It would probably be universally agreed that the Lower Sonoran Zone includes the Estrella River Valley east of San Miguel and north of Paso Robles, eastward to Cholame Valley, and thence southward (excluding the higher hills) through Carrizo Plain and the southern Temblor Range to Cuyama Valley. To the Transition Zone may be assigned the *Pinus ponderosa* forest on Pine Mountain and a few other places in the vicinity, probably the Monterey pine forest near Cambria and the grasslands along the coast to the north, and perhaps the broad-leaf evergreen forest in the Santa Lucia Range as far south as Lopez Canyon. The remainder of the county may be regarded either as Upper Sonoran or, in some districts, as intermediate between Upper Sonoran and one of the other two zones. As a demonstration of the lack of sharp definition of the zones, it may be noted that the flora of Carrizo Plain is distinctively Lower Sonoran as contrasted with the Upper Sonoran flora around Paso Robles; yet temperatures on Carrizo Plain are usually lower than at Paso Robles.

On the basis of their occurrence within our limits and their usual absence from other life-zones, the following may be considered as index species.

LOWER SONORAN

Atriplex polycarpa	*Chorizanthe Xanti*
Atriplex spinifera	*Mentzelia affinis*
Isocoma acradenia	*Phacelia Fremontii*
Frankenia campestris	*Malacothrix glabrata*
Chaenactis stevioides	*Eschscholzia Lemmonii*
var. *brachypappa*	*Coreopsis calliopsidea*

UPPER SONORAN

Pinus Sabiniana	*Arctostaphylos glauca*
Juniperus californica	*Lupinus succulentus*
Quercus Douglasii	*Photinia arbutifolia*
Quercus dumosa	*Eriogonum elongatum*
Adenostoma fasciculatum	*Artemisia californica*
Ceanothus cuneatus	*Salvia mellifera*

TRANSITION

Pinus ponderosa	*Arbutus Menziesii*
Pinus radiata	*Quercus chrysolepis*
Pinus muricata (?)	*Deschampsia holciformis*
Vaccinium ovatum	*Eryngium armatum*
Anaphalis margaritacea	*Iris Douglasiana*
Trillium sessile	*Fragaria californica*
var. *angustipetalum*	*Rubus parviflorus*
Lithocarpus densiflorus	*Heuchera micrantha*

Natural Plant Communities

It would be highly desirable from many points of view to know exactly what the country was like in its primitive state. To attempt to describe "natural" plant communities, in one sense of the word, would be futile, because there is no place where the vegetation has not been subject to some form of human interference or manipulation. Relatively speaking, however, land which has not been used for agriculture or building, and which has not been devastated in "range-improvement" programs, may be regarded as showing fairly well what plant communities would exist in the absence of artificial controls.

The number of ways of naming and classifying plant communities is virtually endless. No matter how they are defined, there are always transitional states between one community and another. The question as to the usefulness of progressively finer distinctions leads to uncertainty of the value of any such classification. For example, how many kinds of "grassland" can profitably be distinguished, or how many kinds of "evergreen forest?" The classification of California plant communities published by Munz and Keck has proved to be a useful guide to the study of the types of vegetation represented in San Luis Obispo County. Because of special local conditions, however, a more accurate description of our plant communities can be devised by making certain modifications in the system of Munz and Keck. The following list of plant communities is offered as a generalized description of the natural vegetation of San Luis Obispo County, but not all details of it will have wider geographic application.

Local designation	*Corresponding designation by Munz and Keck*
1. Beach-Dune	Coastal Strand [1]
2. Coastal Salt Marsh	Coastal Salt Marsh
3. Freshwater Marsh	Freshwater Marsh
4. Riparian Woodland	
5. North Coastal Grassland	Coastal Prairie
6. Northern Coastal Scrub	Northern Coastal Scrub

---

1. The word "strand" ordinarily denotes the very edge of the water; hence the substitution of "Beach-dune" as the name for this community.

7. Coastal Sand-plains and   Probably part of
   Stabilized Dunes   Coastal Sage Scrub
8. Chaparral   Chaparral; also probably
                                    part of Coastal Sage Scrub
9. Evergreen Forest   Mixed Evergreen Forest; Closed-cone
   (several subtypes)   Pine Forest; perhaps small areas of
                                    others
10. Foothill Woodland   Foothill Woodland
11. Interior Herbaceous   Valley Grassland
    (several subtypes)
12. Juniper-Oak Woodland   Part of Pinyon-Juniper Woodland
13. Desert Scrub   Climatically like Shadscale Scrub or
                                    Creosote Bush Scrub, but lacking
                                    most species listed by Munz and
                                    Keck
14. Saline Plains   Alkali Sink; perhaps partly Shadscale
                                    Scrub

1. **Beach-Dune Community.** Most extensively developed from Pismo Beach to mouth of Santa Maria River; also west side of Morro Bay, and scattered small areas northward to Point Sierra Nevada (north of Arroyo de la Cruz). *Carpobrotus chilensis, Atriplex californica* and *leucophylla, Abronia umbellata, maritima,* and *latifolia, Eschscholzia californica* var. *maritima, Malacothrix incana, Franseria Chamissonis* var. *bipinnatisecta, Oenothera cheiranthifolia.* An introduced species, *Cakile maritima,* has become probably the most universally present member of this community.

2. **Coastal Salt Marsh.** Shores of Morro Bay; mouth of San Luis Creek at Avila; scattered patches among dunes south of Pismo Beach, where grading into freshwater marsh; weakly represented at mouths of some streams northward. *Salicornia virginica, Atriplex patula* var. *hastata, Distichlis spicata, Frankenia grandifolia, Limonium commune* var. *mexicanum, Jaumea carnosa, Suaeda californica.*

3. **Freshwater Marsh.** Well represented among the dunes from Oceano (where the plant community has been virtually wiped out) to Oso Flaco Lake; Laguna near San Luis Obispo; small patches elsewhere near coast, and slightly developed in Salinas Valley. *Typha latifolia, Sparganium eurycarpum, Scirpus acutus, S. californicus, Cardamine Gambelii, Epilobium Watsonii* and varieties, *Oenanthe sarmentosa.*

4. **Riparian Woodland.** Mainly along Salinas River and some of its tributaries; also along streams flowing directly into the sea to the west.

The only one of our plant communities in which deciduous trees are well represented. *Salix lasiolepis, S. lasiandra, Populus trichocarpa* (near coast), *P. Fremontii* (inland), *Quercus lobata, Q. agrifolia, Alnus rhombifolia, Platanus racemosa, Acer Negundo.*

5. **North Coastal Grassland.** Poorly represented around San Luis Valley, where the native vegetation has been largely destroyed; well represented on open hills and ocean bluffs from Cambria northward. *Hordeum brachyantherum, Deschampsia holciformis, Danthonia californica, Fritillaria biflora, Mariposa lutea, Oenothera ovata, Eryngium armatum, Lomatium caruifolium, Sanicula arctopoides, S. maritima, Layia platyglossa* var. *platyglossa, Hemizonia corymbosa.*

6. **Northern Coastal Scrub.** Best represented from Cambria northward, although several of the associated species extend considerably farther south. *Ceanothus thyrsiflorus, C. maritimus* (localized), *Rubus ursinus, Lupinus arboreus, Diplacus aurantiacus, Rhus diversiloba, Rhamnus californica, Arctostaphylos tomentosa, A. cruzensis* (localized), *Baccharis pilularis* var. *pilularis, Artemisia californica, Eriophyllum staechadifolium, Iris Douglasiana.* Where the shrubs are more widely spaced, this community merges very gradually into North Coastal Grassland.

7. **Coastal Sand-plains and Stabilized Dunes.** Located just inland from the Beach-Dune Community and not sharply distinguished from it, as many species thrive on both stabilized and shifting sand. South side of Morro Bay, and most extensively developed on Nipomo Mesa. This entire community is rapidly disappearing as a result of the spread of civilization. *Mucronea californica, Eriogonum parvifolium, Lupinus arboreus, L. Chamissonis, Arctostaphylos morroensis* (localized), *A. rudis, Ceanothus impressus, C. cuneatus* var. *fascicularis, Eriastrum densifolium, Phacelia distans, Amsinckia spectabilis* var. *microcarpa, Salvia mellifera, Orthocarpus purpurascens, Corethrogyne filaginifolia, Ericameria ericoides, Cirsium occidentale.*

8. **Chaparral.** Almost reaching the shore on rocky slopes of the San Luis Range; more typically on hills farther inland, extending to La Panza Range. The summits of the northern Temblor Range, and some places on the Caliente Range, have dense stands of shrubs, but these are not typical chaparral. "Chapparal" is, in any case, a sort of catch-all term for a great variety of closely crowded shrubs, nearly all of which have rigid branches and are evergreen. Because of differences in the underlying rocks and local climates, the "chaparrales" of two different localities may consist of two entirely different associations of species. The most constant member of the community is *Adenostoma fascicula-*

tum, which may form extensive pure stands or be mingled with a diversity of other shrubs. *Ceanothus cuneatus* is the second species in order of abundance, and various other species of *Ceanothus* may be abundant in certain localities. *Arctostaphylos* is another very common genus in the chaparral, being represented by one or more species in different climatic or geologic areas. Other species which are often plentiful include *Quercus Wislizeni* var. *frutescens, Q. durata, Q. dumosa, Dendromecon rigida, Garrya Veatchii, Salvia mellifera,* and *Eriodictyon tomentosum.* Under this community may also be listed the remarkable fire-type herbs, mostly annual but a few perennial, which appear in profusion the year after a stand of chaparral is destroyed by fire, but otherwise germinate sparingly if at all. Examples are *Calandrinia Breweri, Silene multinervia, Papaver californicum, Streptanthus heterophyllus, Phacelia brachyloba, Antirrhinum multiflorum, Cryptantha micromeres,* and *Helianthus gracilentus.*

9. **Evergreen Forest.** Irregular, and usually rather small areas near the coast and in the Santa Lucia Range. Close spacing of the trees is a typical condition, but where some species, especially *Quercus agrifolia,* grow farther apart, there is a gradual transition to chaparral or to herbaceous communities. The trees which constitute the evergreen forest differ widely from one place to another, but the associated herbaceous species, for the most part, are not restricted to stands of any particular sort of tree. True coniferous forest is rare in the county and is usually represented by pure stands of a single species, such as *Pinus ponderosa* on Pine Mountain above San Simeon, *P. radiata* around Cambria, *P. muricata* south of Coon Creek in the San Luis Range, *P. attenuata* mixed to a limited extent with *P. Coulteri* southeast of Cuesta Pass, and *Cupressus Sargentii* on Cypress Mountain and (with scattered *Pinus Coulteri*) northwest of Cuesta Pass. Where various dicotyledonous trees are associated with any of these conifers, the community is described as Mixed Evergreen Forest; and, where the conifers thin out or disappear completely, it becomes a Broad-leaf Evergreen Forest. Characteristic trees of the latter include *Quercus agrifolia, Q. chrysolepis, Lithocarpus densiflorus, Umbellularia californica, Arbutus Menziesii,* and in some places the deciduous species *Quercus Kelloggii* and *Acer macrophyllum.* In many places *Quercus agrifolia* is the only tree present, and there the community could appropriately be designated Live-oak Forest. However, because the total area is small, and because the subtypes merge so gradually into one another, it seems more useful to use the inclusive term Evergreen Forest for all these variations.

10. **Foothill Woodland.** From east base of Santa Lucia Range across upper Salinas Valley to hills bordering La Panza Range on both sides; weakly represented in upper Arroyo Grande Valley and in Cuyama Canyon. *Pinus Sabiniana, Quercus lobata, Q. Douglasii, Q. agrifolia, Poa scabrella, Delphinium variegatum, Dodecatheon Clevelandii, Lomatium utriculatum, Collinsia heterophylla.* Most of the herbs in this community are also found in "grassland" areas, but in such places they may be, to a considerable extent, relics from a time before trees had been cleared away.

11. **Interior Herbaceous.** Extensive areas toward the east were clearly treeless in their primitive state, but it is also known that many trees have been, and are being, removed from rangelands. It is therefore sometimes difficult to determine whether an herbaceous community represents the natural condition of a specific locality. I am not convinced that grasses were originally particularly abundant in such communities, and therefore prefer the term "herbaceous" to "grassland." In any case, native grasses are not now a conspicuous element in the flora of the area in question. Herbaceous communities differ conspicuously from one another, because of soil differences and local climates. The following subtypes may be noted:

West of Santa Lucia Range, mainly in rocky clay derived from serpentine: *Trifolium fucatum* var. *flavulum, Allium haematochiton, Phacelia distans, Layia Jonesii, Lasthenia chrysostoma, Linanthus parviflorus, Chorizanthe Palmeri, Hemizonia fasciculata, Astragalus curtipes.*

Yucca grassland, a unique local plant association near San Luis Obispo on hills where serpentine rock has not broken down completely into soil. *Yucca Whipplei* is the dominant scenic feature. *Quercus durata* is often present but is scattered and here grows close to the ground, so that this is not a chaparral formation. Other plants which are often plentiful in this association include *Selaginella Bigelovii, Melica Torreyana, Stipa lepida, Bloomeria crocea, Streptanthus glandulosus, Lotus junceus, Astragalus curtipes, Lomatium parvifolium* var. *pallidum, Cryptantha Clevelandii* var. *hispidissima, Hazardia squarrosa,* and *Eriophyllum confertiflorum.*

Pale, friable (probably calcium-rich) clays in eastern part: *Phacelia ciliata, Monolopia lanceolata, Streptanthus anceps, Madia radiata, Coreopsis calliopsidea, Fritillaria agrestis.*

Sandy or gravelly loams and crumbling shales: *Stipa cernua, Eschscholzia californica* var. *ambigua, Layia platyglossa* var. *breviseta, Lupi-*

nus subvexus, Trifolium albopurpureum, Platystemon californicus, Amsinckia tessellata.

Nearly pure sands: *Layia glandulosa, Gilia tenuiflora, Malacothrix californica, Chaenactis lanosa, Collinsia bartsiaefolia* var. *Davidsonii.*

Vernal pools: *Downingia cuspidata, Eryngium Vaseyi, Boisduvalia glabella, Deschampsia danthonioides, Callitriche marginata, Pilularia americana, Allocarya stipitata.*

12. **Juniper-Oak Woodland.** Around north end of La Panza Range, southward along San Juan River; higher parts of Temblor Range; Caliente Mountain. *Juniperus californica, Quercus Douglasii, Chorizanthe biloba, Eriogonum fasciculatum* var. *polifolium, Gutierrezia bracteata, Stenotopsis linearifolius.*

13. **Desert Scrub.** Best represented in Temblor Range, especially southward, to south slope of Caliente Mountain and (before cleared for agriculture) Cuyama Valley. *Ephedra californica, Atriplex polycarpa, Eriogonum fasciculatum* var. *polifolium, Lycium Andersonii, Eastwoodia elegans, Hymenoclea Salsola, Eurotia lanata, Isomeris arborea* var. *globosa, Stanleya pinnata, Mirabilis laevis.*

14. **Saline Plains.** Cholame Valley; Carrizo Plain around Soda Lake; south to Cuyama Valley. *Distichlis spicata, Atriplex spinifera, A. coronata, Allenrolfea occidentalis* (in strongly alkaline spots), *Frankenia campestris, Cressa truxillensis, Isocoma acradenia, Layia Munzii, Lasthenia Ferrisiae.*

Composition of the Flora

Ecologic conditions, particularly local climates, are reflected in the presence or absence of particular plants. As in all of California, and in the world's floras generally, *Compositae, Leguminosae,* and *Gramineae* are represented by numerous species in San Luis Obispo County. Some groups, such as *Orchidaceae* and some large west-American genera including *Ranunculus, Arabis,* and *Penstemon* are represented by remarkably few species. The lack of high mountains with their forests, meadows, and rocky peaks is an obvious explanation for the absence from the county of many widely distributed Californian species.

On the other hand, some groups of plants are well adapted to the conditions which prevail here. The following genera have the largest numbers of native species in our flora.

Lupinus	23	Chorizanthe [b]	18	Oenothera [c]	13
Eriogonum	23	Juncus [a]	17	Cryptantha	12
Carex [a]	22	Lotus	16	Hemizonia [d]	12
Trifolium	19	Arctostaphylos	15	Elymus [e]	12
Phacelia	18	Atriplex	13	Astragalus	12
				Ceanothus	12

a. The figures for *Carex* and *Juncus* are misleading, because authors dealing with these genera have drawn finer distinctions between species than has been customary in most plant genera. The key-characters used to differentiate the species are in many instances very obscure, and moreover seem to be in some cases inconstant even on a single plant.

b. Excluding *Mucronea* and *Centrostegia*, which would add 5 species to the figure given.

c. In the degree of apparent relationship among the species and their range of diversity, *Oenothera* in the traditional sense is not comparable to most other genera. As bringing the genus concept more nearly in conformity with that in other groups of plants, Raven's recent separation of most of our species under the name *Camissonia* is a step in the right direction. *Camissonia* as defined by Raven, however, still includes some mutually discordant elements.

d. As defined by Keck, including *Centromadia* but excluding *Holocarpha, Calycadenia,* and *Blepharizonia.*

e. Here defined to include *Sitanion* and the native American species which authors have placed in *Agropyron.*

Of equal, if not greater, significance are those genera represented by less numerous species, but by a high proportion of the total number in the genus. San Luis Obispo County is an important center of distribution for these genera.

Genera	Spp. in S. L. O. Co.	Estimated total spp.
Sanicula	10	35 (Calif. total 14)
Lasthenia (including *Baeria*)	10	16
Layia	8	15
Eriastrum	7	14
Amsinckia	7	11
Eschscholzia	6	11
Eriodictyon	5	9

The figures in the following numerical summary will undoubtedly be changed by later discoveries. Any such changes, however, will not affect the general conclusions which may be drawn from these data.

Total native species   1287
Introduced species included   296

The families represented by the greatest numbers of native species in
our flora are the following:

Compositae   203
Leguminosae   84
Gramineae   73
Polygonaceae   56
Scrophulariaceae   52
Polemoniaceae   45
Boraginaceae   44
Cruciferae   40
Cyperaceae   40
Umbelliferae   37
Onagraceae   33
Hydrophyllaceae   32

Species reaching their southern limit in the county (including those extending north only to Monterey Co.)   154

Species of coastal or montane southern California reaching their northern limit in the county (including those extending south only to Santa
Barbara Co.)   66

Desert, Great Basin, or San Joaquin Valley species reaching their western limit in the county   111

Species endemic to San Luis Obispo Co.   21

The number of county endemics would appear considerably larger if
named subspecies, varieties, and formae of localized occurrence were
included. Such a figure would not be accurate, however, because there
are in addition an undetermined number of local races to which botanists have never given names. Some of these unnamed races are more
readily recognizable than many named subspecies. Comparisons of
endemism in different areas, if they are to lead to valid conclusions,
either should take into account all these "kinds" of plants, or else should
be limited to those kinds which are distinct enough to have been classified as species.

Geographical Analysis

Geographically considered, our flora may be regarded as composed chiefly of the following elements.

1. **Endemic.** Found only in San Luis Obispo County and the immediately surrounding areas.

2. **Cosmopolitan or circumboreal.** Species adaptable to a wide range of climates, and usually found in Eurasia as well as over a large part of North America.

3. **Western North American.** Restricted to western North America, and widespread within that extensive region.

4. **General Californian.** Species distributed over a large part of California but not extending far, if at all, beyond the borders of the state.

5. **Great Valley.** Occurring most extensively in the Great Valley of California; not extending far, if at all, south of the Tehachapi and San Emigdio Mountains; extending more or less toward the coast, principally in the upper Salinas basin. The species listed here occur largely north and east of the Kern Endemism Area, which is mentioned in the discussion of the endemic floristic element.

6. **Northern.** Species which are here at, or near, the southern limit of their distribution, occurring more widely in northern California, Oregon, or even farther north.

7. **South coastal.** Species which are here at, or near, the northern limit of their distribution, extending southward in coastal southern California.

8. **Great Basin-Desert.** Species found in the Mohave Desert, and often also far eastward and northward, extending into the southeastern corner of San Luis Obispo County, or in several instances widely distributed through the interior of the county.

1. ENDEMIC FLORISTIC ELEMENT

A. Santa Lucia Range and adjacent coast north to Monterey Bay (Lucian Endemism Area of Jepson). The distribution of the species ranges from strictly coastal (*Allium Hickmanii, Lotus Benthamii*) to the higher and drier interior portions of these mountains (*Lupinus cervinus, Centrostegia Vortriedei*).

<table>
<tr><td>

Abies bracteata
Carex obispoensis
Allium Hickmanii
Chorizanthe Palmeri (genuine)
Centrostegia Vortriedei
Ribes sericeum
Lupinus cervinus
Lotus Benthamii (genuine)

</td><td>

Malacothamnus Palmeri
Lomatium parvifolium
  var. *pallidum*
Arctostaphylos Hooveri
Monardella Palmeri
Diplacus fasciculatus
Galium Hardhamiae
Corethrogyne leucophylla

</td></tr>
</table>

Nearly all of the plants listed above extend into Monterey County. Also within the Lucian Endemism Area, two patterns of more strictly localized distribution are evident. The first of these is centered around Arroyo de la Cruz, and I propose that this small area be called the Cruzian pocket of endemism.

Mariposa clavata var. *recurvifolia*
Bloomeria humilis
Triteleia ixioides var. *Cookii* (at higher altitudes near Pine Mt.)
Ceanothus maritimus
Arctostaphylos Hearstiorum
Arctostaphylos cruzensis (to southern Monterey Co. and Los Osos Valley)
Hemizonia corymbosa subsp. *macrocephala* (to Los Osos Valley)
Cirsium occidentale var. *compacta*
Unnamed local maritime races of several other species.

The other concentration of narrow endemics within the southern part of the Lucian Endemism Area is within a radius of about ten miles around San Luis Obispo, although a few extend for a little distance out from this center. This is here named the Obispoan pocket of endemism.

Cyclobothra obispoensis
Chorizanthe Breweri
Dudleya murina
Dudleya Bettinae
Astragalus didymocarpus subsp. *Milesianus* (at least in typical form)
Sidalcea Hickmanii subsp. *anomala* (nw. of Cuesta Pass)
Arctostaphylos luciana (se. of Cuesta Pass)
Arctostaphylos morroensis (south of Morro Bay)
Arctostaphylos pechoensis, genuine (San Luis Range)
Eriodictyon altissimum (Indian Knob)
Orthocarpus densiflorus var. *obispoensis* (to Arroyo de la Cruz)
Layia Jonesii

Cirsium fontinale var. *obispoense* (to San Simeon Creek)
Stephanomeria carotifera
Local races of *Brodiaea jolonensis*, *Delphinium Parryi*, etc.

B. Salinan Endemism Area. This name is here proposed for the drainage basin of the upper Salinas River. It is approximately the area formerly inhabited by the Salinan Indians. The main centers of endemism within this area are the San Antonio Valley in Monterey County and the La Panza district, with several species showing bicentric distribution and two of these (*Chlorogalum purpureum* and *Calycadenia villosa*) represented by distinguishable races in the two localities. Some of the plants centering in this area extend slightly beyond it, generally to the upper San Benito River or to the high mountains in the northern end of Ventura County. The latter are marked with an asterisk in the following list.

Chlorogalum purpureum
Mariposa simulans
Allium lacunosum
   var. *micranthum*
Chorizanthe biloba
Chorizanthe rectispina
Chorizanthe Douglasii
Centrostegia insignis
*_Astragalus macrodon_
Malacothamnus Jonesii and *M.*
   niveus (probably conspecific)

Clarkia speciosa
   (excluding subspecies)
Eriastrum luteum
*_Amsinckia Douglasiana_ (diploid)
*_Acanthomintha obovata_
Hemizonia pentactis
Calycadenia villosa
Lasthenia leptalea
*_Layia glandulosa_ var. *lutea*

C. Santa Lucia Range (or adjacent coast) to western Santa Barbara County or the vicinity of Santa Barbara.

Mariposa clavata var. *clavata*
Delphinium umbraculorum
Lotus junceus,
   genuine (to w. Ventura Co.)
Astragalus curtipes (excl.
   distinct San Benito Co. form)
Oenothera cheiranthifolia
   var. *nitida*
Sanicula Hoffmanii

Sanicula simulans
Phacelia grisea
   (also La Panza Range)
Monardella villosa var. *obispoensis*
   (to La Panza Range)
Isocoma veneta var. *sedoides*
Lessingia germanorum
   var. *pectinata*
Erigeron sanctarum

D. Santa Maria Endemism Area. Centering about the Santa Maria River, with a few species extending into the upper Salinas Valley. As does the Lucian Area, this includes habitats ranging from the coastline to dry interior hills. This endemism area can be regarded as consisting of three widely overlapping zones, as follows.

Zone 1, belonging largely or entirely to the Beach-Dune Plant Community:

Chorizanthe angustifolia

Lupinus nipomensis
   (highly localized)

Eriastrum densifolium, typical

Monardella crispa

Monardella undulata
   var. *frutescens*

Castilleia mollis

Erigeron Blochmaniae

Senecio Blochmaniae

Cirsium rhothophilum

Malacothrix incana

Zone 2, in coastal area but hardly entering the Beach-Dune Community:

Chorizanthe diffusa var. *nivea*

Prunus punctata

Ceanothus impressus

Ceanothus cuneatus
   var. *fascicularis*

Clarkia speciosa
   subsp. *immaculata*

Arctostaphylos rudis

Amsinckia spectabilis
   var. *microcarpa*

Scrophularia atrata

Zone 3, farther back from the coast; all but one entering the Salinas River watershed:

Agrostis Hooveri

Chorizanthe obovata

Lupinus ludovicianus

Malacothamnus gracilis

Arctostaphylos pilosula
   (to La Panza Range)

E. Inner Coast Range endemics. One of these, typical *Lepidium Jaredii*, is highly localized, but the rest extend northward, in some cases even to Colusa Co. Ecologically associated with the next group, but apparently absent from the San Joaquin Valley plains, even before these were cleared for cultivation.

Chorizanthe ventricosa

Eriogonum Eastwoodianum

Eriogonum Covilleanum

Eriogonum indictum

Eschscholzia Lemmonii

Eschscholzia rhombipetala

Lepidium Jaredii subsp. *Jaredii*

Streptanthus Parryi
   (to e. slope Santa Lucia Range)

Phacelia cryptantha var. *heliophila*

Plagiobothrys infectivus

Antirrhinum ovatum

Madia radiata

Hemizonia Halliana

F. Kern Endemism Area. San Joaquin Valley, mainly in Kern County, but most species extending farther north on the west side; many also present in interior San Luis Obispo County.

Festuca microstachys var. *simulans*
Allium Howellii
Hollisteria lanata
Eriogonum gossypinum
Delphinium gypsophilum
Atriplex vallicola
Streptanthus anceps
Streptanthus californicus
Lupinus nanus var. *Menkerae*

Astragalus oxyphysus
Eriastrum Hooveri
Trichostema ovatum
Layia Munzii
Layia pentachaeta var. *albida*
Hemizonia pallida
Lasthenia Ferrisiae
Eatonella Congdonii
Eastwoodia elegans

2. Cosmopolitan (in all major portions of the earth) and Circumboreal (North America and Eurasia) Floristic Elements

Pteridium aquilinum
Adiantum Capillus-Veneris
Equisetum Telmateia
Typha latifolia
Juncus balticus
Juncus bufonius

Festuca rubra
Suaeda fruticosa
Arabis glabra
Barbarea orthoceras
Achillea Millefolium

3. WESTERN NORTH AMERICAN FLORISTIC ELEMENT

Pityrogramma triangularis
Woodwardia fimbriata
Pinus ponderosa
Bromus carinatus
Poa scabrella
Carex praegracilis
Dichelostemma pulchellum

Montia perfoliata
Linum Lewisii
Epilobium paniculatum
Sambucus mexicana
Solidago occidentalis
Brickellia californica
Lagophylla ramosissima

4. GENERAL CALIFORNIAN FLORISTIC ELEMENT

As this is probably the largest element in our flora, only a few species are listed below as examples.

Pellaea mucronata
Selaginella Bigelovii
Pinus Sabiniana
Stipa cernua
Chlorogalum pomeridianum
Quercus Wislizeni
Ranunculus californicus

Photinia arbutifolia
Lupinus succulentus
Lotus scoparius
Clarkia unguiculata
Phacelia ciliata
Solidago californica
Senecio Douglasii

5. Great Valley Floristic Element

Fritillaria agrestis
Astragalus asymmetricus
Atriplex fruticulosa
Boisduvalia cleistogama
Eryngium Vaseyi
Navarretia nigellaeformis
Gilia tricolor var. *longipedicellata*
Allocarya humistrata

Plagiobothrys fulvus var. *campestris*
Plagiobothrys shastensis
Collinsia sparsiflora var. *solitaria*
Orthocarpus micranthus
Evax caulescens var. *humilis*
Blepharizonia laxa
Lasthenia Fremontii

Also under this heading may be listed species found in the hills on both sides of the San Joaquin Valley but absent, almost or entirely, from the valley plains. These species might be regarded as belonging to the Low Foothill Endemism Area of Jepson, but they do not extend into the northern Sierra Nevada (in no instance north of Placer Co.) or into the North Coast Ranges. In the South Coast Ranges, some of these plants reach only the Temblor Range, while others extend to the Santa Lucia Mountains or northward as far as Contra Costa County.

Mariposa venusta
  (to northern Los Angeles Co.)
Streptanthus Coulteri
  (to northern Los Angeles Co.)
Lupinus Benthamii
  (to northern Los Angeles Co.)
Mentzelia pectinata
Clarkia cylindrica
Eriastrum pluriflorum

Nemophila Fremontii
Mimulus androsaceus
Layia heterotricha (absent
  from Sierras n. of Tehachapi)
Layia pentachaeta var. *Hansenii*
Holocarpha Heermannii
Lasthenia debilis
Chaetopappa fragilis

6. Northern Floristic Element

Of the plants which reach their southern limit in the county, those which extend only to Monterey County have already been listed under the endemic floristic element, and those found principally in the interior of northern California under the Great Valley floristic element. The remainder almost all reflect local conditions of higher rainfall or lower temperatures than the average for San Luis Obispo County—climatic conditions which are duplicated extensively farther north, but are unusual for this part of California. I regard these northern plants as relics from a time (Pleistocene, or immediately following) when this entire area was better supplied with water, and when redwoods, pines, and cypresses were prevalent along our coast. By contrast, the Great Basin and desert plants have been moving into the eastern part of the county, where they are now a numerous element in the flora. Somewhat arbitrarily, the northern species can be listed under three headings.

A. Extending northward only to the San Francisco Bay region (Franciscan Endemism Area of Jepson).

Polystichum Dudleyi	*Sanicula maritima*
Triteleia ixioides (genuine)	*Lomatium parvifolium*
Chorizanthe pungens	*Convolvulus subacaulis*
(incl. var. *Hartwegii*)	*Monardella villosa* var. *subglabra*
Delphinium californicum	*Scorzonella paludosa*
Epilobium Watsonii, typical	*Cirsium fontinale*
Clarkia rubicunda	(incl. var. *obispoense*)

B. Extending to northern California or to southwestern Oregon (not beyond what some writers have called the Californian Biotic Province).

Festuca Elmeri	*Dentaria integrifolia*
Melica Torreyana	var. *integrifolia*
Elymus pacificus	*Heuchera pilosissima*
Deschampsia holciformis	*Ribes glutinosum*
Agrostis californica	*Potentilla californica*
Phalaris californica	*Rosa spithamaea*
Carex Dudleyi	*Sidalcea diploscypha*
Carex mendocinensis	*Oenothera ovata*
Allium unifolium	*Sanicula arctopoides*
Calochortus uniflorus	*Linanthus androsaceus* (genuine)
Disporum Hookeri	*Wyethia glabra*
Castanopsis chrysophylla	*Wyethia helenioides*
var. *minor*	*Layia gaillardioides*
Eriogonum latifolium (genuine)	*Calycadenia truncata*
Montia gypsophiloides	*Lasthenia macrantha*
Delphinium nudicaule	*Senecio aronicoides*
Delphinium variegatum	

C. Extending at least to the vicinity of the Columbia River, and mostly much farther north. A considerable proportion are found around the world in northern regions.

Cheilanthes siliquosa
Melica Harfordii
Elymus mollis
Trisetum canescens
Calamagrostis rubescens
Calamagrostis nutkaensis
Hierochloe occidentalis
Carex obnupta
Carex luzulina
Carex Cusickii
Salix Coulteri
Polygonum Paronychia
Cerastium arvense
Nymphaea polysepala
Heuchera micrantha
Vicia gigantea
Trientalis latifolia
Navarretia squarrosa
Phacelia nemoralis
Cynoglossum grande
Stachys Chamissonis
Orthocarpus castillejoide
Madia madioides

7. South Coastal Floristic Element

This group includes a complete gradation from strictly coastline plants (*Abronia maritima, Coreopsis gigantea*) to those which belong primarily to the mountain ranges bordering the deserts (*Quercus Dunnii, Adenostoma sparsifolium*).

Pellaea andromedaefolia
  var. *pubescens*
Carex triquetra
Allium haematochiton
Quercus Dunnii
Atriplex Watsonii
Abronia maritima
Streptanthus heterophyllus
Cardamine Gambelii
Dudleya pulverulenta
Adenostoma sparsifolium
Opuntia phaeacantha
Sanicula arguta
Tauschia arguta
Leptodactylon californicum
Eriodictyon crassifolium
Penstemon cordifolius
Coreopsis gigantea
Chaenactis glabriuscula var. *curta*
Baccharis Plummerae
Brickellia Nevinii

8. Great Basin-Desert Floristic Element

The strong representation of this element in the eastern part of the county is interpreted as due to increasingly arid conditions since Pleistocene time, resulting in relatively recent migration of desert plants coastward. Annual herbs are represented by numerous species, but desert shrubs, which spread more slowly, are proportionately fewer. The presence of a lone tree of *Pinus monophylla* as a relic near the summit of Caliente Mountain is consistent with this interpretation, as this species inhabits desert mountain ranges which are less arid and is on the point of dying out here. Species which in San Luis Obispo County have been found only on Caliente Mountain are marked with an asterisk.

Ephedra californica	*Asclepias erosa*
Ephedra viridis	*Convolvulus longipes*
Mariposa Palmeri	*Gilia latiflora*
**Eriogonum Heermannii*	**Eriastrum diffusum*
Eriogonum maculatum	*Langloisia Schottii*
Eriogonum Ordii	*Linanthus Parryae*
Chorizanthe Xanti	*Phacelia Fremontii*
Chorizanthe Watsonii	**Phacelia affinis*
Centrostegia Thurberi	*Cryptantha barbigera*
Mucronea perfoliata	*Cryptantha oxygona*
Atriplex polycarpa	*Greeneocharis circumscissa*
Grayia spinosa	*Lycium Andersonii*
Eurotia lanata	**Parishella californica*
Mirabilis Froebelii	*Coreopsis calliopsidea*
Abronia pogonantha	*Coreopsis Bigelovii*
Streptanthus inflatus	*Chaenactis Xantiana*
**Arabis pulchra*	*Chaenactis stevioides*
Lupinus horizontalis	var. *brachypappa*
Oenothera Palmeri	*Malacothrix glabrata*
**Lomatium mohavense*	**Anisocoma acaulis*

Remarks for Botanists

While it is hoped that the notes in the following list will be of general interest, there are certain features which call for explanation.

Any plant classification must be justified by the test of utility. Accordingly, my approach to classification is essentially conservative. Any classification which allows two species on a single plant, as in some groups of *Carex,* or even two genera on a single plant, as in several species of *Elymus,* can serve no useful purpose. Likewise, I can perceive no value in interpreting a group of plants as a "hybrid swarm" in which the great majority of individuals float in a sort of limbo without a specific name. Interspecific hybrids do exist, but when they become more numerous than the parents (or supposed parents) the classification needs to be revised. On the other hand, I have no hesitation about disagreeing with eminent authorities when they have used the same name for two distinct plants. For example, *Chaenactis lanosa* and *C. glabriuscula* are (in San Luis Obispo County, at least) always readily distinguishable and never show any indication of interbreeding when they grow together. In several genera, I have attempted to follow the classification worked out by those botanists who have studied the groups most intensively. The results, it must be admitted, have not been uniformly satisfactory.

Synonyms are listed occasionally, but only when they seem necessary to help the reader know to what plant a name refers. Complete synonymy would add greatly to the bulk of the work without contributing comparable value. Citation of specimens is likewise held to a minimum, and is usually done when the identity of the specimen may be questioned, or when it might be doubted that the report of a plant at a given locality is supported by adequate evidence. Numbers cited without collector's name are my own.

The taxonomic category of "subspecies" is here used very sparingly. According to the rules of nomenclature, it is not a routine replacement for the term "variety" (as many contemporary authors treat it) but indicates a degree of relationship intermediate between variety and species. I have used "subspecies," therefore, only for elements of a species which are geographically isolated, entirely or almost so, from the remainder of the species, and which are separated by small but clearly defined differences that show little if any intergradation.

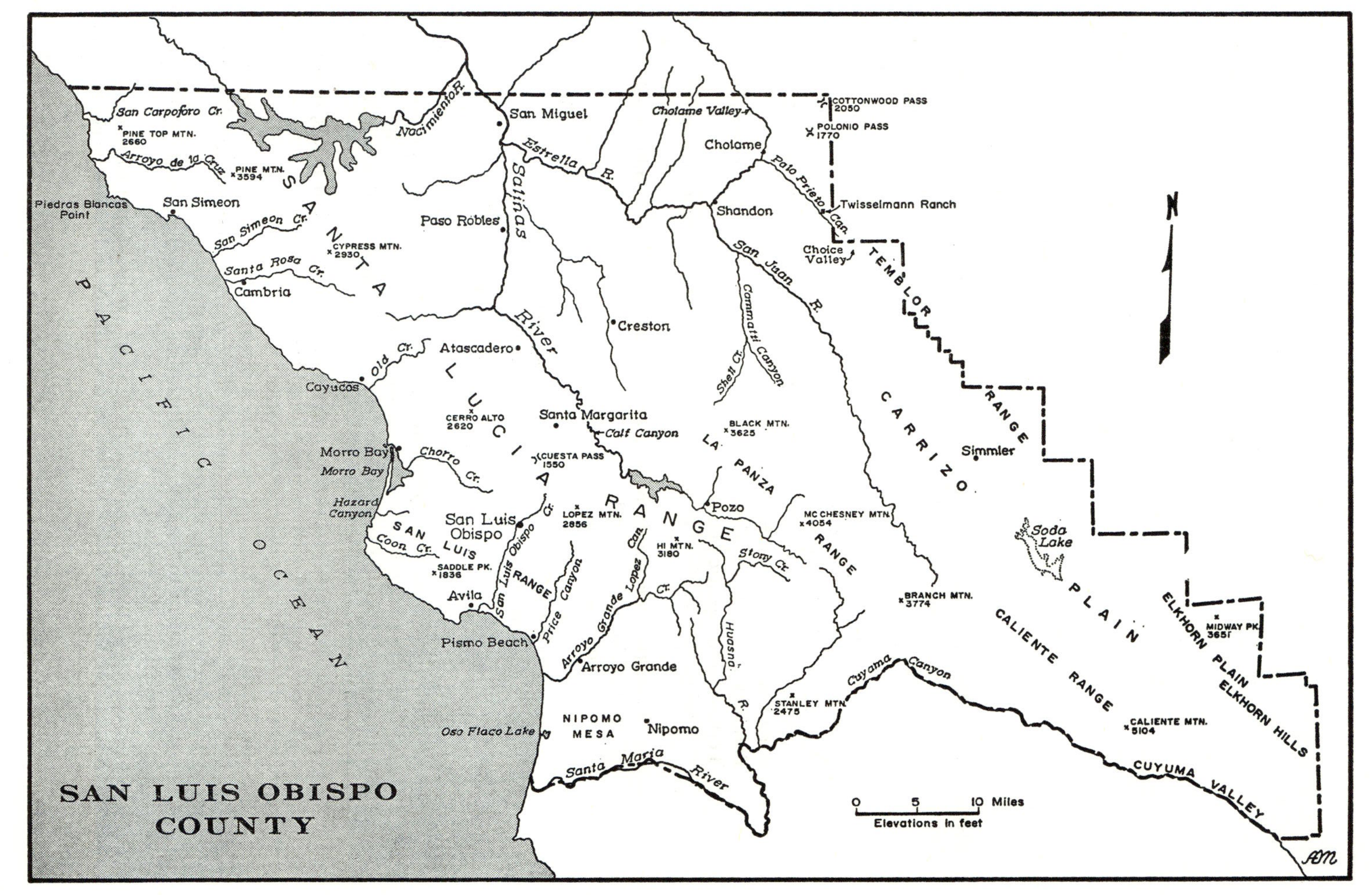

SAN LUIS OBISPO COUNTY
N
PACIFIC OCEAN
San Carpoforo Cr.
PINE TOP MTN. 2660
Arroyo de la Cruz
PINE MTN. 3594
Piedras Blancas Point
San Simeon
San Simeon Cr.
Nacimiento R.
San Miguel
Estrella R.
Salinas
Paso Robles
CYPRESS MTN. 2930
Santa Rosa Cr.
Cambria
SANTA
River
Old Cr.
Atascadero
Cayucos
LUCIA
CERRO ALTO 2620
Santa Margarita
Calf Canyon
CUESTA PASS 1550
Morro Bay
Morro Bay
Chorro Cr.
Hazard Canyon
San Luis Obispo
SAN LUIS
Coon Cr.
San Luis Obispo Cr.
LOPEZ MTN. 2856
RANGE
SADDLE PK. 1836
Avila
Price Canyon
Pismo Beach
Arroyo Grande
Arroyo Grande
Lopez Can.
HI MTN. 3180
Cr.
Stony Cr.
Huasna R.
NIPOMO MESA
Oso Flaco Lake
Nipomo
Santa Maria River
Cholame Valley
Cholame
Shandon
Choice Valley
San Juan R.
Cammatti Canyon
Shell Cr.
Creston
LA PANZA
BLACK MTN. 3625
Pozo
MC CHESNEY MTN. 4054
RANGE
BRANCH MTN. 3774
STANLEY MTN. 2475
Cuyama Canyon
COTTONWOOD PASS 2050
POLONIO PASS 1770
Twisselmann Ranch
Palo Prieto Can.
TEMBLOR
RANGE
CARRIZO
Simmler
Soda Lake
PLAIN
CALIENTE RANGE
CALIENTE MTN. 5104
ELKHORN PLAIN
ELKHORN HILLS
MIDWAY PK. 3651
CUYUMA VALLEY
0 5 10 Miles
Elevations in feet

Key to Artificial Groups of Families

Plants without flowers, with branches (strobili or cones) bearing specialized spore-bearing leaves, or with sporangia on backs or margins of leaves.
    Herbaceous plants, not bearing seeds (pteridophytes)................**Group 1.**
    Woody plants, bearing seeds (gymnosperms) ....**Group 2.**
Plants with flowers (in a few species flowers rarely produced) (angiosperms).
    Flowers without corolla, or with perianth in which none of the segments resemble typical petals (in *Polygonaceae* the "inner sepals" may differ markedly from the outer; in *Nyctaginaceae, Clematis,* and *Fremontodendron* the calyx is showy and resembles a typical corolla).
        Parasites on branches of trees ....*Loranthaceae.*
        Terrestrial to aquatic plants.
            Submerged or floating aquatic plants (see also *Salviniaceae* in group 1)
                Group 3.
            Terrestrial or marsh plants, or sometimes the base submerged and rooted in bottom mud.
                Trees, shrubs, or woody vines (see also *Myrtaceae* in group 11)
                    Group 5.
                Herbaceous plants, or stems woody only at base.
                    Leaves long and parallel-veined, or narrow and "grass-like" without evident veins (monocotyledons in part) ....**Group 4.**
                    Leaves either with or without evident veins, but not parallel-veined or "grass-like" ....**Group 6.**
Flowers with corolla, or with perianth-segments resembling typical petals (in *Polygonaceae* the perianth, although often showy, is conventionally called a "calyx"; in *Eucalyptus* the calyptra or "cap" on the flower-bud is described as consisting of fused petals).
    Perianth-segments 6, at least the outer 3 resembling typical petals, or all 6 essentially alike and often united toward base.
        Herbs with a tap-root, or shrubs ....*Polygonaceae.*
        Plants neither shrubby nor with a tap-root ....**Group 4.**
Calyx and corolla markedly unlike, their parts various in number, the sepals or calyx-lobes usually green, rarely petal-like.
    Petals distinct to the base, usually falling separately from the old flower (sometimes persistent).
        Stems markedly fleshy; leaves greatly reduced (the flattened joints of the stem may suggest leaves); plants with axillary clusters of spines ....*Cactaceae.*
        Stems not markedly fleshy; spines, if present, not axillary in position.

Trees, shrubs, or woody vines ....**Group 7.**
Herbaceous, or woody only at base of stem.
Stamens more numerous than petals (sometimes the same number in *Elatinaceae* and *Frankeniaceae* .....**Group 8.**
Number of stamens equalling number of petals or fewer (in *Mesembryanthemaceae* both petals and stamens are numerous, probably about the same number... **Group 9.**
Petals united at least at base, the whole corolla usually falling as a unit as fruit develops.
Leaves opposite or whorled, or some upper ones alternate
Group 10.
Leaves alternate, or crowded in a basal rosette or dense tuft
Group 11.

Group 1. (pteridophytes)

Small floating plants with overlapping scale-like leaves ....*Salviniaceae.*
Plants not floating.
Moss-like in appearance, the stems repeatedly branched and bearing numerous small narrowly lanceolate leaves ....*Selaginellaceae.*
Not moss-like in appearance.
Stems jointed, hollow; leaves reduced, whorled, those of each whorl united into a sheath ....*Equisetaceae.*
Stems not jointed or hollow; leaves well developed.
Leaves slender, grass-like, attached to a short solid stem (corm)
Isoetaceae.
Leaves not grass-like; or, if so, not attached to a corm.
Leaves resembling either a four-leaf clover or very fine grass; sporangia in hard basal structures (sporocarps) .....*Marsileaceae.*
Leaves pinnate or pinnately lobed; sporangia on backs or margins of leaves ....*Polypodiaceae.*

Group 2. (gymnosperms)

Internodes all or mostly shorter than leaves; leaves green.
Leaves needle-like or narrow and flattened, at least 1 cm. long.
Leaves when old falling from branchlets; cone-scales not peltate, each bearing 2 seeds ....*Pinaceae.*
Leaves persistent, the branchlets in age falling entire; cone-scales peltate, each bearing 3 to 7 seeds ....*Taxodiaceae.*
Leaves very small, scale-like, densely crowded and overlapping ..*Cupressaceae.*
Internodes many times as long as leaves; leaves reduced to brown pointed scales, opposite or in whorls of 3 ....*Ephedraceae.*

Group 3. (submerged or floating aquatics)

Leaves and stems clearly differentiated.
Plants of fresh or brackish water; inflorescence neither one-sided nor flattened.
Leaves limp, mostly over 2.5 cm. long, entire ....*Potamogetonaceae.*
Leaves rather stiff, mostly less than 2.5 cm. long, usually with spine-like divisions or teeth.
Leaves coarsely toothed to nearly entire ....*Najadaceae.*
Leaves pinnate, with slender prickly-margined divisions
Ceratophyllaceae.
Plants of salt water; flowers borne on one side of a flattened axis ..*Zosteraceae.*
Very small floating or submerged plants without clear differentiation into stems and leaves ....*Lemnaceae.*

Group 4. (monocotyledons in part)

Perianth green or brown, consisting of scales or bristles, or absent.
    Plants acaulescent, the leaves all in a basal tuft and the flowers in a slender raceme or spike borne on a scape . . . .*Juncaginaceae.*
    Plants not acaulescent, though the leaves may be crowded toward the base (a few species of *Juncus* are nearly acaulescent but the flowers are not in a raceme or spike).
        Perianth of bristles, of fewer than 6 scales, or none; fruit indehiscent, 1-seeded (rarely 2-seeded in *Sparganiaceae*).
            Flowers in a dense brown-velvety spike or in globose heads, the upper staminate and the lower pistillate.
                Flowers in a dense cylindrical spike . . . .*Typhaceae.*
                Flowers in globose heads . . . .*Sparganiaceae.*
            Flowers not in dense brown-velvety spikes or in globose heads, perfect or unisexual.
                Leaves 2-ranked; stems terete or a little flattened; fruit a caryopsis (grain) which is never 3-angled and not usually flattened
                                                Gramineae.
                Leaves 3-ranked or in some species reduced to basal sheaths; stems usually sharply or obtusely 3-angled, rarely terete (*Scirpus acutus*); achene 3-angled or lenticular . . . .*Cyperaceae.*
        Perianth-segments 6, scale-like; fruit a capsule . . . .*Juncaceae.*
Perianth white or bright-colored, corolla-like; green in a few species, but the segments with the texture of typical petals, not scale-like.
    Leaves in a large basal rosette, fleshy at base or throughout, rigid, tipped with a stout spine . . . .*Agavaceae.*
    Leaves variously arranged, not fleshy or rigid, not spine-tipped.
        Flowers not in umbels.
            Ovary superior; stamens 6.
                Fruit a capsule; plants usually with a bulb (except *Asphodelus*)
                                          Liliaceae.
                Fruit a berry; plants with a rhizome or a tough cluster of fibrous roots . . . .*Convallariaceae.*
            Ovary inferior; stamens 3 or 1.
                Inner perianth-segments (petals) alike . . . .*Iridaceae.*
                Inner perianth-segments unlike, two resembling the sepals, the third ("lip") markedly different . . . .*Orchidaceae.*
        Flowers in an umbel, the umbel subtended by one or more bracts
                                          Amaryllidaceae.

Group 5. (woody Apetalae)

Plant with simple trunk topped by a cluster of very large leaves (palm) . . . .*Palmae*
Plant with branched stems (not palms).
    Leaves opposite.
        Leaves simple.
            Flowers in catkins . . . .*Garryaceae.*
            Flowers not in catkins.
                Leaves silvery with appressed scales . . . .*Elaeagnaceae.*
                Leaves green, glabrous . . . .*Oleaceae.*
        Leaves compound.
            Vines; sepals 4, white, petal-like in appearance . . . . . .*Ranunculaceae.*
            Tree; sepals inconspicuous . . . .*Aceraceae.*

Leaves alternate.
    Flowers unisexual, at least the staminate in catkins.
        Pistillate flowers in catkins or in cone-like woody spikes.
            Fruit a capsule containing minute seeds, each with a tuft of hair
                Salicaceae.
            Fruit one-seeded, an achene or samara.
                Pistillate flowers in cone-like woody spikes; staminate catkins
                long and drooping . . . .*Betulaceae.*
                Pistillate flowers in short catkins; catkins not drooping
                  Myricaceae.
            Pistillate flowers in small clusters or solitary, with an involucre which
            develops into a husk, bur, or acorn-cup.
                Leaves simple . . . .*Fagaceae.*
                Leaves pinnate . . . .*Juglandaceae.*
    Flowers not in catkins, perfect or unisexual.
        Calyx white to rose-pink or yellow, appearing corolla-like.
            Flowers in umbels, small (less than 1 cm. wide).
                Small shrubs, not aromatic; umbels with an involucre
                  Polygonaceae.
                Tree or large shrub, strongly aromatic; umbels with early-
                deciduous bracts . . . .*Lauraceae.*
            Flowers solitary, large (over 2 cm. wide), bright yellow
                Sterculiaceae.
         Calyx green or yellow-green, or none.
            Leaves pinnately veined or at least with a midrib, entire or
            toothed, or fleshy and without evident veins.
                Leaves entire, often more or less fleshy, or reduced to minute
                fleshy scales . . . .*Chenopodiaceae.*
                Leaves toothed, not fleshy.
                    Leaves coarsely serrate; receptacle tubular; fruit an
                  achene with plumose persistent style . . . .*Rosaceae.*
                    Leaves serrulate or denticulate to spinose-dentate; recep-
                  tacle cup-shaped; fruit a berry . . . .*Rhamnaceae.*
            Leaves palmately veined and palmately lobed.
                Large deciduous tree; flowers in globose heads which become
                woody in fruit . . . .*Platanaceae.*
                Evergreen shrub; flowers in racemes . . . .*Euphorbiaceae.*

Group 6. (herbaceous Apetalae)

Flowers in a dense stout erect spike, with one or more showy white bracts at base.
    Inflorescence subtended by a single large bract (spathe) . . . .*Araceae.*
    Inflorescence with several petal-like bracts at base . . . .*Saururaceae.*
Flowers not in stout erect spikes, without petaloid bracts.
    Calyx white to pink, deep purple, or yellow, corolla-like in appearance.
        Leaves alternate or in a basal rosette (a few species of *Eriogonum* have op-
        posite or whorled leaves as well as a basal rosette).
            Leaves simple, entire . . . .*Polygonaceae.*
            Leaves compound . . . .*Ranunculaceae.*
        Leaves opposite, not forming a basal rosette.
            Flowers in an involucre, usually in heads, sometimes solitary
                Nyctaginaceae.
            Flowers without involucre, solitary in the leaf-axils . . . . .*Primulaceae.*
    Calyx, if present, green and not corolla-like.

Calyx present, though often minute or present only in staminate flowers; stamens usually more than one.

Flowers hypogynous (calyx or stamens attached at base of ovary).

Flowers not in bractless racemes.

Leaves simple.

Fruit an achene or utricle.

Calyx usually of 6 segments (if fewer, the flowers in an involucre); achene 3-angled . . . . *Polygonaceae.*

Calyx of fewer than 6 segments; flowers not in an involucre; fruit not 3-angled.

Calyx with 4 lobes or sepals; plant usually with stinging hairs (except one species) . . *Urticaceae.*

Calyx with 3 or 5 lobes or sepals (or none in pistillate flowers).

Leaves without stipules.

Flowers without bractlets; calyx not scarious (in *Salsola* developing scarious wings) . . . . *Chenopodiaceae.*

Flowers with bractlets; calyx more or less scarious . . . . *Amaranthaceae.*

Leaves with scarious stipules. . *Illecebraceae.*

Fruit a capsule.

Flowers perfect; capsule many-seeded

Caryophyllaceae.

Flowers unisexual; capsule 3-seeded. . *Euphorbiaceae.*

Leaves compound . . . . *Ranunculaceae.*

Flowers in bractless racemes . . . . *Cruciferae.*

Flowers epigynous or perigynous (ovary inferior, with calyx-lobes or stamens attached above it; or stamens attached at the mouth of a calyx-tube).

Flowers perigynous . . . . *Rosaceae.*

Flowers epigynous.

Low spreading annual; leaves somewhat fleshy, entire

Tetragoniaceae.

Tall perennial; leaves not fleshy, deeply lobed or compound and toothed . . . . *Datiscaceae.*

Calyx absent; flower reduced to one pistil or one stamen.

Terrestrial plants, usually in dry places; flowers in a cup-shaped structure (cyathium) . . . . *Euphorbiaceae.*

Aquatic or on wet soil; flowers not in a cyathium . . . . . *Callitrichaceae.*

Group 7. (woody Choripetalae)

Leaves alternate.

Leaves, or some of them, simple (*Stanleya* has some lower leaves pinnate; *Pickeringia* has leaves either simple or 3-foliolate).

Leaves represented on mature plants by parallel-veined phyllodes

Mimosaceae.

Leaves not parallel-veined.

Leaves pinnately veined or very narrow and without evident veins, never palmately veined.

Leaves not reduced to scales.

Petals 4 (rarely 6), yellow.

Stamens many; calyx falling off as flower opens

Papaveraceae.

Stamens 6; calyx of 4 distinct sepals, not falling before
petals *Cruciferae.*
Petals 5, not yellow.
Corolla irregular, pea-like (papilionaceous) *Leguminosae.*
Corolla regular.
Stamens more than 5 ....*Rosaceae.*
Stamens 5 only ....*Rhamnaceae.*
Leaves scale-like, triangular, appressed to branchlets
Tamaricaceae.
Leaves palmately veined.
Shrubs, small trees, or vines without tendrils.
Leaves toothed, palmately lobed, or both.
Calyx-lobes larger and more showy than petals; ovary inferior ....*Grossulariaceae.*
Calyx-lobes inconspicuous; ovaries 3 or more, superior
Rosaceae.
Leaves cordate at base, otherwise entire .....*Cesalpiniaceae.*
Vines with tendrils ....*Vitaceae.*
Leaves all compound (or in *Isomeris* sometimes a few simple).
Corolla strongly irregular, pea-like (papilionaceous) or of one petal only
Leguminosae.
Corolla regular to slightly irregular, the petals of about equal size.
Flowers 1 cm. or more in diameter.
Petals 4; fruit an inflated capsule; shrub strongly odorous
Capparidaceae.
Petals 5; fruit not inflated; plants not markedly odorous.
Corolla white or rose-pink, regular ....*Rosaceae.*
Corolla yellow, slightly irregular ....*Cesalpiniaceae.*
Flowers small, less than 1 cm. in diameter.
Leaflets spinose-dentate ....*Berberidaceae.*
Leaflets not spinose-dentate.
Leaves pinnate, with many leaflets ....*Simarubaceae.*
Leaves with 3 leaflets only ....*Anacardiaceae.*
Leaves opposite.
Leaves simple.
Leaves palmately lobed ....*Aceraceae.*
Leaves entire or toothed.
Flowers large (about 5 cm. wide); stamens many ....*Cistaceae.*
Flowers small (less than 1 cm. wide); stamens 4 or 5.
Petals 5; leaves firm; fruit a capsule ....*Rhamnaceae.*
Petals 4; leaves thin and flexible; fruit fleshy ....*Cornaceae.*
Leaves compound.
Leaves palmate ....*Hippocastanaceae.*
Leaves pinnate ....*Oleaceae.*

Group 8. (herbaceous Choripetalae with more stamens than petals)

Petals 3; stamens 6.
Erect plants; leaves well over 1 cm. long.
Leaves all basal; pistils many ....*Alismataceae.*
Leaves, or some of them, borne above base of stem; pistil 1.
Leaves narrow, parallel-veined, alternate and basal ....*Liliaceae.*
Leaves broad, net-veined, in a whorl or 3 around the solitary flower
Trilliaceae.

Small plants forming mats on wet soil; leaves less than 1 cm. long..*Elatinaceae*.
Petals 4 or more (sometimes fewer in *Ranunculaceae*); stamens more than 4, usually 8 or more.
  Ovary superior.
    Pistils 2 or more, or the carpels only lightly joined toward base.
      Carpels more than 2.
        Leaves succulent, simple and entire ....*Crassulaceae*.
        Leaves not succulent, usually toothed, lobed, or compound.
          Stamens hypogynous (attached below ovaries)
                                      Ranunculaceae.
          Stamens attached on the margin of a perigynous disk or cup
                                       Rosaceae.
      Carpels 2 only ....*Saxifragaceae*.
    Pistil 1, sometimes with 2 or more styles.
      Leaves simple (sometimes deeply lobed).
        Leaves alternate or basal.
          Stamens many (more than 15).
            Sepals more than 5.
              Small tufted plant with fleshy terete leaves
                                  Portulacaceae.
              Large aquatic plant with round floating leaves
                                Nymphaeaceae.
            Sepals 5 or fewer.
              Calyx falling off as flower opens .....*Papaveraceae*.
              Sepals persistent in fruit ....*Cistaceae*.
           Stamens fewer than 15.
             Sepals 2; petals 5; stamens variable ......*Portulacaceae*.
            Sepals and petals 4 or 5.
              Flowers regular to slightly irregular, without conspicuous spur.
                Sepals 4, deciduous; petals 4; stamens 6
                                    Cruciferae.
                Sepals 5, persistent; petals 5; stamens 10
                                    Geraniaceae.
              Flowers irregular; upper sepal with conspicuous spur
                                Tropaeolaceae.
        Leaves opposite (sometimes crowded at base of stem).
          Calyx falling as flower opens; petals 6 (rarely 4)
                                  Papaveraceae.
          Calyx persistent in fruit; petals normally 5.
           Calyx of distinct sepals.
            Stamens many ....*Guttiferae*.
            Stamens 10 or fewer.
              Leaves all entire; flowers mostly terminal on branches, axillary in a few species
                                  Caryophyllaceae.
              Leaves toothed; flowers in axils of leaves
                                  Elatinaceae.
          Calyx tubular.
           Capsule with free central placentae; plants not woody at base or scarcely so ......*Caryophyllaceae*.
           Capsule with parietal placentae; woody-based perennials of salt marshes or alkaline depressions
                                  Frankeniaceae.

Leaves compound.
Fruit a berry; petals of variable number, small, white, and inconspicuous ....*Ranunculaceae.*
Fruit dry; petals regularly either 4 or 5.
Petals 4.
Petals all alike.
Flowers not in racemes; stamens many *Papaveraceae.*
Flowers in bractless racemes; stamens normally 6
Cruciferae.
Petals in 2 unlike pairs ....*Fumariaceae.*
Petals 5.
Corolla irregular, with banner, wings, and keel
Leguminosae.
Corolla regular.
Leaves clover-like, with 3 leaflets .....*Oxalidaceae.*
Leaves with more than 3 leaflets.
Leaves pinnately divided, the primary divisions mostly incised or dissected into narrow segments
Limnanthaceae.
Leaves simply pinnate, the leaflets entire
Zygophyllaceae.
Ovary inferior or half-inferior.
Petals 5.
Leaves fleshy; sepals 2 ....*Portulacaceae.*
Leaves not fleshy; calyx-lobes 5.
Stamens 10.
Leaves palmately lobed or crenate; flowers in a raceme, white
Saxifragaceae.
Leaves entire; flowers solitary in the leaf-axils, yellow
Onagraceae.
Stamens many ....*Loasaceae.*
Petals 4 ....*Onagraceae.*

Group 9. (herbaceous Choripetalae with stamens of same number as petals or fewer)

Ovary superior (in *Lythraceae* closely invested by a perigynous tube but not fused with it; in *Plumbaginaceae* enclosed in a tubular calyx).
Leaves opposite, or some also basal.
Petals and stamens 5 or 6.
Petals and stamens 5.
Sepals 2 ....*Portulacaceae.*
Sepals 5 ....*Caryophyllaceae.*
Petals and stamens 6 ....*Papaveraceae.*
Petals and stamens 4 or fewer.
Leaves succulent; carpels distinct ....*Crassulaceae.*
Leaves not succulent; pistil 1, compound ....*Elatinaceae.*
Leaves alternate or basal only.
Calyx of distinct sepals.
Sepals 2; petals 4 or fewer ....*Portulacaceae.*
Sepals 5; petals 5.
Flowers regular or nearly so.
Leaves simple, entire, narrow; capsule not beaked..*Linaceae.*
Leaves lobed, crenate, or compound; fruit beaked
Geraniaceae.

       Flowers irregular .... *Violaceae.*
    Calyx tubular or funnelform.
       Leaves alternate .... *Lythraceae.*
       Leaves basal only .... *Plumbaginaceae.*
Ovary inferior or half-inferior.
    Leaves succulent; petals and stamens many .... *Mesembryanthemaceae.*
    Leaves not succulent; petals and stamens 5 or 4.
       Flowers not in umbels or heads.
          Petals and stamens 5; flowers rather small, in cymes or panicled cymes
                                                   Saxifragaceae.
          Petals and stamens 4; flowers over 3 cm. wide, in axils of upper leaves
                                                     Onagraceae.
       Flowers in umbels, compound umbels, panicled umbels, or heads, rarely
       solitary.
          Flowers in panicled umbels; fruit a berry .... *Araliaceae.*
          Flowers otherwise arranged, most commonly in compound umbels;
          fruit a schizocarp .... *Umbelliferae.*

Group 10. (Sympetalae with opposite or whorled leaves)

Ovary superior.
    Corolla regular.
       Pistil 1; fruit a capsule or of nutlets.
          Stamens attached at middle of base of each corolla-lobe .. *Primulaceae.*
          Stamens alternating with corolla-lobes.
             Ovary one-celled; corolla usually withering-persistent
                                                       Gentianaceae.
             Ovary 2- to 4-celled; corolla usually promptly deciduous from de-
             veloping fruit.
                Ovary 3-celled; style 3-branched .... *Polemoniaceae.*
                Ovary 2- or 4-celled; style 2-branched.
                    Leaves pinnately lobed or compound; fruit a capsule
                                            Hydrophyllaceae.
                  Leaves entire; fruit of 4 nutlets .... *Boraginaceae.*
       Ovaries distinct, becoming 2 follicles in fruit.
          Flowers not in umbels; stamens not united with stigmas.. *Apocynaceae.*
          Flowers in umbels; stamens united into a column with 5 hoods
                                                 Asclepiadaceae.
    Corolla 2-lipped to slightly irregular.
       Fruit of 4 or fewer nutlets.
          Flowers in usually dense terminal spikes, not whorled ... *Verbenaceae.*
          Flowers variously arranged, not in spikes except when spikes are inter-
          rupted (flowers in whorls) .... *Labiatae.*
       Fruit a capsule.
          Capsule 1-celled .... *Portulacaceae.*
          Capsule 2-celled .... *Scrophulariaceae.*
Ovary inferior.
    Stamens distinct; flowers variously arranged.
       Stamens 4 or 5.
          Shrubs or woody vines; fruit a berry .... *Rubiaceae.*
          Herbs (one species with woody but weak slender, sprawling stems).
             Leaves whorled; stems weak and slender .... *Rubiaceae.*
             Leaves opposite; stems strong, erect .... **Dipsacaceae.**

Stamens 1 to 3 ....*Valerianaceae.*
Anthers united into a tube around the style; flowers in heads .....*Compositae.*

Group 11. (Sympetalae with alternate or basal leaves)

Stamens 8 or more, more numerous than corolla-lobes or petals.
  Filaments of stamens united, adherent to base of corolla, which is divided into almost distinct petals.
    Leaves palmately veined and palmately lobed or crenate; stipules present
                Malvaceae.
    Leaves not palmately veined, entire; stipules none.
      Small woody-based herb; flowers irregular ....*Polygalaceae.*
      Shrub; flowers regular ....*Styracaceae.*
  Filaments of stamens distinct.
    Herbs with succulent leaves in a basal rosette ....*Crassulaceae.*
    Woody plants.
      Corolla represented by a calyptra ("cap") on flower-bud, falling off as flower opens; stamens numerous ....*Myrtaceae.*
      Corolla urn-shaped to short-tubular or campanulate; stamens 10.
        Ovary superior ....*Ericaceae.*
        Ovary inferior ....*Vacciniaceae.*
Stamens 7 or fewer, equalling the number of corolla-lobes or fewer.
  Parasites without chlorophyll; leaves reduced to scales or absent.
    Root-parasites; stem fleshy, bearing scale-like leaves.
      Corolla regular ....*Lennoaceae.*
      Corolla irregular ....*Orobanchaceae.*
    Stems hair-like, rootless and leafless ....*Convolvulaceae.*
  Green plants.
    Ovary superior, or attached only to base of flower-tube.
      Corolla regular or nearly so.
        Leaves basal only.
          Flowers not in spikes; corolla soft and not scarious.
             Flowers in an umbel ....*Primulaceae.*
             Flowers in a head or panicle ....*Plumbaginaceae.*
          Inflorescence a spike; corolla dry and scarious *Plantaginaceae.*
        Stems with some cauline leaves, often with a basal rosette also.
          Stamens standing directly in front of corolla-lobes
                Primulaceae.
          Stamens alternating with corolla-lobes.
            Style 3-branched or stigma 3-lobed (4-lobed in a few species); ovary 3-celled (1-celled in a few species)
                Polemoniaceae.
            Style 2-branched or stigma 2-lobed, sometimes entire; ovary 2-celled or 4-celled.
              Flowers not in an erect raceme.
                Ovary 2-celled; fruit a capsule or berry.
                  Stems twining, trailing, or creeping; flowers usually solitary on axillary peduncles, sometimes 2 or 3 on an axillary branch
                    Convolvulaceae.
                  Stems usually erect or ascending, sometimes trailing but never twining or creeping; flowers variously arranged, almost never solitary and axillary.

Calyx divided nearly to base; style 2-branched, or styles 2 . . *Hydrophyllaceae.*
Calyx parted about half way to base or less (except *Petunia*); style not branched (stigma 2-lobed in *Datura*) . *Solanaceae.*
Ovary 4-celled; fruit of 4 or fewer one-seeded nutlets . . . .*Boraginaceae.*
Flowers in a long erect raceme . . .*Scrophulariaceae.*
Corolla irregular.
Capsule not beaked . . . .*Scrophulariaceae.*
Capsule long-beaked, the beak splitting into 2 stout hooks at maturity . . . .*Martyniaceae.*
Ovary inferior.
Flowers not in heads.
Herbaceous vines with tendrils . . . .*Cucurbitaceae.*
Herbs without tendrils.
Corolla regular; anthers distinct . . . .*Campanulaceae.*
Corolla irregular; anthers united into a tube . . . .*Lobeliaceae.*
Flowers in heads . . . .*Compositae.*

Identifying Characteristics of Families

Succulent plants:

Agavaceae
Chenopodiaceae
Tetragoniaceae
Nyctaginaceae
Mesembryanthemaceae
Caryophyllaceae (Spergularia)
Cruciferae (Cakile)
Crassulaceae
Cactaceae
Solanaceae (Lycium)
Compositae (several species with slightly fleshy leaves)

Parasites without chlorophyll:

Loranthaceae (Arceuthobium)
Lennoaceae
Convolvulaceae (Cuscuta)
Orobanchaceae

Plants with twining stems:

Convolvulaceae
Compositae (Senecio mikanioides)

Plants with tendrils:

Ranunculaceae (petioles of *Clematis*)
Leguminosae (terminal leaflets of *Vicia* and *Lathyrus*)
Tropaeolaceae (petioles of *Tropaeolum*)
Vitaceae
Cucurbitaceae

Leaves markedly odorous or aromatic:

Conifers (*Pinaceae, Taxodiaceae, Cupressaceae*)
Gramineae (Hierochloe)
Lauraceae
Capparidaceae
Tropaeolaceae
Simarubaceae
Anacardiaceae (Schinus)
Myrtaceae
Umbelliferae
Labiatae
Solanaceae
Martyniaceae
Compositae

Leaves reduced to inconspicuous scales or sheaths:

Equisetaceae
Cupressaceae
Ephedraceae
Cyperaceae (few species)
Juncaceae (few species)
Chenopodiaceae
Tamaricaceae
Compositae (few species)

Flowers in umbels:

Amaryllidaceae
Polygonaceae
Lauraceae
Geraniaceae
Oxalidaceae
Araliaceae
Umbelliferae
Primulaceae
Asclepiadaceae

Flowers in heads:

Sparganiaceae
Juncaceae

Nyctaginaceae
Platanaceae
Mimosaceae
Leguminosae (Trifolium)
Umbelliferae (Eryngium, Sanicula)
Plumbaginaceae
Polemoniaceae
Labiatae (Monardella)
Rubiaceae
Dipsacaceae

Flowers conspicuously irregular:

Ranunculaceae (Delphinium)
Leguminosae
Tropaeolaceae
Polygalaceae
Violaceae
Onagraceae (Clarkia concinna)
Labiatae
Scrophulariaceae
Martyniaceae
Orobanchaceae
Valerianaceae (Plectritis)
Lobeliaceae
Compositae (Cichorieae, and ray-flowers in other tribes)

PTERIDOPHYTES

Polypodiaceae. FERN FAMILY

Sporangia on backs of leaves, not marginal.
  Backs of leaves not powdery; sporangia in sori.
    Leaves deeply pinnately lobed, the divisions broadened at base and confluent on rachis ....1. *Polypodium.*
    Leaves compound, the divisions (pinnae or pinnules) narrowed at base and usually with petiolules.
      Petioles tough, fibrous; plant 3 dm. or more tall.
        Sori circular or nearly so.
          Indusium not notched, attached at center, often absent from sori on older leaves ....2. *Polystichum.*
          Indusium deeply notched on one side, attached at base of notch ....3. *Dryopteris.*
        Sori more or less elongate.
          Sori mostly oval, very small, less than 1.5 mm. long
                4. *Athyrium.*
          Sori narrowly oblong or linear, mostly more than 2 mm. long
                5. *Woodwardia.*
      Petioles fragile; plant less than 3 dm. tall ....6. *Cystopteris.*
  Backs of leaves with yellow or white waxy powder (this sometimes sparse); sporangia scattered, not in sori ....7. *Pityrogramma.*

Sporangia on margins of leaves, in a continuous band or in marginal sori.
  Petioles tan to dark brown or black, brittle.
    Pinnules widened from base upward, the sporangia along upper margin only, covered by the closely reflexed margin ....8. *Adiantum.*
    Pinnules widest near middle or near base, or uniformly narrow, the sporangia along both edges.
      Leaves when mature without scales or hairs.
        Pinnules entire; sporangia in a marginal band.
          Petioles tan or light brown; pinnules obtuse, mostly 2 mm. wide or more ....9. *Pellaea.*
          Petioles dark brown; pinnules mucronate, less than 2 mm. wide.
            Leaf-blades dull or gray-green, much longer than broad
                                  9. *Pellaea.*
            Leaf-blades bright green, not much longer than broad
                                10. *Cheilanthes.*
        Pinnules deeply incised or coarsely toothed; sporangia in small marginal sori or even solitary ....10. *Cheilanthes.*
      Leaves bearing scales or hairs ....10. *Cheilanthes.*
  Petioles green to light brown, tough, fibrous ....11. *Pteridium.*

1. Polypodium L.

Leaves thin; terminal pinna or lobe less than 1/4 total length of blade; largest sori about 2.5 mm. in diameter ....1. *P. californicum.*
Leaves thick, firm-textured; terminal pinna usually at least 1/3 total length of blade; largest sori about 4 mm. in diameter ....2. *P. Scouleri.*

  1. **P. californicum** Kaulf. Common in rocky places from coast eastward to La Panza Range.

  2. **P. Scouleri** H. & G. North slope of Morro Rock (*R. M. Lloyd 3378*). Hybrids with *P. californicum* were also found there.

2. Polystichum Roth

Leaves simply pinnate, the leaflets evenly serrate ....1. *P. munitum.*
  Petioles and rachises covered with conspicuous brown membranous scales
                                    subsp. *munitum.*
  Scales, except at base of petioles, much reduced and inconspicuous
                                    subsp. *curtum.*
Leaves twice pinnate, the secondary divisions incised ....2. *P. Dudleyi.*

  1. **P. munitum** (Kaulf.) Presl subsp. **munitum.** Sword Fern. Shaded, rather moist places: San Luis Range, especially in the Coon Creek watershed, north slope of Morro Rock, and near coast from near Cambria northward; densely wooded or sheltered rocky places in Santa Lucia Mts.

  Subsp. **curtum** Ewan. Back from the coast, often in drier sites: Lopez Canyon; near Rocky Butte. Perhaps most of the sword ferns in the Santa Lucia Range proper, as distinguished from the coastal hills, will prove to be this subspecies. Insufficient attention has so far been given to the plants in their various habitats. Possibly subsp. *curtum* is distinct enough to be classified as a species.

  2. **P. Dudleyi** Maxon. Damp shaded rocky banks: upper part of Lopez Canyon (*8807*); coast just south of Monterey Co. line (*6672*).

3. Dryopteris Adans.

1. **D. arguta** (Kaulf.) Watt. CALIFORNIA WOOD-FERN. Common in woods near coast and throughout Santa Lucia Range; more sparingly eastward to La Panza Range.

4. Athyrium Roth. LADY FERN

Primary pinnae widely divergent, the fronds well over 15 cm. wide
A. Filix-femina var. *californicum.*
Primary pinnae ascending, the fronds less than 10 cm. wide . . . . var. *angustum.*

1. **A. Filix-femina** (L.) Roth var. **californicum** Butters. Hazard Canyon, in moist thicket (*7368*). The plant is placed in var. *californicum* because the branches of the rachis are puberulent rather than scaly, although the occurrence is outside the normal area of the variety as given by Munz.

Var. **angustum** (Willd.) Farwell. In wet place near Piedras Blancas Point (*7327*). (Reported as var. *sitchense* in Am. Fern Journ. 56: 20. 1966.)

5. Woodwardia Sm. CHAIN FERN

1. **W. fimbriata** Sm. Wet places around springs and along small streams, mostly in shaded sheltered positions, near coast and in Santa Lucia Mts. Notably plentiful in upper part of Lopez Canyon, where there are magnificent natural ferneries. Rarely found where there is no surface indication of moisture.

6. Cystopteris. BLADDER FERN

1. **C. fragilis** (L.) Bernh. Rare locally, in shaded moist places: between Rocky Butte and Pine Mt. above San Simeon (*8014*); Garcia Mt. south of Pozo; trail from Stoney Creek to Colwell Mesa (*7968*).

7. Pityrogramma Link

Backs of leaves yellow or white with copious waxy granules.
Backs of leaves yellow . . . .*P. triangularis* var. *triangularis.*
Backs of leaves white to pale yellow (apparently often darkening in drying)
var. *semipallida.*
Backs of leaves green, with sparse waxy granules . . . . var. *viridis.*

1. **P. triangularis** (Kaulf.) Maxon var. **triangularis.** GOLDBACK FERN. Wooded hills and rocky places, our commonest fern: throughout Santa Lucia, San Luis, and La Panza Ranges and the hills between; even in exceptionally sheltered and shaded spots in northern Temblor Range. The following less common varieties are noticeable when found but do not have much geographic significance.

Var. **semipallida** J. T. Howell. SILVERBACK FERN. Occasionally found associated with var. *triangularis,* locally rare: Tassajera Creek, Lopez Canyon, See Canyon, and probably elsewhere. Much more common than the white-powdered form are intergrades in which the backs of the leaves are various shades of pale yellow.

Var. **viridis** Hoover. GREENBACK FERN. Rare in Santa Lucia and La Panza Ranges, mixed with the other varieties. Tassajera Creek is the type locality of this variety.

8. Adiantum L. MAIDENHAIR FERN

Petioles divergently 2 to 4 times forked at summit; leaflets (except lowest) attached at one side . . . . .1. *A. pedatum.*

Petioles continuing into a rachis with shorter lateral branches; leaflets attached near middle of a cuneate base.

Fertile leaflets narrowly fan-shaped, the angle at base usually less than 90 degrees, with short portions (not over 4 mm.) of the upper margin reflexed to cover sporangia . . . . .2. *A. Capillus-Veneris.*
Fertile leaflets broadly fan-shaped, the angle at base usually more than 90 degrees, with longer portions (mostly over 4 mm.) of the margin reflexed to cover sporangia . . . . .3. *A. Jordanii.*

1. **A. pedatum** L. FIVE-FINGERED FERN. Cool, permanently moist, more or less shaded banks: upper Lopez Canyon (plentiful); Coon Creek and Diablo Canyon in San Luis Range; coast north of San Carpoforo Creek. The name var. *aleuticum,* used in recent references for all California plants of this species, is not here used, because our plants seem essentially identical with specimens from the eastern states.

2. **A. Capillus-Veneris** L. MAIDENHAIR FERN. Rocky banks kept permanently moist by seepage: first ravine north of San Carpoforo Creek; apparently also in upper Lopez Canyon (plants very few, sterile, and in poor condition).

3. **A. Jordanii** C. Muell. CALIFORNIA MAIDENHAIR. Wooded or rocky slopes in summer-dry places, usually in shade: common from near coast eastward to La Panza Range. Grows best when rooted in a soil composed largely of decomposed leaves. In contrast with *A. Capillus-Veneris,* this species is normally summer-dormant. The only exception noted was close to the stream, although not in wet soil, in upper Lopez Canyon. Properly treated, *A. Jordanii* will become evergreen in cultivation.

9. **Pellaea** Link. CLIFF-BRAKE

Petioles light brown; leaflets obtuse at apex . . . . .1. *P. andromedaefolia.*
Rhizomes extensively creeping, the leaves not crowded; rachis and its branches glabrous . . . . .var. *andromedaefolia.*
Rhizomes shortly branched, the leaves tufted; rachis and branches usually minutely hairy . . . . .var. *pubescens.*
Petioles dark brown leaflets mucronate . . . . .2. *P. mucronata.*
Leaves partly tripinnate, some of the secondary divisions bearing 3 to 9 pinnules . . . . .var. *mucronata.*
Leaves bipinnate only, the secondary divisions consisting of simple pinnules var. *californica.*

1. **P. andromedaefolia** (Kaulf.) Fee var. **andromedaefolia.** COFFEE FERN. Openly wooded or rocky slopes, common in western part and occasional, except for extremely arid localities, in eastern part.

Var. **pubescens** D. C. Eaton. On serpentine rock east of Morro Bay and around San Luis Obispo. These occurrences are the farthest north known for the variety.

2. **P. mucronata** D. C. Eaton var. **mucronata.** BIRD'S-FOOT FERN. Rocky places, commonest in central part of county: rare near coast. A local variant found on sandstone hills north of Arroyo Grande and east of Pismo Beach is very vigorous with many leaves, has leaflets almost twice as long as usual in the species, and its leaves remain greener in age. Herbarium specimens resembling this variant have been seen from Santa Barbara and Ventura Counties.

Var. **californica** (Lemmon) M. & J. *P. compacta* Maxon. Upper San Juan River (*Rodin 7127*).

10. **Cheilanthes** Sw.

Leaf-segments acute or mucronate, without scales or hairs.
    Leaflets incised or coarsely toothed, with small sori in the sinuses
                                                  1. *C. californica.*
    Leaflets of fertile leaves finely toothed to entire, the sori confluent into a marginal band, or at least closely spaced ....2. *C. siliquosa.*
        Leaflets nearly all with a continuous marginal indusium ......f. *siliquosa.*
        Leaflets, or some of them, with the marginal indusium interrupted
                                        f. *Carlotta-Halliae.*
Leaf-segments rounded, obtuse, bearing scales or hairs.
    Leaves finely glandular-hairy ....3. *C. Cooperae.*
    Leaves with scales on backs of leaflets, not glandular.
        Scales ovate-acuminate, sparingly if at all fringed, usually completely concealing sporangia ....4. *C. Clevelandii.*
        Scales lanceolate or narrower, copiously long-fringed, not concealing sporangia ....5. *C. intertexta.*

1. **C. californica** (Hook.) Mett. Occasional in sheltered rocky places, mostly on sandstone or granite and not ordinarily on serpentine: scattered through western half of county, seldom in large numbers.

2. **C. siliquosa** Maxon f. **siliquosa.** Upper Chorro Creek, among serpentine rocks (*6569* in part); Cypress Swamp, Cypress Mt. (*Twisselmann 3230*).

Forma **Carlotta-Halliae** (Wagner & Gilbert) Hoover. *C. Carlotta-Halliae* W. & G. With typical *C. siliquosa* in the region of upper Chorro Creek. The Marin County plants originally described under this name are said by Wagner to have originated by hybridization between *C. siliquosa* and *C. californica,* followed by a doubling of the chromosomes. It seems rather doubtful that the local plants, as well as similar plants from the Sierra Nevada, are genetically identical with the Marin County form.

3. **C. Cooperae** D. C. Eaton. Crevices in limestone or calcareous sandstone: Franklin Creek, Camp Natoma in Adelaida district, according to Clare Hardham (Leafl. West. Bot. 9: 129).

4. **C. Clevelandii** D. C. Eaton. *C. Covillei* Maxon. Rocky places, most commonly on sandstone or granite: summit of Mt. Bishop near San Luis Obispo (*R. J. Rodin*); hills near upper Salinas River; more frequent in La Panza Range but not common even there; upper San Juan River (*Rodin 7126*). Of the plants found in this county, only a collection from the Pine Mt. of the La Panza Range (*6584*) shows the creeping rhizomes with comparatively widely spaced leaves which, according to authors, characterize "typical" *C. Clevelandii.* The rest are more closely tufted and so represent *C. Covillei.* Because I find a continuous range of variation and no geographic separation of the plants into distinguishable groups, both forms are included under the earlier published name.

5. **C. intertexta** (Maxon) Maxon. East of Middle Branch of Huerhuero Creek, 6 miles south of Creston, tightly wedged in crevices of granite (*6578* in 1946). This only known occurrence of the species in the county was obliterated by road-building operations about 1950.

11. **Pteridium** Scop. BRACKEN

1. **P. aquilinum** (L.) Kuhn. BRACKEN. Common in coastal woodlands, less often in exposed places (where dwarfed); woods in Santa Lucia Mts.; occasional in shel-

tered shaded places east of Salinas River (Rocky Canyon near Atascadero). The western American representation of this species has been called var. *pubescens* Underw. and var. *lanuginosum* (Bong.) Fernald, but wild plants observed by me in Great Britain were pubescent also and differed in no evident way from the bracken of California.

Marsileaceae. CLOVER-FERN FAMILY

Leaves with 4 leaflets, resembling a "four-leaf" clover ....1. *Marsilea.*
Leaves very slender, resembling a fine-textured grass ....2. *Pilularia.*

1. Marsilea L. CLOVER-FERN

1. **M. vestita** H. & G. Depressions which are flooded during growing season but often dry in summer; rare in upper Salinas Valley: Atascadero Lake; 7 miles southeast of Santa Margarita. At the latter locality, where both *Marsilea* and *Pilularia* were collected on May 11, 1952, no trace of either could be found on the same date in 1964.

2. Pilularia L. PILLWORT

1. **P. americana** A. Br. Beds of vernal pools, which become very dry in summer, in interior: near Estrella; 7 miles southeast of Santa Margarita; north of Soda Lake on Carrizo Plain.

Salviniaceae. FLOATING-FERN FAMILY

1. Azolla Lam.

1. **A. filiculoides** Lam. DUCKWEED FERN. Ponds, pools, and sluggish streams: noticed particularly in Los Osos Valley and on lakes among the dunes south and west of Arroyo Grande; Trout Creek east of Santa Margarita. In late summer the surface of the water may become red from solid masses of this plant.

Equisetaceae. HORSETAIL FAMILY

1. Equisetum L. HORSETAIL

Erect shoots of 2 distinct sorts: one unbranched, pale, and ending in a strobilus; the other with whorls of slender branches, bright green, and without strobilus.
    Sterile stems mostly over 5 mm. thick (except near tip), over 30 cm. tall at maturity; strobilus 4 to 10 cm. long ....1. *E. Telmateia.*
    Sterile stems 3 mm. thick or less, rarely over 30 cm. tall at maturity; strobilus about 2 to 3 cm. long ....2. *E. arvense.*
Erect shoots essentially alike, simple or branched, all bright green.
    Main erect shoots with about 15 to 25 longitudinal ridges, dying down to base in winter ....3. *E. laevigatum.*
    Main erect shoots with about 25 to 42 longitudinal ridges, evergreen at least in lower portion.
        Lower part of shoots remaining green in winter, with about 25 to 34 ridges
                                                    4. *E. Ferrissii.*
        Shoots wholly green in winter, with 35 to 42 ridges ....5. *E. hiemale.*

1. **E. Telmateia** Ehrh. Common near coast in moist ground, and extending into canyons of the Santa Lucia Mts. When once established, it may spread by its creeping rhizomes into drier ground or persist when the soil becomes drier.

2. **E. arvense** L. In a low swampy place just north of Piedras Blancas Point (*7768*).

3. **E. laevigatum** A. Br. *E. kansanum* Schaffn. *E. Funstonii* A. A. Eaton. Occasional in moist places in or near the Santa Lucia Range: between Rocky Butte and Pine Mt.; forks of San Simeon Creek; Santa Rita Creek; Morro Creek; Serrano Canyon; Alamo Creek near Cuyama River.

4. **E. Ferrissii** Clute. Moist places near coast from Villa Creek between Cayucos and Cambria to Morro Creek, and probably overlooked elsewhere. These plants, according to R. L. Hauke, are hybrids between *E. laevigatum* and *E. hiemale*. In general appearance, some individuals resemble more closely one of the presumed parents, some the other.

5. **E. hiemale** L. var. **affine** (Engelm.) A. A. Eaton. Low moist places near Oceano (*6422*) and Arroyo Grande, and probably elsewhere. Insufficiently collected and readily confused with robust plants of the preceding. R. L. Hauke has included all North American plants of the species in var. *affine*, although there is complete intergradation with the typical Eurasian form.

Selaginellaceae. Spike-Moss Family

1. Selaginella Beauv.

1. **S. Bigelovii** Underw. Common in rocky places, on sandstone, serpentine, granite, etc., from coast eastward through La Panza Range.

Isoetaceae. Quillwort Family

1. Isoetes L. Quillwort

Plant at bases of leaves over 1.5 cm. in diameter; leaves about 25 to 50 per plant, sometimes more or fewer; membranous leaf-margins extending about 1.5 to 3 cm. above sporangia ....1. *I. Nuttallii.*
Plant at bases of leaves 1.5 cm. in diameter or less; leaves 5 to 25 per plant; membranous leaf-margins extending about 1 to 1.5 cm. above sporangia ..2. *I. Orcuttii.*

1. **I. Nuttallii** A. Br. In damp soil of meadow between Rocky Butte and Pine Mt., Santa Lucia Range (*7897*).

2. **I. Orcuttii** A. A. Eaton. Cambria, in moist swales in sandy soil under pines (*6948, 7855*). These plants, although identified by Clyde F. Reed as *I. Nuttallii,* are readily distinguishable by the characters given in the above key.

GYMNOSPERMS

Pinaceae. Pine Family

Needles enclosed in a sheath at base, usually in bundles of 2 or more, rarely solitary
1. *Pinus.*
Needles without sheath at base ....2. *Abies.*

1. Pinus L. Pine

Needles in groups of 3.
    Needles mostly over 15 cm. long; cone-scales tipped either with a prickle or with a stout hooked sharp-pointed spur.
        Cones less than 15 cm. long, each scale with a small prickle..1. *P. ponderosa.*
        Cones over 15 cm. long, each scale with a heavy woody spur.

Cones elongate; branchlets over 5 mm. thick . . . . .2. *P. Coulteri.*
Cones hardly longer than thick; branchlets 5 mm. thick or less
3. *P. Sabiniana.*
Needles mostly less than 15 cm. long; cone-scales tipped either with a rounded
knob or with a straight or slightly curved pointed knob.
Cone-scales facing away from branches with pointed knobs . .4. *P. attenuata.*
Cone-scales facing away from branches with rounded knobs . . .5. *P. radiata.*
Needles in groups of 2 or solitary.
Needles in groups of 2.
Needles slender, flexible; cone-scales rounded at apex . . . . . .5. *P. radiata.*
Needles stout, stiff; cone-scales with a spur or prickle . . . . . .6. *P. muricata.*
Cones reflexed, strongly asymmetrical, the scales with stout woody
spurs . . . . .f. *muricata.*
Cones spreading, nearly symmetrical, the scales each with a prickle
f. *remorata.*
Needles solitary . . . . .7. *P. monophylla.*

1. **P. ponderosa** Dougl. Yellow Pine. Forming almost a pure stand on Pine Mt.
above San Simeon, extending southeast along the ridge toward Rocky Butte and,
mixed with other trees, in small groves and as scattered individuals down both
flanks of the main ridge of the Santa Lucia Mts. to about the 2500-foot level. The
trees in the past have been logged at times, but fortunately not in such a way as to
destroy the forest. Also on and near Pine Top Mt., on the coast just south of San
Carpoforo Creek. In the latter locality the cones persist on the trees at least until
the year following release of the seeds, and do not detach readily even then. This
unusual feature, which apparently has not been mentioned in the literature dealing
with *P. ponderosa,* could possibly be explained by past hybridization with one of the
so-called "closed-cone" pines, of which *P. radiata* is the only one now occurring reasonably near. In all structural features, however, the trees in question seem to be
typical *P. ponderosa.*

2. **P. Coulteri** D. Don. Big-Cone Pine. Nearly always at altitudes above 1500 feet,
and usually above 2000 feet: Santa Lucia Mts. from the Cambria-Templeton road
northward and on both sides of Cuesta Pass; Pine Ridge in Huasna River watershed; highest ridges in middle portion of La Panza Range. There is also an isolated
stand at a lower altitude in the vicinity of Yaro Creek north of Pozo. The cones
vary in size, but the largest are the heaviest of those of any pine. Oddly, this species grows on the ridge north of San Carpoforo Creek, while *P. ponderosa* occupies
a similar site to the south.

3. **P. Sabiniana** Dougl. Digger Pine. Our most widespread pine, well distributed
through much of the Santa Lucia Range, the upper Salinas Valley from Atascadero
and near Creston southward, and eastward to east base of La Panza Range. Largely
absent from the somewhat moister habitats favored by *P. Coulteri,* but growing near
that species in several localities. On the west slope of the Santa Lucia Mts. it is less
frequent but is known in the watersheds of Chorro Creek and San Simeon Creek,
and in southern Monterey Co. grows on slopes directly above the ocean. Digger
pine has frequently been planted about houses in towns and on farms. Trees with
unusually large cones occur notably in hills adjacent to the La Panza Range, and
are also known in other parts of California. These cones are as stout as those of
P. Coulteri but markedly shorter.

4. **P. attenuata** Lemmon. Knob-Cone Pine. In this county restricted to an area of

probably several square miles on the higher ridges of the Santa Lucia Range east of San Luis Obispo. The trees here grow on siliceous shale and in places form a dense forest. *Pinus Coulteri* grows in the same area but mostly in less rocky, somewhat moister spots; the two pines mix only to a limited extent. The seeds of *P. attenuata* are commonly believed to be released only during fires, but trees which showed no indication of burning have been seen with the cone-scales partly separated.

5. **P. radiata** D. Don. MONTEREY PINE. Forming an extensive forest over a considerable area around Cambria, and in an isolated stand to the north near Pico Creek. Branches on mature trees apparently form bundles of two needles in years following a season of deficient rainfall, while bundles of three needles are usually formed if the rainfall was adequate in the preceding year. Planted trees, which probably are nearly all descended from the form found at Monterey, nearly always have the needles in threes. There is some uncertainty as to whether three trees at Ontario Grade (east of Avila) represent a last remnant of a former Monterey pine forest in that area. The upper one of these trees has been destroyed by the State Division of Highways, but the two near the base of the hill remain. It was observed in 1963 that young trees were coming up spontaneously near them, and their position, not bordering a road but on a steep hillside mixed among live oaks, suggests the possibility that they were not planted. If this conclusion is correct, then this tiny and vanishing "grove" represents the southernmost truly native occurrence of Monterey pine on the mainland, although it is known farther to the south on Guadalupe Island and thrives in cultivation in coastal southern California.

6. **P. muricata** D. Don. BISHOP PINE. Local distribution centered in western parts of San Luis Range, where notably plentiful in Coon Creek watershed, extending from its summit to the sea. The stand which is closest to a main travelled road is easily visible on the hillside above Sycamore Mineral Springs, near Avila. Locally the trees grow mainly on siliceous shale. A few small groves and isolated trees extend into the sandy country south of Morro Bay. To the east a few trees occur on a sandstone hill near the edge of San Luis Valley, a short distance east of the Carpenter Canyon road to Arroyo Grande. At this last locality the trunks divide into several ascending branches, as is usual in *P. Sabiniana* but not normal in our other pines. Bishop Pine has special local historical importance as having been discovered here by Dr. Thomas Coulter in 1831. Jepson in "The Silva of California" quotes Don, who named the species, as stating that it was "found by Dr. Coulter at San Luis Obispo . . . at an elevation of 3000 feet above the sea." Although the highest summits of the San Luis Range do not actually reach this altitude, there is no question but that Bishop Pine was first collected in this vicinity. The cones are variable. In at least some of the groves, a portion of the trees bear cones corresponding to f. *remorata* (Mason) Hoover, or intermediate between the "muricata" and "remorata" types. Both forms are true closed-cone pines to an even greater degree than *P. attenuata*. The cone-scales remain tightly pressed together until subjected to considerable heat, so that the seeds are normally released only during a fire.

7. **P. monophylla** Torr. & Frém. ONE-LEAF PIÑON. A solitary full-grown tree, discovered by Eben McMillan, grows on the north slope of Caliente Mt. near the highest peak, among small oaks (*8276*). This can be interpreted as the last vestige of a piñon-juniper-oak woodland, of which elsewhere only the junipers and oaks remain. On the opposite side of Cuyama Valley, in the mountains of northeastern

Santa Barbara Co. and toward Mt. Abel, the species is abundant. An exhaustive search all over Caliente Mt. might reveal additional trees, but even so it is very odd that such a short distance should involve such a conspicuous difference in the composition of the plant community, at the same altitude and under seemingly identical climatic conditions.

2. **Abies** Mill. Fir

1. **A. bracteata** (D. Don) D. Don. Bristle-Cone Fir. Santa Lucia Fir. Restricted to the Santa Lucia Mts., mainly in Monterey Co. but also present in the more remote mountains of northwestern San Luis Obispo Co. Because the trees are on property of the Hearst Corporation, exact information about their occurrence is difficult to obtain. Donald Hemphill, in a letter to P. A. Munz, has provided the following information. "By driving into the back country several miles inland from Hearst Castle, we did find a very unique grove of firs in association with Digger Pine and oak woodland. I think the elevation was about 1500 feet and certainly less than 2000 feet. One grove was on serpentine soil, and we noted there were some seedlings in the dense shade of the large trees coming up through the litter. . . . On the drainage of Arroyo de la Cruz they were downstream from Marmolejo Flat. One group of about 100 large trees may actually be less than 1500 feet in elevation." Another locality is documented by a specimen in the herbarium of the California Academy of Sciences labelled "upper Las Tablas Creek" (*Chester Dudley* in 1927). It is hoped that this latter record can be confirmed by more recent observation.

Taxodiaceae. Bald Cypress Family

1. **Sequoia** Endl.

1. **S. sempervirens** (D. Don) Endl. Redwood. The southernmost natural stand of redwood is in a steep coastal ravine a short distance north of Salmon Creek in Monterey Co. (not in the Salmon Creek canyon itself, as seems to be widely believed). The species is included here because of the occasional occurrence of seedling trees which were not planted. Around trees which were planted many years ago in San Carpoforo Creek Canyon near the Monterey Co. line, a few young trees have come up spontaneously. According to Robert J. Rodin, a single redwood grows in a steep canyon on the coast just south of the Monterey Co. line.

Cupressaceae. Cypress Family

Cones woody, with distinct peltate scales and many seeds. . . . . .1. *Cupressus.*
Cones becoming fleshy and berry-like, with one or few seeds . . . . .2. *Juniperus.*

1. **Cupressus** L. Cypress

Cones nearly always over 2.5 cm. in diameter . . . . .1. *C. macrocarpa.*
Cones 2 cm. or less in diameter . . . . .2. *C. Sargentii.*

1. **C. macrocarpa** Hartw. Monterey Cypress. Although native only on the coast of Monterey Co., this species has been very extensively planted. If seeds germinate near planted trees, the seedlings seldom survive the first summer. However, a grove north of San Carpoforo Creek (*8351*) is actually spreading, and thriving young trees from spontaneous seeding have been seen on the hill south of San Carpoforo Creek and elsewhere near the coast. At the south end of Morro Bay a few Monterey cypress trees have come up, apparently spontaneously, in the eucalyptus plantations.

2. **C. Sargentii** Jepson. SARGENT CYPRESS. Areas of serpentine rock along main summit of Santa Lucia Range, forming three extensive but well separated stands: northwest of Cuesta Pass; Cypress Mt.; reported by Dr. Carl B. Wolf (El Aliso 1: 231. 1948) from "northeast [actually northwest?] end of the Pine Mountain ridge on the slopes above Tobacco and Little Burnett Creeks."

2. **Juniperus** L. JUNIPER

1. **J. californica** Carr. CALIFORNIA JUNIPER. Widespread through the less extremely arid hills of the interior: eastern foothills of Santa Lucia Mts. from about Paso Robles northward; Cottonwood Pass; Cholame Hills; Shandon Hills; hilly country north and east of La Panza Range; higher parts of Temblor Range; Caliente Range, including low hills on north side of Cuyama River. The trees have been destroyed in some places where they formerly occurred, but the species still forms a characteristic woodland, mixed with small oaks or sometimes growing alone. The largest trees seen were growing on shale in a canyon near Mariannas Ranch in the Temblor Range.

Ephedraceae. EPHEDRA FAMILY

1. **Ephedra** L.

Mostly over 1 m. tall; branches dull green ....1. *E.californica.*
Mostly less than 1 m. tall; branches bright grass-green ....2. *E. viridis.*

1. **E. californica** Wats. DESERT TEA. Gravelley or rocky soils, dry hills on east side of Carrizo Plain from Panorama Hills southward, extending to Cuyama Valley, where now largely destroyed. In the southern Temblor Range, this is the most plentiful shrubby species in the plant community which I call "Desert Scrub." The branchlets when broken up were used by pioneers in making a beverage, and to a limited degree are even yet an article of commerce for this purpose.

2. **E. viridis** Cov. Caliente Mt., plentiful all along the summit ridge in clay, sand, or rocky soil; extending down into canyons on north side of lower Cuyama Valley. In the Temblor Range found in Cedar Canyon, Kern Co., but not yet detected in the San Luis Obispo Co. portions of that range.

ANGIOSPERMS

Typhaceae. CAT-TAIL FAMILY

1. **Typha** L. CAT-TAIL.

Pistillate portion of spike at least 2 cm. thick, the staminate portion contiguous above it ....1. *T. latifolia.*
Pistillate portion of spike about 1 cm. thick or less, the upper or staminate portion separated from it by a short interval of bare stem ....2. *T. angustifolia.*

1. **T. latifolia** L. CAT-TAIL. Common in wet places near coast; less common in Salinas Valley, as at Santa Margarita.

2. **T. angustifolia** L. Narrow-leaved cat-tails are of frequent occurrence, mainly in canyons of the hills rather than in the lowland marshes where *T. latifolia* abounds. Probably only one species is involved, but its identity is uncertain, both because of inadequate collections and because our plants do not consistently show the key-characters used by different authors to distinguish either *T. angustifolia* or *T. do-*

mingensis Pers. Twisselmann has used the name *T. domingensis* for plants of the Temblor Range. Because of the uncertainty of identification, it seems best for now to include our plants under the earlier published name, *T. angustifolia,* until a convincing reason appears for doing otherwise.

Sparganiaceae. Bur-Reed Family

1. Sparganium L. Bur-Reed

1. **S. eurycarpum** Engelm. *S. californicum* Greene. *S. Greenei* Morong. Occasional in coastal fresh-water marshes: Baywood Park; Edna; Small Twin Lake south of Arroyo Grande to Oso Flaco Lake. Also upper Salinas Valley near Santa Margarita.

Potamogetonaceae. Pondweed Family

Leaves opposite; flowers sessile or subsessile, whorled in the leaf-axils 1. *Zannichellia.*
Leaves alternate; flowers in spikes, heads, small umbels, or pairs, borne on peduncles.
  Fruits solitary or in pairs or apparent umbels; perianth absent . . . . .2. *Ruppia.*
  Fruits in spikes or heads; perianth present (though very inconspicuous)
                                                                3. *Potamogeton.*

1. Zannichellia L.

1. **Z. palustris** L. Horned Pondweed. In pools and slow streams in the interior: Cottonwood Pass; Twisselmann Ranch south of Cholame (in water-tank) ; upper Arroyo Grande.

2. Ruppia L.

1. **R. maritima** L. Submerged in brackish water near coast: estuary of Osos Creek; Little Pico Creek. Some of the plants occurring locally may be *R. spiralis* L., but at least the one specimen now available (San Simeon, *7406*) does not have the peduncles noticeably coiled in fruit.

3. Potamogeton L. Pondweed

Flowers in an elongate spike . . . . .1. *P. pectinatus.*
Flowers crowded in a very short head-like spike . . . . .2. *P. foliosus.*

1. **P. pectinatus** L. Fennel-Leaved Pondweed. Locally abundant in Oso Flaco Lake; probably elsewhere near coast.
2. **P. foliosus** Raf. Submerged in running water: San Luis Creek; Rinconada Creek; Alamo Creek near Cuyama River.

Zosteraceae. Eel-Grass Family

Plants of sheltered bays; leaves (in our form) about 5 to 10 mm. wide . . .1. *Zostera.*
Plants of exposed rocky surf; leaves 3 mm. wide or less . . . . .2. *Phyllospadix.*

1. Zostera L.

1. **Z. marina** L. Eel-Grass. In salt water, Morro Bay. All the plants found here are evidently var. *latifolia* Morong. The plants at high tide are often deposited on the salt-marshes, where the leaves after drying look like narrow strips of paper.

2. **Phyllospadix** Hook. Surf-Grass

Leaves about 1 to 1.5 mm. wide; flowering stems bearing usually more than 2 spikes
1. *P. Torreyi.*
Leaves about 2 to 3 mm. wide; flowering stems bearing 1 spike, or sometimes 2
2. *P. Scouleri.*

1. **P. Torreyi** Wats. On rocks in the surf, often in tide-pools. *Phyllospadix* is abundant on the rocky parts of our coast but is not often observed in flower or fruit. Where both species occur together, *P. Torreyi* can be readily distinguished at any stage by having narrower and (on the average) much longer leaves than *P. Scouleri*. Without doubt *P. Torreyi* is the more plentiful of the two along the San Luis Obispo Co. coast.

2. **P. Scouleri** Hook. Definitely identified as present at Leffingwell Landing near Cambria, growing mixed with *P. Torreyi.* The detailed local distribution of the two species remains yet to be worked out.

Najadaceae. Naiad Family

1. Najas L.

Leaves coarsely toothed .... 1. *N. marina.*
Leaves minutely serrulate, sometimes entire .... 2. *N. guadalupensis.*

1. **N. marina** L. Oso Flaco Lake, where locally abundant.

2. **N. guadalupensis** Morong. Submerged in fresh water: Atascadero Lake (*Nobs & Smith 870* in 1949).

Juncaginaceae. Arrow-Grass Family

Flowers all in racemes; carpels 6 or 3 .... 1. *Triglochin.*
Plants with sessile pistillate flowers at base as well as flowers in spikes; carpel 1
2. *Lilaea.*

1. Triglochin L.

Fruit 5 to 6 mm. long, much longer than thick; carpels 6 .... 1. *T. concinna.*
Fruit about 2 mm. long, about as thick as long; fertile carpels 3 .... 2. *T. striata.*

1. **T. concinna** Davy. Found locally in moist to comparatively dry saline soil on the mud-flats bordering Morro Bay (*6919*) and at Avila (*9015*).

2. **T. striata** R. & P. Edge of salt lagoon bordering beach west of Oceano. (*7558*).

2. Lilaea H. & B.

1. **L. scilloides** (Poir.) Haum. Occasional in low places which are wet in winter, both coastal and interior: Piedras Blancas Point (*6964*); Cambria (*6939*): 1 mile north of Creston (*8473*).

Alismataceae. Water-Plantain Family

1. Alisma L.

1. **A. triviale** Pursh. Pools along Salinas River at Atascadero; Osos Creek estuary; Laguna near San Luis Obispo. Recent authors agree in separating American plants, under the name *A. triviale,* from the European *A. Plantago-aquatica* L., but the differences are not exactly clear to me. If our plants are not native here, as could

well be the case, analogy with many other genera suggests that they may have been introduced from Europe as probably as from elsewhere in America.

Gramineae. GRASS FAMILY

KEY TO TRIBES

Spikelets with 2 or more well-developed florets, at least the lowest 2 florets fertile (spikelets dimorphic in *Lamarckia* and *Cynosurus,* some of them entirely sterile).
    Spikelets with pedicels (these very short in some genera).
        Second glume shorter than lowest lemma .... 1. *Festuceae.*
        Second glume at least as long as lowest lemma.
            Sheaths closed; glumes thin-papery .... *Festuceae (Melica).*
            Sheaths split down to base; glumes firm .... 3. *Aveneae.*
    Spikelets sessile, in a simple spike or in a more complex inflorescence in which the branches are spikes.
        Spikelets in a simple spike or sometimes in a narrowly branched spike, alternating on opposite sides of the rachis, thus showing an obviously 2-ranked arrangement .... 2. *Hordeae.*
        Spikelets in a widely branched panicle, not obviously 2-ranked
                                     5. *Chlorideae.*
Spikelets with 1 well-developed floret, often with additional reduced or rudimentary ones.
    Perfect floret without staminate or sterile florets below it, sometimes with 1 or more vestigial florets above it (see also *Phalaris paradoxa*).
        Spikelets in simple or digitate spikes.
            Spikelets in simple spikes .... 2. *Hordeae.*
            Spikelets in digitate spikes .... 5. *Chlorideae.*
        Spikelets pedicelled, in a panicle.
            Fertile spikelets 1-flowered, surrounded by sterile spikelets containing many empty lemmas .... *Festuceae (Lamarckia).*
            Spikelets all alike.
                Spikelets without rudimentary lemmas .... 4. *Agrostideae.*
                Spikelets with a rudiment consisting of reduced sterile lemmas above fertile floret .... *Festuceae (Melica imperfecta).*
    Perfect floret with 1 or 2 staminate florets or empty (sometimes vestigial) lemmas below it.
        Glumes persistent after fall of florets, folded along midrib and laterally compressed .... 6. *Phalarideae.*
        Spikelets falling entire; glumes rounded on back or dorsally compressed.
            Fertile lemma firmer in texture than glumes or sterile lemma
                                     7. *Paniceae.*
        Glumes of fertile spikelet firmer in texture than fertile lemma
                                 8. *Andropogoneae.*

Tribe 1. Festuceae

Plants rarely over 1 m. tall; stem not over 5 mm. thick.
    Spikelets all alike on same plant.
        Plants with fragile rhizomes or (usually) none, not dioecious; vegetative shoots not showing a conspicuous 2-ranked arrangement of leaves.
            Lemmas with 5 or more nerves, or not obviously nerved.
                Florets appressed to rachilla; lemmas longer than broad.
                Spikelets in panicles.

Spikelets not in 1-sided clusters.
Leaf-sheaths split down one side to base; glumes firm, like lemmas in texture; lemmas 5-nerved.
Lemmas awned from a bifid or obtuse apex

1. *Bromus.*

Lemmas awnless, or tapering into an awn at tip (apex minutely bifid in *Festuca Elmeri*).
Lemmas usually awned, or at least acuminate ....2. *Festuca.*
Lemmas neither awned nor acuminate.
Pedicels slender ....3. *Poa.*
Pedicels short and thick. . 4. *Scleropoa.*
Leaf-sheaths closed; glumes thin-papery; lemmas usually 7-nerved ....5. *Melica.*
Spikelets in 1-sided clusters at end of panicle-branches

6. *Dactylis.*

Spikelets in racemes.
Pedicels stout, 0.5 to 1 mm. long ......7. *Brachypodium.*
Pedicels slender, over 1 mm. long ......8. *Pleuropogon.*
Florets widely divergent from rachilla; lemmas as broad as long

9. *Briza.*

Lemmas prominently 3-nerved ....10. *Eragrostis.*
Plants with tough rhizomes, dioecious; vegetative shoots with many conspicuously 2-ranked leaves ....11. *Distichlis.*
Spikelets dimorphic, sterile and fertile in the same inflorescence.
Fertile spikelets with 2 or 3 flowers; panicle green ....12. *Cynosurus.*
Fertile spikelets 1-flowered; panicle yellowish or pale purplish

13. *Lamarckia.*

Plants mostly over 2 m. tall; stems over 5 mm. thick in lower part.
Plant with creeping rhizomes, forming extensive thickets; leaves evenly distributed along stem ....14. *Phragmites.*
Plant forming very large clumps of long rigidly serrulate leaves; leaves on upper part of stem few and reduced ....15. *Cortaderia.*

Tribe 2. **Hordeae**

Spikelets with 2 or more flowers.
Spikelets with 2 glumes.
Glumes narrowly oblong or linear-lanceolate to subulate or long-bristly.
Spikelets erect or slightly spreading, not markedly compressed

16. *Elymus.*

Spikelets crowded, divergent, strongly compressed. ....17. *Agropyron.*
Glumes ovate ....18. *Triticum.*
Spikelets, except terminal one, with 1 glume ....19. *Lolium.*
Spikelets 1-flowered.
Spikelets in groups of 3 at each node of rachis, the 2 lateral spikelets in most species reduced and sterile ....20. *Hordeum.*
Spikelets solitary at each node of rachis (*Scribneria* rarely has branched spikes).
Glumes on sides of floret; lemma awned ....21. *Scribneria.*
Glume or glumes in front of floret; lemmas not awned.
Glume 1, except on terminal spikelet ....22. *Monerma.*
Glumes 2 ....23. *Parapholis.*

Tribe 3. **Aveneae**

Spikelets more than 1 cm. long.
    Annuals; spikelets many, pendent . . . . .24. *Avena.*
    Perennials; spikelets 5 or fewer, erect or spreading . . . . .25. *Danthonia.*
Spikelets less than 1 cm. long.
    Leaves glabrous to (rarely) short-pilose; flowers 2 or more per spikelet, at least
    the lower 2 both perfect; awns if present straight or bent and twisted.
        Lemmas awnless or awned from a bifid apex.
            Second glume equalling or shorter than lemma of second floret
                        26. *Koeleria.*
            Glumes distinctly longer than second floret . . . . .27. *Schismus.*
        Lemmas awned from back below apex.
            Lemmas rounded or truncate and erose-dentate at apex
                        28. *Deschampsia.*
            Lemmas with 2 sharply pointed or short-awned teeth at apex.
                Small delicate annuals; spikelets less than 5 mm. long; awn aris-
                ing from below middle of lemma . . . . .29. *Aira.*
                Perennial at least 3 dm. tall; spikelets over 5 mm. long; awn aris-
                ing from upper part of lemma .30. *Trisetum.*
    Leaves velvety; spikelets 2-flowered, the lower floret perfect, awnless, the upper
    staminate, the lemma bearing a short hooked awn on back . . . . .31. *Holcus.*

Tribe 4. **Agrostideae**

Lemma shorter than glumes (except for its awn, if any).
    Glumes persistent, the fruiting lemmas falling from them.
        Lemma soft, not closely enveloping the grain, awnless or bearing a dorsal
        awn.
            Glumes not ventricose at base; our species perennial.
            Spikelets less than 6 mm. long; blades rather short and soft.
                Lemma glabrous or with a very short inconspicuous tuft of
                hairs at base . . . . .32. *Agrostis.*
                Lemma with a tuft of hairs at least 1 mm. long at base.
                    Lemmas awnless . . . .*Agrostis Hallii.*
                    Lemmas awned . . . . .33. *Calamagrostis.*.
            Spikelets 8 to 10 mm. long; blades very long and firm
                        34. *Ammophila.*
             Glumes slightly ventricose at base; annual . . . . .35. *Gastridium.*
         Lemma firm, closely adherent to the grain and not readily separable from
        it, bearing a terminal awn.
            Awn single . . . . .36. *Stipa.*
            Awn 3-forked from base . . . . .37. *Aristida.*
    Spikelets falling entire, the pedicels disarticulating just below them
                        38. *Polypogon.*
Lemma longer than glumes . . . . .39. *Muhlenbergia.*

Tribe 5. **Chlorideae**

Spikelets several-flowered, in panicles . . . . .40. *Leptochloa.*
Spikelets 1-flowered, in digitate spikes.
    Tufted annual; spikes appearing feathery from long hairs on lemmas
                        41. *Chloris.*
    Sod-forming perennial with tough rhizomes; spikes not feathery . .42. *Cynodon.*

Tribe 6. **Phalarideae**

Lemmas of 2 lower florets as large as that of the terminal perfect floret.
Lower 2 florets staminate ....43. *Hierochloe.*
Lower 2 florets represented by empty lemmas ....44. *Ehrharta.*
Lower 2 florets reduced to narrow sterile lemmas not more than half as long as fertile lemma, sometimes 1 or none ....45. *Phalaris.*

Tribe 7. **Paniceae**

Spikelets many, borne above the leaves in various inflorescences.
Spikelets imbedded in one side of a flattened rachis ....46. *Stenotaphrum.*
Spikelets not imbedded in rachis.
Spikelets on one side of branches of inflorescence.
Branches of inflorescence digitate or nearly so ....47. *Digitaria.*
Branches of inflorescence geminate or racemosely arranged
48. *Paspalum.*
Spikelets in a panicle (often a narrow spike-like one), its branches not obviously one-sided.
Spikelets not surrounded or subtended by bristles; panicle with divergent branches.
Spikelets on slender pedicels in a loose panicle .....49. *Panicum.*
Spikelets subsessile, crowded on the short panicle-branches
50. *Echinochloa.*
Spikelets surrounded or subtended by bristles; panicle narrow and spike-like.
Bristles persistent, the spikelets deciduous ....51. *Setaria.*
Bristles falling with the spikelets ....52. *Pennisetum.*
Spikelets few, hidden in upper leaf-sheaths ....*Pennisetum clandestinum.*

Tribe 8. **Andropogoneae**

Only one genus represented ....53. *Sorghum.*

1. **Bromus L.**

Lemmas strongly keeled; spikelets laterally compressed.
Lemmas sharp-pointed or with awn less than 4 mm. long.
Lemmas 12 to 17 mm. long, with 11 or 13 nerves ......1. *B. Willdenowii.*
Lemmas 10 to 12 mm. long, mostly 9-nerved ....2. *B. unioloides.*
Lemmas with awn over 4 mm. long.
Panicle-branches short, erect or ascending; awns mostly less than 1 cm. long. Tufted perennials.
Stems erect; leaves covered with long soft hairs
3. *B. breviaristatus.*
Stems widely spreading or prostrate; leaves glabrous or nearly so
4. *B. maritimus.*
Annual; stems 1 or few, scarcely tufted ....5. *B. arizonicus.*
Panicle-branches widely spreading; awns mostly 1 cm. long or more
6. *B. carinatus.*
Lemmas rounded on back; spikelets not compressed or only slightly so.
Perennials.
Panicle-branches short, showing little or no tendency to droop: pedicels straight, stiff ....7. *B. Orcuttianus.*
Panicle-branches elongate, drooping; pedicels flexuous.

    Lemmas rather evenly pubescent over the back.
     Blades pilose . . . . .8. *B. grandis.*
     Blades glabrous or merely scabrous . . . . .9. *B. frondosus.*
    Lemmas markedly hairy near margins and base, otherwise less pubescent or subglabrous.
     Ligule of upper leaves 2 to 4 mm. long . . . . .10. *B. laevipes.*
     Ligule of upper leaves not over 1 mm. long . .11. *B. pseudolaevipes.*
  Annuals.
    Lemma acute or rounded toward apex, not acuminate, the apical teeth hardly 1 mm. long.
     Spikelets erect or ascending.
      Lemmas pubescent; panicle usually dense, with spikelets much overlapping.
       Spikelets hardly compressed . . . . .12. *B. mollis.*
       Spikelets somewhat compressed . . . . .13. *B. molliformis.*
      Lemmas glabrous; panicle loose . . . . .14. *B. commutatus.*
     Spikelets drooping, on long capillary pedicels . . . . . .15. *B. arenarius.*
    Lemma gradually acuminate toward apex, the apical teeth 2 to 5 mm. long.
     Awns straight or slightly flexuous.
      Spikelets drooping; lemmas usually soft-pubescent 16. *B. tectorum.*
      Spikelets spreading to erect; lemmas scabrous but scarcely pubescent.
       Some or all of pedicels over 5 mm. long, the panicle open or less compact; spikelets mostly more or less spreading.
        Lemmas 2.5 to 3 cm. long, the awn 3.5 to 5 cm. long
               17. *B. diandrus.*
        Lemmas 14 to 20 mm. long, the awn 16 to 30 mm. long.
         First glume 7 to 9 mm. long; second glume 11 to 13 mm. long; panicles green to slightly purple-tinged
              18. *B. sterilis.*
         First glume 9 to 12 mm. long; second glume 14 to 16 mm. long; panicles usually dark reddish purple
              19. *B. madritensis.*
       Pedicels rarely over 2 mm., and never over 4 mm. long, the panicle compact; spikelets crowded, mostly erect 20. *B. rubens.*
    Awns twisted below and geniculate . . . . .21. *B. Trinii.*

   1. **B. Willdenowii** Kunth. Rescue Grass. Cultivated for forage and occasionally becoming a roadside and garden weed in western part: Cambria; San Luis Obispo; Santa Maria Valley. Prior to 1960, all references listed this species as *B. catharticus* Vahl.

   2. **B. unioloides** H. B. K. *B. Haenkeanus* (Presl) Kunth. Adventive at Twisselmann Ranch (*Twisselmann 1177*). Specimens listed under this name and the preceding are distinguished on the basis stated by Raven (Brittonia 12: 219–221. 1960).

   3. **B. breviaristatus** Buckl. Occasional, sometimes locally plentiful, in open woods or on open north slopes: Shandon Hills and Santa Lucia Mts. The species is here defined on the basis of its densely white-hairy leaves and short appressed panicle-branches. The name *B. marginatus* Nees has been used by some botanists for the species here called *B. breviaristatus,* and by others for plants which differ in no evident way from *B. carinatus.*

   4. **B. maritimus** (Piper) Hitchc. Common on ocean bluffs, sand-dunes, and open hills along coast south of Islay Creek, and from near Cambria northward. These plants are not typical *B. maritimus,* to judge from descriptions of that species, and

perhaps should be called by some other name. Characteristically in our plants the stems radiate out from a central tuft of narrow leaves, and the panicle-branches are short and appressed.

5. **B. arizonicus** (Shear) Stebbins. Occasional in more arid portions of interior, usually in sandy soil.

6. **B. carinatus** H. & A. Common throughout western part in woods, sparsely brushy areas, or on open uncultivated hills in loose soil. I have been unable to detect *B. marginatus* Nees as a distinguishable entity in this area, unless indeed it and *B. breviaristatus* are the same (see note under that species). Because of the confusion which surrounds the application of the name *B. marginatus*, I intentionally exclude it from our flora.

7. **B. Orcuttianus** Vasey. Open woods, Santa Lucia Mts.: near Rocky Butte Fire Lookout (*9056*); Chris Flood Creek (*9001*). In these specimens the panicle shows little or no tendency to droop. However, in our area the names *B. Orcuttianus, grandis, frondosus, laevipes,* and *pseudolaevipes* represent specimens which can be distinguished only by the use of arbitrarily selected "key-characters," rather than natural populations which are geographically or ecologically distinct. According to Wagnon (Brittonia 7: 415–480. 1952), all of these "species" have the same chromosome number. These facts support a suspicion that previous authors have attempted to draw too fine a line of distinction among species in this group.

8. **B. grandis** (Shear) Hitchc. Open woods and partly shaded banks, fairly common in Santa Lucia Mts. and San Luis Range.

9. **B. frondosus** (Shear) Woot. & Standl. Open woods, Santa Lucia Mts.: 6 miles west of Templeton on road to Cambria (*7987*). The key and descriptions in Hitchcock's "Manual of the Grasses of the United States," ed. 2, do not permit these plants to be referred to any other species, despite their being outside the main area of *B. frondosus* (Colorado, Utah, Arizona, and New Mexico). Although this species is given full status in Wagnon's revision of the group (Brittonia 7: 145–480. 1952), Wagnon unaccountably omitted it from his key to the species.

10. **B. laevipes** Shear. Open woods and partly shaded banks, Santa Lucia Mts. and San Luis Range.

11. **B. pseudolaevipes** Wagnon. To the extent that this name represents a distinguishable group of plants, it seems to belong more to the semi-arid interior than do its close relatives. One collection agrees well with Wagnon's original description and is cited by him: Pilitas Creek district, on shaded slope under oaks (*7197*). Other available material seems to correspond as closely to descriptions of *B. laevipes, B. grandis,* or *B. frondosus* as to typical *B. pseudolaevipes*. To this species may belong a coastal collection which seems identical with *B. laevipes* except that it has a very short ligule: See Canyon, San Luis Range (*8861*). The short ligule, whatever may be its theoretical taxonomic importance, does not seem (in San Luis Obispo County, at least) to mark a group of plants which is morphologically coherent otherwise, or which grows in a separate area from long-liguled plants or in a different environment.

12. **B. mollis** L. SOFT CHESS. Abundant in open places, especially disturbed ground, both coastal and interior (but generally absent from rocky hills, chaparral, and dense woods). Most, if not all, of the plants here called by this name have twisted, somewhat divergent awns and in that respect correspond to *B. scoparius* L. That species, however, is described as having glabrous lemmas, unlike our plants.

13. **B. molliformis** Lloyd? Garden weed at San Luis Obispo (*8711*). This collec-

tion corresponds to published descriptions of *B. molliformis* and is a close match for herbarium specimens so labelled, but on the other hand it has all the appearance of immature, smaller-than-average plants of *B. mollis*. Whether *B. molliformis* can be distinguished at all from *B. mollis* is another unanswered question, apart from the identity of this one collection.

14. **B. commutatus** Schrad. East end of Los Osos Valley, in moist meadow, clay soil (*9024*).

15. **B. arenarius** Labill. Common in sandy soil, often in decomposed granite, from Salinas River eastward through La Panza Range and to Cholame Hills.

16. **B. tectorum** L. Occasional in sandy or clay soils of the interior: Middle Branch of Huerhuero Creek; Twisselmann Ranch.

17. **B. diandrus** Roth. *B. rigidus* Roth. Ripgut Grass. Abundant everywhere in sandy soils or disturbed places (roadsides, gardens, etc.). In a sandy area south and west of Nipomo, it has formed a dense stand in which no native plant could grow even if any seeds yet survive.

18. **B. sterilis** L. A garden weed, apparently not yet very plentiful, at San Luis Obispo; Los Berros Canyon, and doubtless elsewhere.

19. **B. madritensis** L. Recently disturbed ground, mostly on road-banks, Santa Lucia Mts.: Rocky Butte Fire Lookout; south fork of Old Creek; Tassajera Peak. Apparently uncommon, but more probably overlooked because of its resemblance to *B. rubens*.

20. **B. rubens** L. Red Brome. Very abundant in a wide variety of environments throughout the county. No other introduced species has established itself so completely in so many different places. Serpentine rock seems to inhibit its growth, however, and it is also absent from shifting sand-dunes, salt or fresh-water marshes, and deeply shaded sites.

21. **B. Trinii** Desv. Occasional in eastern part of county, usually in sandy soil: Cholame Hills to Temblor Range and Caliente Mt. Also in barren serpentine area, East Fork of Corral de Piedra Creek near San Luis Obispo.

<h3 align="center">2. Festuca L. Fescue</h3>

Perennials.
    Blades flat, mostly more than 3 mm. wide.
        Panicle-branches at maturity widely spreading and flexuous; awns of lemmas mostly over 3 mm. long . . . . 1. *F. Elmeri.*
        Panicle-branches short and ascending; awns 3 mm. long or less, or absent
                                  2. *F. elatior.*
    Blades either becoming involute or very narrow from the first.
        Lemmas with awn much shorter than the body, or merely acuminate.
            Densely tufted; stems not decumbent at base; basal sheaths not reddish
                               3. *F. californica.*
            Loosely tufted; stems decumbent at base; basal sheaths reddish-tinged
                               4. *F. rubra.*
        Lemmas with awn at least as long as the body . . . . 5. *F. occidentalis.*
Annuals.
    Panicle narrow, its branches strictly erect or but slightly spreading.
        Awns rarely over 4 mm. long; florets mostly 5 or more . . . . . . 6. *F. octoflora.*
            Lemmas glabrous . . . . var. *octoflora.*
            Lemmas appressed-pubescent . . . . var. *hirtella.*
        Awns 8 to 15 mm. long; florets 6 or (mostly) fewer.

> Lemmas not ciliate.
>> First glume 4 mm. long; second glume 6 to 7 mm. long
>>> 7. *F. dertonensis.*
>> First glume 1 to 1.5 mm. long; second glume 4 to 4.5 mm. long
>>> 8. *F. Myuros.*
> Lemmas, except the lowest one in each spikelet, long-ciliate
>> 9. *F. megalura.*
Branches of panicle widely divergent at maturity ....10. *F. microstachys.*
> Spikelets appressed to panicle-branches or slightly spreading ...var. *ciliata.*
> Spikelets at maturity widely divergent or reflexed.
>> Florets 1 to 3 ....var. *microstachys.*
>> Florets 3 to 6 ....var. *simulans.*

1. **F. Elmeri** Scribn. & Merr. Common on wooded slopes in San Luis Range: See Canyon (*7511, 8860*), perhaps the southernmost locality for the species. Santa Lucia Mts.: South Fork of Old Creek; forks of San Simeon Creek, and undoubtedly more widespread. Rare east of Salinas River: 2 miles east of Templeton (*7717*).

2. **F. elatior** L. Meadow Fescue. Sown for pasture and often becoming established in low moist places: Los Osos Valley (*9023*). Also found in dry woods near Rocky Butte Fire Lookout (*9055*), where perhaps sown after fire. The collections cited here differ from published descriptions of *F. elatior* in having awns up to 3 mm. long on the lemmas, but available references include no other species which they resemble more closely.

3. **F. californica** Vasey. Frequent in wooded or rocky canyons in San Luis Range and in Santa Lucia Mts. Around San Luis Obispo it usually is found in areas of serpentine (Reservoir Canyon; Steiner Creek; Perfumo Canyon), but elsewhere grows on other substrata. Most of our plants differ from descriptions of *F. californica* in having the collars of the sheaths not "villous" but only inconspicuously pubescent or nearly glabrous. These San Luis Obispo Co. plants key out in current references to *F. idahoensis* Elmer, which differs from them in having filiform blades, a narrow panicle, and much shorter glumes. They have entirely the aspect of typical *F. californica*.

4. **F. rubra** L. Red Fescue. Campus of California Polytechnic College at San Luis Obispo (*Robert Page*), where probably introduced as a component of a lawn or pasture seed mixture. Apparently native along the coast northward, on open hills or among low shrubs: near Piedras Blancas Point (*7662*); Arroyo de la Cruz (*8058*).

5. **F. occidentalis** Hook. Cambria, on shaded slope in Monterey pine forest (*Howell 40,844*).

6. **F. octoflora** Walt. var. **octoflora.** Widespread in sandy or loose soil, usually among chaparral, especially in burned areas.

Var. **hirtella** Piper. At scattered localities (e.g., south of Price Canyon, *6753*), usually not growing with the typical form of the species.

7. **F. dertonensis** (All.) Asch. & Graebn. Common in western part, mostly in or near cultivated ground and by roadsides.

8. **F. Myuros** L. Apparently rarely established, but perhaps overlooked for *F. megalura,* a presumably native species which resembles it very closely.

9. **F. megalura** Nutt. Common in both coastal and interior regions, often very plentiful in cultivated ground, by roadsides, and in sandy pastures. One of the very few native species which have been favored rather than harmed by agricultural activities.

10. **F. microstachys** Nutt. var. **microstachys.** Rather common in western and central portions of county in places where not crowded out by ranker-growing plants, such as coarse sandy soils and crumbling serpentine or shale. Also in Temblor Range and near east end of Caliente Range. To var. *microstachys* I refer all forms of the species with few-flowered spikelets which are divergent or reflexed. The commonest form has glabrous spikelets (the so-called *F. reflexa* Buckley). Plants with hairy lemmas have been called "true" *F. microstachys*; with hairy glumes, *F. Tracyi*; and with hairs on both glumes and lemmas, *F. Eastwoodiae.* These various forms often grow in mixed stands. A collection from 9 miles east of Creston (*8097*) consists mainly of the "*F. Tracyi*" form, but mixed in the same stand were plants having glabrous spikelets (hence *F. reflexa*) but otherwise identical. Conversely, a collection from the San Juan River south of Shandon (*7473*) consists mostly of plants with glabrous spikelets, with a slight admixture of plants with sparsely long-hairy glumes.

Var. **ciliata** Gray. Here I include the group of forms in which only the main panicle-branches are divergent. There is complete intergradation with var. *microstachys,* but often var. *ciliata* grows where the other is absent. Exactly parallel variations exist in the pubescence of the spikelets: the form called *F. pacifica* Piper has glabrous spikelets; *F. confusa,* hairy glumes with glabrous lemmas; and *F. Grayi,* hairy lemmas. A collection from the road to Hi Mountain (*8783*) represents an unmixed stand of the "*F. confusa*" form. One from Fernandez Creek (*7483*) is mainly the "*F. pacifica*" form, but some of the plants have scabrous-hirtellous lemmas, thus approaching "*F. Grayi.*" It is notable, however, that some of the lemmas on these same plants are glabrous.

Var. **simulans** (Hoover) Hoover. Loose or sandy soils in eastern part: San Juan River valley and Carrizo Plain and bordering hills. This variety, which is like var. *microstachys* in having divergent or reflexed spikelets but has more numerous flowers in the spikelet (up to 6), is the only one of the components of the species which has much geographical significance.

3. **Poa** L. Bluegrass

Most of spikelets replaced by dark purple bulblets . . . . .1. *P. bulbosa.*
Plants without bulblets.
    Lemmas with long cobwebby hairs at base.
        Tufted annual or perennial, without rhizomes . . . . .2. *P. Howellii.*
        Plants not tufted, spreading by creeping rhizomes . . . . .3. *P. pratensis.*
    Lemmas more or less puberulent, without cobwebby hairs.
        Weedy annual, mostly in towns and in cultivated ground; usually less than
        2 dm. tall . . . . .4. *P. annua.*
        Perennial bunchgrass, mostly 3 dm. tall or more . . . . .5. *P. scabrella.*

1. **P. bulbosa** L. Sparingly established in an ungrazed pasture at Twisselmann Ranch near north end of Temblor Range (*Twisselmann 6942*).

2. **P. Howellii** Vasey & Scribner. Very common in hilly country, usually in shade but sometimes in partly sunny exposures, from coast eastward to the hills south of Pozo. Not yet detected in La Panza Range.

Most plants of *Poa Howellii* are annual as described. However, a collection from Lopez Canyon (*8288*) is clearly perennial, showing in spring the dead bases of the preceding year's stems. In addition, what seems to be either an unnamed species or an aberrant form of *P. Howellii* has been found repeatedly along the ridge southeast of Cuesta Pass (*7669, 7923, 8817, 8917*), growing in disintegrated shale. Not only are these last mentioned plants evidently perennial, but also the panicle is narrow with erect branches. The spikelets are in all details like those of typical *P. Howellii.*

3. **P. pratensis** L. Kentucky Bluegrass. Occasional in moist places near coast and in Santa Lucia Mts.: Oak Park district north of Arroyo Grande; Santa Rita Creek; between Rocky Butte and Pine Mt. These occurrences may represent escapes from cultivation, as the species is commonly planted for lawns, although it is generally supposed to be native in North America as well as in Europe.

4. **P. annua** L. Annual Bluegrass. Very common in temporarily moist places, especially in lawns and gardens.

5. **P. scabrella** (Thurb.) Benth. Common in uncultivated ground on open hills and plains or in open woods throughout the county. Highly variable: the most vigorous plants with many-flowered spikelets grow in gypseous clay soils of the interior, whereas plants with short leaves, narrow panicles, and few-flowered spikelets are found on rocky hills. Formerly the species was undoubtedly more plentiful than now, because it is often drastically grazed before seeds can ripen.

4. Scleropoa Griseb.

1. **S. rigida** (L.) Griseb. Chorro Creek, *Eastwood & Howell 2221* in 1936; not since detected.

5. Melica L.

Spikelets mostly with 3 to 5 perfect florets.
    Glumes much shorter than spikelet; lemmas narrowed toward apex or short-awned.
        Lowest internodes not swollen; panicle-branches erect; lemmas pubescent on lower part ....1. *M. Harfordii.*
        Lowest internodes swollen; panicle-branches spreading; lemmas glabrous or nearly so ....2. *M. Geyeri.*
    Glumes nearly as long as spikelet; lemmas broadly rounded at apex; lowest internodes slightly swollen ....3. *M. californica.*
Spikelets with 1 or 2 florets in addition to the rudiment.
    Second glume 4 to 6 mm. long, equalling or exceeding the lemma; rudiment obovoid, 1 mm. long, on a stalk 2 mm. long ....4. *M. Torreyana.*
    Second glume 3 to 4 mm. long, distinctly shorter than lemma; rudiment oblong, 2 mm. long, on a stalk 0.5 mm. long ....5. *M. imperfecta.*
        Panicle-branches erect or closely ascending ....var. *imperfecta.*
        Panicle-branches widely spreading ....var. *flexuosa.*

1. **M. Harfordii** Bol. Occasional in dry woods, Santa Lucia Mts.: North Fork San Simeon Creek to Rocky Butte (*7674*); Pine Top Mt. (*8885*). The plants were more plentiful and more vigorous following a fire. Our plants are here classified as belonging to var. *Harfordii*, although, because of having lemma awns up to 4 mm. long, they could be interpreted as intermediate toward var. *aristata* (Thurb. ex Bol.) Hoover.

2. **M. Geyeri** Munro. Shady woods in Santa Lucia Mts.: Pine Mt. (*7901*), the southernmost locality known for the species and the only one in the county.

3. **M. californica** Scribn. Hilly country in open woods or treeless areas back from coast, extending eastward even to Temblor Range.

4. M. Torreyana Scribn. Frequent in rocky, wooded, or brushy areas in Santa Lucia Mts., plentiful in areas of serpentine rock and scarce elsewhere. Also on serpentine in San Luis Range. Of the collections seen from the county, only one is strictly "typical" *M. Torreyana* with appressed-hairy lemmas: road to Hi Mt. from Pozo-Arroyo Grande summit (*8778*). This, incidentally, may be the southernmost locality for the species. The rest of the plants have the lemmas scaberulous to glabrous. A report by Munz of *M. frutescens* in San Luis Obispo Co. is evidently based on an immature collection of this form with glabrous lemmas (*Rose 33,114*).

5. M. imperfecta Trin. var. **imperfecta.** Very common in the hills from coast eastward to La Panza Range, in both shaded and sunny sites; rare in Temblor Range.

Var. **flexuosa** Bol. Shady woods, Santa Lucia Mts. A variety of little geographic significance, perhaps environmentally rather than genetically induced.

6. Dactylis L.

1. D. glomerata L. Orchard Grass. Occasionally found in moist soil, as at San Luis Obispo, but perhaps not permanently established here.

7. Brachypodium Beauv.

1. B. distachyon (L.) Beauv. Weed in gardens and along streets, becoming well established in recent years at San Luis Obispo and extending north to Cambria.

8. Pleuropogon R. Br.

1. P. californicus (Nees) Benth. Winter-wet depressions near northwest end of Laguna, San Luis Obispo (*7917* in 1950, *8880* in 1964).

9. Briza L. Quaking Grass

Spikelets 12 to 22 mm. long, 10 to 14 mm. wide ....1. *B. maxima.*
Spikelets about 4 mm. long and as wide ....2. *B. minor.*

1. B. maxima L. Quaking Grass. Common near coast from vicinity of Cambria northward, both in woods and in the open, mostly in sandy soils which are moist in winter.

2. B. minor L. Small Quaking Grass. Frequent in coastal region, mostly in sandy but firm soils, from hills on west side of San Luis Valley northward.

10. Eragrostis Beauv. Lovegrass

Spikelets 2.5 to 3 mm. wide ....1. *E. cilianensis.*
Spikelets 1 to 1.5 mm. wide.
    Blades mostly 4 to 8 mm. wide ....2. *E. Orcuttiana.*
    Blades 2 mm. wide or less ....3. *E. diffusa.*

1. E. cilianensis (All.) Lutati. Stink-Grass. Twisselmann Ranch, "in occasionally watered adobe soil of alfalfa pasture," *Twisselmann 1672*; Salinas River at Paso Robles.

2. E. Orcuttiana Vasey. Weed in cultivated ground near Creston and in Cuyama Valley.

3. E. diffusa Buckl. Widespread, but not a common weed, in places which are seasonally flooded or artificially watered: Twisselmann Ranch; Salinas River bridge east of Santa Margarita; San Luis Obispo. Hitchcock in Jepson, Fl. Calif. 1: 143, gives *"E. pilosa"* as occurring in San Luis Obispo Co.; but, since the name *E. diffusa* was not mentioned in that reference, probably the same species was meant as is called *E. diffusa* in more recent references.

11. **Distichlis** Raf.

1. **D. spicata** (L.) Greene. Salt-Grass. Abundant in moist saline soil on bluffs all along coast, in salt-marshes around Morro Bay, at Avila, and bordering dunes south of Pismo Beach. Occasionally also in quite dry sand, or in moist spots which are not obviously saline. Also in interior from Cholame Valley to Cuyama Valley. The forms of salt-grass in San Luis Obispo Co. do not fit well into the varieties distinguished by A. A. Beetle (Bull. Torr. Club 70: 638–650. 1943). Plants in the salt-marsh at Avila have erect stems, as in typical *D. spicata* of the Atlantic Coast, combined with the short ascending leaves of var. *nana* Beetle. On sand-dunes at the mouth of Morro Creek the plants form wide-spreading stolons (hence presumably var. *stolonifera* Beetle) but have the short divaricate leaves of var. *divaricata* Beetle. Plants corresponding to var. *divaricata* and to var. *nana* occur in mixed stands on Carrizo Plain. The variety *stricta* (Torr.) Beetle, defined as having longer leaves, has been found in Cuyama Valley.

12. **Cynosurus** L.

1. **C. echinatus** L. Open woods, Santa Lucia Mts.: near Rocky Butte Fire Lookout (*9054*). Probably introduced in a seed mixture sown after fire.

13. **Lamarckia** Moench

1. **L. aurea** (L.) Moench. Golden-Top. Frequent on rocky hills from coast to summits of Santa Lucia Range.

14. **Phragmites** Trin. Reed

1. **P. communis** Trin. Common Reed. Forming extensive "brakes" along the river in Cuyama Valley.

15. **Cortaderia** Stapf

1. **C. Selloana** (Schult.) Asch. & Graebn. Pampas Grass. Planted for ornament; escaping in moist places from Price Canyon to vicinity of Arroyo Grande and occasionally elsewhere near coast.

16. **Elymus** L.

Glumes straight, subulate to narrowly lanceolate or oblong.
    Plants with well developed rhizomes.
        Glumes subulate or narrowly linear-acuminate.
            Rhizomes short, thick, scaly; blades mostly over 15 mm. wide; spikes branched, over 30 cm. long .... 1. *E. condensatus.*
            Rhizomes slender, extensively creeping, not scaly; blades 10 mm. wide or less; spikes usually simple, sometimes branched, not over 25 cm. long.
                Stems 30 cm. tall or more, not overtopped by the leaves; spike 6 cm. long or more .... 2. *E. triticoides.*
                    Spike simple .... var. *triticoides.*
                    Spike branched .... var. *multiflorus.*

> Stems less than 20 cm. tall, mostly overtopped by the leaves;
>   spike 3 to 5 cm. long ....3. *E. pacificus.*
> Glumes oblong to lanceolate.
>   Spikes not disjointing; spikelets mostly in pairs, hairy; glumes obscurely
>   nerved ....4. *E. mollis.*
>   Spikes readily disjointing at maturity; spikelets solitary at the nodes,
>   glabrous; glumes prominently several-nerved ....5. *E. multinodus.*
> Plants without rhizomes (sometimes rooting at a decumbent base).
>   Glumes and lemmas truncate or rounded at apex *Agropyron intermedium.*
>   Glumes and lemmas acuminate or awned.
>   Spikelets mostly solitary at the nodes (paired at less than half of the
>   nodes on the most robust culms).
>   Spikes 10 cm. long or more, nearly always with 10 or more joints;
>   lower internodes of spike more than 10 mm. long.
>   Lemmas merely acuminate, or tipped with an awn hardly 1
>   mm. long; blades involute, less than 2 mm. wide
>   6. *E. trachycaulus.*
>   Lemmas with an awn at least 3 mm. long; blades mostly flat,
>   the larger ones over 3 mm. wide ....7. *E. laevis.*
>   Spikes up to about 8 cm. long, with 10 or fewer joints; lower in-
>   ternodes of spike less than 10 mm. long ....8. *E. subsecundus*
>   Spikelets in pairs (except sometimes at a few of the uppermost nodes
>   or in drought-dwarfed plants).
>   Lemmas with awn at least as long as the body .....9. *E. glaucus.*
>   Lemmas awnless, or with an awn up to about 2 mm. long
>   10. *E. virescens.*
> Glumes having the form of long divergent bristly awns, or branching into such awns.
>   Annual; spike not disjointing at maturity ....11. *E. Caput-Medusae.*
>   Perennials; spike readily disjointing.
>   Spike, including awns, longer than broad ....12. *E. Hansenii.*
>   Spike, including awns, as broad as long.
>   Glumes entire or divided into 2 awns ....13. *E. elymoides.*
>   Glumes divided into 3 or more awns ....14. *E. multisetus.*

1. **E. condensatus** Presl. Giant Rye-Grass. Common on hillsides and in canyons from coast inland through Santa Lucia Range, often among trees or dense brush.

2. **E. triticoides** Buckl. var. **triticoides.** Common in low ground and along streams in both coastal and interior regions, usually in sandy soil.

Var. **multiflorus** (Gould) Hoover, n. comb. *E. triticoides* subsp. *multiflorus* Gould, Madroño 8: 46. 1945. Of scattered occurrence in western portion: Tar Spring Canyon (*7551*). Possibly originating as a hybrid between typical *E. triticoides* and *E. condensatus.* It is regularly sterile, but so are all the rhizome-forming sorts of *Elymus* in our region, so that sterility in itself does not demonstrate hybrid origin. For example, I have searched in vain at the proper season for fertile seeds on spikes of *E. mollis* and of typical *E. triticoides.*

3. **E. pacificus** Gould. Loose sand near coast from south end of Morro Bay (*7866*) to Hazard Canyon (*Lee McHenry* in 1947) and Spooner's Cove (*7527*). Plants of *E. triticoides* in dry, sterile sand sometimes become dwarfed and approach *E. pacificus* rather closely.

4. **E. mollis** Trin. American Dunegrass. Loose sand of beaches, dunes, or cliffs in vicinity of Morro Bay: north end of Morro Bay (*6617*); south of Hazard Canyon.

5. **E. multinodus** Gould. Shifting sand-dunes, Oceano Beach (*7309* in 1947). It is

unknown whether it was planted earlier for a sand-binder or was introduced accidentally.

6. **E. trachycaulus** (Link) Hoover. Steiner Creek ("Serrano Canyon") near San Luis Obispo *(7286)*, in dry black clay in area of serpentine rock, among shrubs or in open places.

7. **E. laevis** (Scribn. & Sm.) Hoover. *Agropyron Parishii* Scribn. & Sm., not *E. Parishii* Davy & Merrill. *A. Parishii* var. *laeve* Scribn. & Sm. *A. laeve* Hitchc. *E. Stebbinsii* Gould. Occasional in open woods, or sometimes in brushy areas, through the Santa Lucia Mts. Variation in this species is discussed in Leafl. West. Bot. 10: 339. Plants with pubescent or glabrous nodes, and long-awned or short-awned plants, do not constitute entities which are distinct geographically, ecologically, or genetically.

8. **E. subsecundus** (Link) Hoover. On outcrop of white volcanic rock, Mallagh's Landing near Avila *(9017)*; on serpentine, San Bernardo Creek. Local representatives of this species differ markedly from *E. laevis* in their shorter spikes and smaller spikelets. Their appearance suggests a starved form of *E. glaucus*, from which they differ in having spikelets solitary at all the nodes of the spike.

9. **E. glaucus** Buckl. Very common, though scattered rather than in dense stands, in wooded areas from coast eastward to La Panza Range; Temblor Range, according to Twisselmann. There are occasional indications of hybridization with *E. laevis*, but ordinarily *E. glaucus* is distinguishable by its short dense spikes with comparatively small spikelets.

10. **E. virescens** Piper. Steiner Creek ("Serrano Canyon"), on brushy slopes and open flat in area of serpentine *(7167)*.

11. **E. Caput-Medusae** L. Open range on hills at California Polytechnic College, *K. E. Wade* in 1948 and 1949. The species has not been noticed recently and may not have become permanently established.

12. **E. Hansenii** Scribn. Ridge west of upper Lopez Canyon, among sagebrush and bush lupine *(8936)*; west slope of Temblor Range *(8312)*; reported from Templeton by Hitchcock in Jepson, Fl. Calif. 1: 187. This plant is often stated to be a hybrid between *E. elymoides* and some other species such as *E. glaucus*. Its scarcity and sporadic occurrence tend to confirm this supposition. A similar plant, but with spikelets solitary at most of the nodes, was found along Steiner Creek near San Luis Obispo. Because *E. trachycaulus* was present nearby, it seems likely that this plant was a hybrid between that species and *E. elymoides*.

13. **E. elymoides** (Raf.) Swezey. Either this species or the following, perhaps both, is frequent in rocky soils in both coastal and interior districts, being especially common in serpentine areas. In proportion to the abundance of the plants, few collections have been made. This and *E. multisetus* are not clearly distinguishable in our area.

14. **E. multisetus** (J. G. Smith) Jones. See note under preceding species. Choice Valley *(E. McMillan 136)*; Templeton *(Davy 7601,* cited by Hitchcock as *Sitanion jubatum)*; Steiner Creek near San Luis Obispo; ridge northwest of Cuesta Pass, on serpentine *(8942,* seemingly intermediate toward *E. elymoides)*; Salsipuedes Creek west of Pozo. Reported from Temblor Range by Twisselmann, as *Sitanion jubatum*.

17. Agropyron Gaertn.

Spikes short and dense; spikelets divergent ....1. *A. desertorum.*
Spikes elongate; spikelets erect ....2. *A. intermedium.*

1. **A. desertorum** (Fisch.) Schult. Planted for pasture in arid regions, and rarely appearing spontaneously in eastern part: Twisselmann Ranch (*Twisselmann 1255 in 1954*).

2. **A. intermedium** (Host) Beauv. Abundant along roadsides in hills west of Templeton (Bert Noble's ranch), presumably escaped from cultivation. This species probably should be included in *Elymus*, but its correct name under that genus can not now be determined with certainty. Available references make no clear statement that this species is, or is not, the same as *E. intermedius* Bieb. In view of the rather broad and truncate glumes and lemmas of *A. intermedium*, there also seems to be no compelling reason for excluding it from the genus *Triticum*, under which it was originally published.

18. **Triticum** L. Wheat

1. **T. aestivum** L. Wheat. Extensively cultivated in eastern part but seldom coming up spontaneously, and never becoming established: bed of Estrella River near Estrella (*8748*).

19. **Lolium** L. Rye-Grass

Glume shorter than spikelet.
    Perennial; lemmas awnless or nearly so . . . . 1. *L. perenne.*
    Usually annual; lemmas awned . . . . 2. *L. multiflorum.*
Glume as long as spikelet (except for awns of lemmas) . . . . 3. *L. temulentum.*

1. **L. perenne** L. Perennial Rye-Grass. Commonly included in seed mixtures for low-quality or temporary lawns and often persisting or escaping, especially in places which are at least temporarily moist. This is probably also the species which is extensively sown, under the erroneous name of "rye," after forest and brush fires. This practice can not be too strongly condemned, because seeds of native plants are abundantly present to grow under favorable conditions after fires, and those desirable native species are too easily crowded out by the worthless *Lolium*.

2. **L. multiflorum** Lam. Italian Rye-Grass. A weed especially in clay soils, abundant near coast, occasional in interior. To susceptible persons, the pollen of this species is a major cause of the agonies of hay fever in May and June.

3. **L. temulentum** L. Darnel. A widely distributed weed, but not very plentiful here, in fertile soils, both in cultivated ground and in openly wooded areas in the hills.

20. **Hordeum** L.

Perennials.
    Blades usually glabrous, 3 mm. wide or more; anthers 1 to 1.5 mm. long
                                            1. *H. brachyantherum.*
    Blades usually pubescent, less than 3 mm. wide; anthers 1.5 to 3 mm. long
                                               2. *H. californicum.*
Annuals.
    Middle spikelet at each node sessile, fertile, the lateral pedicelled and sterile or staminate.
        Glumes not ciliate.
            Spike 4 to 7 cm. long; fertile lemma 7 to 8 mm. long; lemmas of sterile spikelets awnless . . . . 3. *H. depressum.*

Spike 1.5 to 3 cm. long; fertile lemma about 5 mm. long; lemmas of
sterile spikelets short-awned ....4. *H. marinum.*
Inner glumes of lateral spikelets slightly broadened just above the
base, then tapering into the awn ....var. *marinum.*
All glumes setaceous from the base, not dilated   var. *Gussonianum.*
Glumes of fertile spikelet long-ciliate.
Internodes of rachis about 3 mm. long; florets of lateral spikelets
larger than that of middle one ....5. *H. leporinum.*
Internodes of rachis about 2 mm. long; lateral florets not larger than
middle one ....6. *H. glaucum.*
All 3 spikelets at a node sessile, usually all fertile ....7. *H. vulgare.*

1. **H. brachyantherum** Nevski. Common along coast from Cambria northward.
On exposed maritime bluffs the stems are prostrate or nearly so. In sheltered sites,
as in ravines or in hollows of the pine woods, the plants grow more nearly erect.
Some of the plants on ocean bluffs near Cambria have somewhat larger anthers,
thus tending to break down the distinction between this species and *H. californicum.*

2. **H. californicum** Covas & Stebbins. Frequent in western part, extending east to
La Panza Range, mostly along small temporary streams or in clay soils which dry
out slowly.

3. **H. depressum** (Scribn. & Sm.) Rydb. Occasional in winter-wet depressions in the
interior: Creston, San Juan Valley, Temblor Range, and Carrizo Plain. Seldom
found near coast: salt marsh on east side of Morro Bay (*6920*).

4. **H. marinum** Huds. var. **marinum.** Weedy roadside, Cambria (*Howell 40,811
in 1964*). Perhaps many of the plants which have been assumed to belong to the
following variety are really typical *H. marinum.* The difference is slight and, in
many specimens, subject to intergradation. Several collections of var. *marinum* have
been seen from farther north in California. In every instance they had been identified as var. *Gussonianum* or a synonym of that name.

Var. **Gussonianum** (Parl.) Thell. *H. Gussonianum* Parl. *H. Hystrix* Roth. *H. gen-
iculatum* All. Mediterranean Barley. Clay soils of open range or pastures, common near coast, less so in interior.

5. **H. leporinum** Link. Foxtail. Very abundant in cultivated ground, as well as
better soils in the wild areas, in both interior and coastal districts. Some of the
plants assumed to be of this species may actually belong to the following. The two
are identical in superficial appearance and are distinguishable only by careful measurements under magnification. Although "foxtail" is the time-honored vernacular
name for the genus *Alopecurus,* Californians rarely use the name for anything other
than *H. leporinum.*

6. **H. glaucum** Steudel. Probably a common weed, but definite records of its occurrence are few, because of its close resemblance to the preceding species: Twisselmann Ranch (*Twisselmann 844*); Santa Maria River (*Eastwood 350*).

7. **H. vulgare** L. Barley. Extensively planted; often coming up spontaneously in
disturbed soil but probably never long persistent outside of cultivation.

21. **Scribneria** Hack.

1. **S. Bolanderi** (Thurb.) Hack. In gravelly or sandy soil which is moist early in
the season, uncommon or overlooked: northwest of Cuesta Pass (*8477*); Navajo
Creek, La Panza Range (*7878*); 18 miles east of Creston.

22. **Monerma** Beauv.

1. **M. cylindrica** (Willd.) Coss. & Dur. Rarely established along the coast: Piedras Blancas; north end of Morro Bay.

23. **Parapholis** C. E. Hubbard

1. **P. incurva** (L.) C. E. Hubbard. Common on sea-cliffs and borders of salt-marshes near coast.

24. **Avena** L. OATS

Lemmas long-hairy; awn geniculate and twisted below.
    Teeth of lemma not awned . . . . .1. *A. fatua.*
    Teeth of lemma awned . . . . .2. *A. barbata.*
Lemmas glabrous; awn usually straight or absent .3. *A. sativa.*

1. **A. fatua** L. WILD OATS. A very plentiful weed of cultivated fields, city lots, roadsides, open hillsides, and fertile plains. In many places this, together with other introduced species, has crowded out almost the last remnant of native plants.

2. **A. barbata** Brot. SLENDER WILD OATS. Like the preceding, plentiful and widespread. Generally speaking, *A. barbata* grows in drier, more rocky places than does *A. fatua,* but often the two are found together.

3. **A. sativa** L. CULTIVATED OATS. Seldom spontaneous in or near fields where it was formerly grown; never spreading into uncultivated areas.

25. **Danthonia** Lam. & DC.

1. **D. californica** Bol. Frequent on open hills and plains near coast, in firm but often sandy soil, from San Luis Range and Los Osos Valley northward; moist openings in pine woods at Cambria. Since grazing animals seek it out, often before seeds can mature, and since it is crowded out by rank-growing introduced weeds, the species is gradually vanishing.

26. **Koeleria** Pers.

Perennial bunch-grass; lemmas awnless or barely awn-tipped . . . . .1. *K. cristata.*
Annual; lemmas with short but clearly defined awn . . . . .2. *K. phleoides.*

1. **K. cristata** (L.) Pers. JUNEGRASS. Common and widely distributed in western part of county, on rocky or fertile clay hills, sand-dunes, and sandy coastal plains. Extending inland to La Panza Range and hills to the north.

2. **K. phleoides** (Vill.) Pers. Locally established in sandy soil: Gillis Canyon southeast of Shandon (*Howell 24,376* in 1948).

27. **Schismus** Beauv.

Glumes 5 to 6 mm. long; lemmas 2.5 to 3 mm. long, with 2 short acute lobes at apex; palea acute, shorter than lemma . . . . .1. *S. arabicus.*
Glumes 4 to 5 mm. long; lemmas about 2 mm. long, rounded and emarginate at apex; palea rounded, as long as lemma . . . . .2. *S. barbatus.*

1. **S. arabicus** Nees. Common in sandy soil in eastern part: Shell Creek to San Juan Valley, southern Temblor Range, and Cuyama Valley.

2. **S. barbatus** (L.) Thell. Sandy soil, apparently sparingly established: hills near Elkhorn Plain to east end of Caliente Range and Cuyama Valley. Additional occur-

rences of this species may have been overlooked because of its close resemblance to the preceding.

28. **Deschampsia** Beauv.

Tufted perennials with many basal leaves.
    Panicle pyramidal to oblong in outline, less than 1⁄3 total height of plant; leaves of basal tuft mostly over 15 cm. long, of firm texture.
        Panicle open, some of its lower branches over 5 cm. long, not spikelet-bearing toward the base . . . . .1. *D. caespitosa.*
        Panicle dense, its branches not over 4 cm. long, with spikelets crowded throughout . . . . .2. *D. holciformis.*
    Panicle very slender, with erect branches, about 1⁄2 total height of plant; leaves of basal tuft 15 cm. long or mostly shorter, filiform and flexuous   3. *D. elongata.*
Annual with few basal leaves, hardly tufted . . . . .4. *D. danthonioides.*

1. **D. caespitosa** (L.) Beauv. Tufted Hairgrass. Moist swales in pine woods near Cambria. Regarding the form of the species occurring here, Agnes Chase (in Hitchcock's Man. Grasses U.S. ed. 2, 295) states: "Large plants from Oregon and California have been described under *Deschampsia caespitosa* subsp. *beringensis* (Hultén) Lawr., but are not *D. beringensis* Hultén, of the Aleutians."

2. **D. holciformis** Presl. Plentiful locally on ocean bluffs and in open fields along the coast north of San Simeon.

3. **D. elongata** (Hook.) Munro. Slender Hairgrass. Frequent in moist places, usually in woods, from coast eastward through Santa Lucia Mts.

4. **D. danthonioides** (Trin.) Munro. Annual Hairgrass. Common in interior, and occasional near coast, in places wet in winter, such as beds of streamlets, vernal pools, etc.

29. **Aira** L.

Spikelets with 2 exserted awns of about equal length . . . . .1. *A. caryophyllea.*
Spikelets with only 1 exserted awn; lemma of lower floret awnless or with a very short awn . . . . .2. *A. elegans.*

1. **A. caryophyllea** L. Silver Hairgrass. Well established near coast from around Cambria northward, generally in moist sandy soil. Rare in sandy places in the interior: Calf Canyon (*8346*).

2. **A. elegans** Willd. Open hillside in San Luis Range above head of Perfumo Canyon (*8873*).

30. **Trisetum** Pers.

1. **T. canescens** Buckl. Moist shady woods near coast and in Santa Lucia Mts.: oak woods in Los Osos Valley; Santa Rita Creek west of Templeton (*8467*); San Carpoforo Creek (*8990*). The latter collection differs from descriptions of *T. canescens* in having glabrous leaves.

31. **Holcus** L.

1. **H. lanatus** L. Velvet Grass. Grown for pasture; sometimes spreading as a weed in moist places, as along roadsides and in lawns: San Simeon (*7956*).

32. Agrostis L. Bent-Grass.

Plants with extensively creeping rhizomes and stolons.
  Panicle pyramidal, with divergent branches; palea more than half as long as
  lemma . . . . 1. *A. alba.*
  Panicle narrow, with ascending branches (or somewhat spreading at anthesis);
  palea absent.
      Hairs at base of lemma 1 mm. long or less . . . . 2. *A. diegoensis.*
      Hairs at base of lemma 1 to 2 mm. long . . . . 3. *A. Hallii.*
Bunch-grasses without rhizomes or stolons.
  Panicle narrow, with erect or ascending branches bearing spikelets from near
  the base.
      Stems (in the form occurring locally) prostrate or widely spreading, radiat-
      ing from a central leaf-tuft; blades of upper leaves about 2 or 3 cm. long
                                                                    4. *A. californica.*
      Stems erect; blades of upper leaves nearly all over 5 cm. long . . 5. *A. exarata.*
          Lemma awnless, or with awn not exceeding the glumes   var. *exarata.*
          Lemma with awn exceeding the glumes.
              Panicle 10 cm. long or more, with some branches at least 3 cm.
              long . . . . var. *pacifica.*
              Panicle less than 10 cm. long, with branches not over 2 cm. long
                                                                    var. *monolepis.*
  Panicle open, its branches bearing spikelets only toward the ends   6. *A. Hooveri.*

1. **A. alba** L. Redtop. Twisselmann Ranch, "occasional escape from lawns;"
Santa Margarita Creek near Santa Margarita.

2. **A. diegoensis** Vasey. San Diego Bent-Grass. Plentiful in woods and thickets,
or rarely on open hills, near coast; more sparingly inland to La Panza Range. This
species is potentially of great value as a lawn grass. It is obviously drought-resistant,
is at least equal in appearance to the exotic grasses ordinarily used, would not be
difficult to keep under control, and probably would stand mowing well.

3. **A. Hallii** Vasey. Woods and thickets near coast: Cambria (*Howell 40,866*);
Arroyo de la Cruz (*7942*). Doubtfully distinct from *A. diegoensis.*

4. **A. californica** Trin. Ocean bluffs from near Cambria northward. The plants
here referred to *A. californica* have prostrate stems radiating out from the center of
the tuft. This feature does not seem to be characteristic of the species as it occurs
farther north.

5. **A. exarata** Trin. var. **exarata.** Occasional in moist places in western part: San
Simeon Creek (*8030*); Los Osos (*7174*).

Var. **pacifica** Vasey. Widely distributed through western part of county in moist
places, apparently not mixed with typical (awnless) *A. exarata.* This form with
awned lemmas seems to be the commoner of the two here.

Var. **monolepis** (Torr.) Hitchc. Steiner Creek near San Luis Obispo; pine woods
at Cambria; just north of Piedras Blancas Point. This may be only a seasonal form
of var. *pacifica.* Local specimens corresponding to the description of var. *pacifica*
have been collected in May and June, whereas smaller plants corresponding to var.
monolepis have been collected from late July to January.

6. **A. Hooveri** Swallen. Dry soil derived from sandstone or siliceous shale: low
ridge on northeast side of Los Osos Valley (*8591*); hills bordering San Luis Valley
on the south, eastward to east slope of Santa Lucia Mts. (Rinconada Mine), upper
Arroyo Grande, and southward to Purisima Hills, Santa Barbara Co. The distribu-

tion of this species is centered in the Arroyo Grande watershed, where in several places it is locally plentiful.

33. **Calamagrostis** Adans.

Plants with extensively creeping rhizomes, forming a loose sod; sheaths pubescent on the collar . . . . .1. *C. rubescens.*
Plants with short rhizomes, forming dense clumps; sheaths glabrous on the collar.
 Growing in wet soil; blades when flat 6 to 14 mm. wide, often becoming involute; panicle 20 cm. long or more, with some of its lower branches 6 cm. long or more . . . . .2. *C. nutkaensis.*
 Growing in dry soil; blades mostly 2 to 3 mm. wide; panicle less than 20 cm. long, none of its branches over 3.5 cm. long . . . . .3. *C. koelerioides.*

 1. **C. rubescens** Buckl. PINEGRASS. Oak and pine woods in Santa Lucia Mts.: 7 miles west of Templeton (*7991*); between Rocky Butte and Pine Mt. (*8022*). This grass may be more common than it appears to be, as extensive stands of it flower only in certain years, and then only sparingly.

 2. **C. nutkaensis** (Presl) Steud. PACIFIC REEDGRASS. Rare in wet places near coast: bog in Black Lake Canyon (*7339*), the southernmost known locality; in swampy area on hillside north of Piedras Blancas Point (*7331*).

 3. **C. koelerioides** Vasey. Rocky hills, known definitely at only one place in the county: top of sandstone hill near summit on Carpenter Canyon road from San Luis Obispo to Arroyo Grande. (*6192*).

34. **Ammophila** Host. BEACH-GRASS

 1. **A. arenaria** (L.) Link. EUROPEAN BEACH-GRASS. Planted on dunes as a sand-binder, and now well established on west side of Morro Bay southward to near Hazard Canyon, and from Pismo Beach southward.

35. **Gastridium** Beauv.

 1. **G. ventricosum** (Gouan) Schinz & Thell. NIT-GRASS. Common on open hills in western part, sometimes a weed in gardens and agricultural land.

36. **Stipa** L. NEEDLE-GRASS

Panicle erect, with short ascending branches; lemma or its awn with conspicuous hairs 3 mm. long or more.
 Plant 3 to 6 dm. tall; blades very narrow and involute; body of lemma minutely pubescent; basal portion of awn long-hairy . . . . .1. *S. speciosa.*
 Plant 8 to 18 dm. tall; blades flat or subinvolute; body of lemma densely long-hairy toward apex; awn scabrous, not hairy . . . . .2. *S. coronata.*
Panicle with spreading or nodding branches; lemma and its awn with appressed hairs not over 1 mm. long, or glabrous.
 Glumes 6 mm. long or more; lemma 5 mm. long or more, appressed-pubescent; awn twice geniculate.
  Glumes 12 to 20 mm. long; lemma more than 7 mm. long.
   Awn mostly 6 to 7 cm. long, its terminal segment mostly 3 to 4 cm. long
                      3. *S. pulchra.*
   Awn mostly 7 to 9 cm. long, its terminal segment mostly 5 to 6 cm. long
                      4. *S. cernua.*
  Glumes 6 to 12 mm. long; lemma not more than 6 mm. long . .5. *S. lepida.*
 Glumes about 3 mm. long; lemma 2 mm. long, glabrous; awn slightly flexuous or obscurely geniculate . . . . .6. *S. miliacea.*

1. **S. speciosa** Trin. & Rupr. Desert Needle-Grass. Frequent on dry rocky hills in eastern part, extending westward to 5 miles east of Creston.

2. **S. coronata** Thurb. On rocky slopes, widely scattered, the plants never numerous, Santa Lucia Mts. and from La Panza Range southward.

3. **S. pulchra** Hitchc. Purple Needle-Grass. Common on open hills and plains near coast and in Santa Lucia Range, in soils varying from heavy clay to firm sandy loam. Undoubtedly less plentiful than formerly, but still abundant in spots. This species and *S. cernua* have been described as varying more widely in the length of the awn than is indicated in the above key, but in San Luis Obispo Co. the length of the awn and the proportional length of its terminal segment furnish the most satisfactory basis for distinguishing between plants near the coast and those in the interior.

4. **S. cernua** Stebbins & Love. Common throughout the interior except in the most arid parts. Some ranchers believe that this species was the main constituent of the primitive grassland of eastern San Luis Obispo. Both it and *S. pulchra* are excellent forage, to judge from the fact that animals graze them drastically as long as they are available. Although *S. cernua* reaches the coast in some parts of California, there is no available record of its occurrence in coastal San Luis Obispo Co. The summits of the Santa Lucia Range seem to mark the approximate boundary between *S. cernua* to the east and *S. pulchra* to the west. However, much remains to be learned about the exact local distribution of both species. It may be significant that plants growing on a serpentine outcrop on the road to Hi Mt. from the Pozo-Arroyo Grande summit (*8775*), on about the geographic boundary between the two, combine the characteristic lemmas and awns of *S. pulchra* with the pale, very narrow and involute leaves which have been regarded as a distinguishing mark of *S. cernua*.

5. **S. lepida** Hitchc. Foothill Needle-Grass. Common in openly wooded and brushy areas, less often on open rocky hills, from the coast inland at least to La Panza Range. Notably abundant and vigorous after chaparral fires. As a quick-growing natural replacement for burned stands of chaparral, this species seems preferable by far to the weedy species which are often sown. When the chaparral has not burned for several years, plants of *S. lepida* gradually become dwarfed. This ecologic form evidently is the basis for the name *S. lepida* var. *Andersonii* (Vasey) Hitchc.

6. **S. miliacea** (L.) Hoover. *Oryzopsis miliacea* (L.) B. & H. Smilo. Introduced in dry ground, frequent as a weed near coast, and at widely scattered places in interior.

37. Aristida L.

1. **A. adscensionis** L. Six-Weeks Three-Awn. Occasional on rocky hills around San Luis Obispo and southward.

38. Polypogon Desf.

Glumes awnless ....1. *P. semiverticillatus.*
Glumes awned.
    Awns of glumes longer than the body.
        Glumes tapering into awns; awns slightly longer than the body
                                2. *P. interruptus.*
        Glumes rounded and more or less lobed above attachment of awns; awns 3 or 4 times as long as the body.

    Glumes slightly lobed; lemma awned ....3. *P. monspeliensis.*
    Glumes with 2 long ciliate lobes; lemma awnless ....4. *P. maritimus.*
  Awns of glumes shorter than the body ....5. *P. elongatus.*

1. **P. semiverticillatus** (Forsk.) Hyl. Common in wet places, widespread, both coastal and interior regions.

2. **P. interruptus** H. B. K. In wet places, often associated with the preceding, from which it is readily distinguishable by the presence of awns. Not yet reported from eastern part of county.

3. **P. monspeliensis** (L.) Desf. Rabbitfoot Grass. Very common in wet soil. The plants are clearly annual in temporarily wet places. In permanently wet places they often become perennial, forming dense tough clumps and growing taller than described in current references.

4. **P. maritimus** Willd. In moist (or formerly moist) soil of stream-beds: upper Buckhorn Canyon, La Panza Range (*Twisselmann 2171*); Stoney Creek (*7962*). Plants of *P. monspeliensis* were inadvertently mixed in both collections. This fact suggests that *P. maritimus* may have been overlooked in other localities because of its superficial resemblance to the other species.

5. **P. elongatus** H. B. K. Moist spots among sand-dunes, Oceano Beach (*7314*); Oso Flaco Lake. The inflorescence is more compact than in Mexican specimens of this species. This difference has been noted also by H. L. Mason in "A Flora of the Marshes of California."

39. Muhlenbergia Schreb.

Annual; lemma long-awned ....1. *M. microsperma.*
Perennial; lemma awnless.
    Plants with extensively creeping rhizomes and stolons, not usually over 3 dm. tall.
        Panicle slender, with rather few spikelets ....2. *M. utilis.*
        Panicle open, about as broad as long, with many spikelets 3. *M. asperifolia.*
    Plants forming large dense clumps, mostly 10 to 15 dm. tall ......4. *M. rigens.*

1. **M. microsperma** (DC.) Kunth. Dry hills in southwestern part (Price Canyon, *6750*).

2. **M. utilis** (Torr.) Hitchc. Aparejo Grass. Uncommon in moist sandy soil near coast: Los Osos (*6602*); Black Lake Canyon; upper Arroyo Grande. Used for stuffing pack saddles, according to A. S. Hitchcock.

3. **M. asperifolia** (Nees & Mey.) Parodi. Scratchgrass. Low-lying, often alkaline ground, known in the county definitely only from Cuyama Ranch (*10,134*). To be expected in Salinas Valley.

4. **M. rigens** (Benth.) Hitchc. Deergrass. Occasional in dry soil in Salinas River watershed: Palo Prieto Canyon; Creston to Santa Margarita. Also in Arroyo Grande watershed. This species was used by California Indians in making baskets.

40. Leptochloa Beauv.

Spikelets 7 to 9 mm. long; lemmas awned ....1. *L. fascicularis.*
Spikelets 5 to 7 mm. long; lemmas awnless ....2. *L. uninervia.*

1. **L. fascicularis** (Lam.) Gray. Recently adventive in wet places near San Luis Obispo and by roadsides in Santa Maria Valley.

2. **L. uninervia** (Presl) Hitchc. & Chase. Roadside and garden weed in the interior: Choice Valley and Estrella Plains, according to Twisselmann.

41. **Chloris** Swartz.

1. **C. virgata** Swartz. Feather Fingergrass. Becoming established in irrigated fields around Shandon (Twisselmann).

42. **Cynodon** L. Rich.

1. **C. Dactylon** (L.) Pers. Bermuda Grass. An aggressive weed, common and becoming increasingly established in towns and in agricultural lands, spreading into moist places on open range. Plants which can survive in competition with this pest are few.

43. **Hierochloe** R. Br.

1. **H. occidentalis** Buckl. Vanilla Grass. California Sweetgrass. In shady woods and on their borders, locally common in watershed of San Carpoforo Creek (*8980*), the southern limit of the known distribution of the species.

44. **Ehrharta** Thunb.

1. **E. calycina** J. E. Smith. Well established in sandy soil near the coast, around the south end of Morro Bay and on Nipomo Mesa.

45. **Phalaris** L. Canary Grass

Perennials, forming robust clumps.
    Basal internodes swollen; panicle cylindric, over 5 cm. long . . . . . 1. *P. aquatica.*
    Basal internodes not swollen; panicle ovate or short-oblong in outline, 3 to 5 cm. long . . . . 2. *P. californica.*
Annuals.
    Spikelets in lower part of panicle deformed, the glumes reduced and ending in hard knobs . . . . 3. *P. paradoxa.*
    Spikelets all alike.
        Panicle about 13 to 17 mm. wide, not more than 4 times as long as wide; folded glumes about 1.5 mm. wide, winged on the keel . . . . 4. *P. minor.*
        Panicle about 8 to 14 mm. wide, usually more than 4 times as long as wide; folded glumes barely 1 mm. wide, very narrowly if at all winged.
        Fertile lemma acuminate into a slender glabrous point
                                                  5. *P. Lemmonii.*
      Fertile lemma acute.
        Spikelets 5 to 6 mm. long . . . . 6. *P. caroliniana.*
        Spikelets 3.5 to 4 mm. long . . . . 7. *P. angusta.*

1. **P. aquatica** L. *P. tuberosa* L. Harding Grass. Sown for forage and in recent years becoming established by roadsides, in low fields, and even on dry hills. The plants introduced in California are usually called *P. tuberosa* var. *stenoptera* or *P. stenoptera.* According to Hitchcock's "Manual of the Grasses of the United States," ed. 2, p. 556, var. *stenoptera* differs from typical *P. tuberosa* (i.e., *P. aquatica*) in having "the base of the culms little or not at all swollen." The plants found here have the base of the culms distinctly swollen and therefore are included in *P. aquatica.*

2. **P. californica** H. & A. Frequent in moist places near coast, usually in sandy soil: Oak Park district (*7520*) and northward. The pollen of this and probably of all other species of *Phalaris* is a cause of hay fever.

3. **P. paradoxa** L. Common in clay soils near coast, mostly in cultivated ground or abandoned fields. The var. *praemorsa* (Lam.) Coss. & Dur. may also be present, but the only local specimens now at hand belong to typical *P. paradoxa*.

4. **P. minor** Retz. In clay soils near coast, often associated with *P. paradoxa*.

5. **P. Lemmonii** Vasey. Frequent in clay beds of vernal pools, Salinas Valley, Los Osos Valley, and San Luis Valley.

6. **P. caroliniana** Walt. Serrano Canyon (*Eastwood & Howell 2265*).

7. **P. angusta** Nees. In moist sandy soil, Los Osos Valley (*7182*); Oak Park district north of Arroyo Grande (*7522*). Reported from San Luis Obispo by Hitchcock in Jepson's "Flora of California." The differentiation of our local plants into the three species *P. angusta, P. Lemmonii,* and *P. caroliniana* is not so clear as could be desired.

46. **Stenotaphrum** Trin.

1. **S. secundatum** (Walt.) Kuntze. St. Augustine Grass. Introduced as a lawn grass, seldom spontaneous: Cambria (*Howell 40,807*). A coarse sod-former resembling *Pennisetum clandestinum*; vegetative shoots are distinguishable from it by being wholly glabrous.

47. **Digitaria** Heister.

1. **D. sanguinalis** (L.) Scop. Crabgrass. Frequent in places which receive some watering during summer, such as lawns and gardens.

48. **Paspalum** L.

Forming extensively creeping rhizomes; inflorescence with nearly always 2 branches
1. *P. distichum.*
Forming dense tough clumps; inflorescence with several racemosely arranged branches
2. *P. dilatatum.*

1. **P. distichum** L. Knotgrass. Rarely if ever collected, but probably of frequent occurrence in wet places: San Luis Obispo; San Juan River.

2. **P. dilatatum** Poir. Dallis Grass. Moist soil: near Creston; San Luis Obispo; Oceano Beach, locally abundant in low moist ground along railway (*7322*); probably of more general occurrence as an escape from irrigated pastures.

49. **Panicum** L.

Perennial; panicle about 5 to 10 cm. long in vernal phase, smaller in autumnal phase . . . . .1. *P. pacificum.*
Annual; panicle about 20 to 30 cm. long . . . . .2. *P. capillare.*

1. **P. pacificum** Hitchc. & Chase. Cambria, in rather dry to moist sandy soil in shade of pines (*7222, 7260*).

2. **P. capillare** L. var. **occidentale** Rydb. Witch Grass. Summer-moist ground, Atascadero Lake (*Hardham 2655*); Salinas River near Santa Margarita; to be expected elsewhere in Salinas Valley.

50. **Echinochloa** Beauv.

Stems erect or nearly so; blades mostly 5 to 10 mm. wide; branches of inflorescence in robust plants compound . . . . .1. *E. Crus-galli.*
Stems widely spreading; blades mostly 3 to 6 mm. wide; branches of inflorescence simple . . . . .2. *E. colonum.*

1. **E. Crus-galli** (L.) Beauv. Barnyard Grass. Frequent in summer-irrigated places. Most of the plants seen are probably referable to var. *zelayensis* (H. B. K.) Hitchc., with awns reduced or virtually absent.

2. **E. colonum** (L.) Link. Not clearly separable from *E. Crus-galli* var. *zelayensis* in our area. Plants growing as a weed at Twisselmann Ranch (*Twisselmann 1463*) are referred here, largely because of having spreading rather than erect stems. The leaves in these particular plants have transverse purple bands.

51. **Setaria** Beauv.

Perennial with short knotty rhizomes ....1. *S. geniculata.*
Annuals with cluster of fibrous roots.
    Bristles below each spikelet 5 or more ....2. *S. lutescens.*
    Bristles below each spikelet 1 to 3.
        Bristle 1 below each spikelet, 1 to 3 times as long as the spikelet, retrorsely scabrous ....3. *S. verticillata.*
        Bristles 1 to 3, 3 to 4 times as long as the spikelet, antrorsely scabrous
                                          4. *S. viridis.*

1. **S. geniculata** (Lam.) Beauv. In places which are watered in summer, such as lawns and gardens, at San Luis Obispo and probably elsewhere.

2. **S. lutecsens** (Weigel) Hubbard. "A scarce garden weed at the Twisselmann Ranch," according to Twisselmann. The various species of *Setaria* are probably more widely distributed as weeds than the very scanty collections would indicate.

3. **S. verticillata** (L.) Beauv. Garden weed at Twisselmann Ranch (*Twisselmann 1314*).

4. **S. viridis** (L.) Beauv. San Miguel (*Chester Dudley* in 1936); 2 miles south of Creston, bordering irrigated field (*10,064* in 1966).

52. **Pennisetum** L. Rich.

Spikelets 2 to 4, hidden in the upper leaf-sheaths ....1. *P. clandestinum*
Spikelets many, in an exserted plumose panicle.
    Panicle yellowish, oblong or oval in outline, 3 to 10 cm. long ...2. *P. villosum.*
    Panicle purplish, long-cylindric, 15 to 35 cm. long ....3. *P. setaceum.*

1. **P. clandestinum** Hochst. Kikuyu Grass. Well established in moist places in the hills north of Arroyo Grande, and probably elsewhere near the coast.

2. **P. villosum** R. Br. Feathertop. Originally introduced for cultivation as an ornamental, it has for many years grown wild along streets in San Luis Obispo.

3. **P. setaceum** (Forsk.) Chiov. Fountain Grass. Cultivated as an ornamental; recently escaped along roadsides at San Luis Obispo (*9189* in 1964).

53. **Sorghum** Moench

1. **S. halepense** (L.) Pers. Johnson Grass. Occasionally established in or near cultivated ground: San Luis Obispo; Choice Valley and near Shandon, according to Twisselmann.

Cyperaceae. Sedge Family

Flowers all, or at least some, perfect; achene not enclosed in a saccate structure.
    Spikelets compressed, the scales in 2 rows ....1. *Cyperus.*
    Spikelets not compressed, the scales spirally arranged.

Leaf-margins smooth or merely scabrous; spikelets several-flowered, with perfect flowers only.

Spikelet solitary, exactly terminal, not subtended by an involucral bract; base of style persisting as a tubercle on the achene. 2. *Eleocharis*
Spikelets subtended by one or more involucral bracts: involucral bract, if solitary, usually appearing like a continuation of the stem, the inflorescence thus apparently lateral; style wholly deciduous, or its base persistent on the achene as a minute point only ...3. *Scirpus.*
Leaf-margins rigidly serrulate (cutting); spikelets with staminate flowers in addition to a single terminal perfect flower ....4. *Cladium.*
Flowers unisexual; achene enclosed in a saccate structure (perigynium) ...5. *Carex*

1. Cyperus L.

Tufted annuals or perennials less than 1 m. tall, with a cluster of fibrous roots; lower leaves with blades; involucral leaves mostly 8 or fewer.

Spikelets in compound umbels, green to medium brown.

Perennial; spikelets green; scales not decurrent on rachis ...1. *C. Eragrostis.*
Annual; spikelets light or medium brown; scales decurrent on rachis, their bases persistent as membranous wings ....2. *C. erythrorhizos.*
Spikelets in a small dense head, dark brown ....3. *C. niger.*
Perennial with rhizomes, not tufted, 1 m. tall or more; lower leaves represented by sheaths only; involucral leaves mostly 10 or more ....4. *C. alternifolius.*

1. **C. Eragrostis** Lam. Common in western part in drying streambeds, along permanent streams, and around springs; scarce toward the interior.

2. **C. erythrorhizos** Muhl. In wet soil along upper Arroyo Grande.

3. **C. niger** R. & P. var. **capitatus** (Britt.) O'Neill. Baywood Park, in brackish marsh (*6359*).

4. **C. alternifolius** L. Umbrella Plant. Well established along a stream-course at San Luis Obispo.

2. Eleocharis R. Br. Spike-Rush

Stems capillary, barely 0.5 mm. in diameter or mostly less; spikelets usually with fewer than 10 flowers.

Achenes with longitudinal ridges connected by fine transverse lines
1. *E. acicularis.*
Achenes without longitudinal ridges ....2. *E. parvula.*
Stems firm, not capillary, over 0.5 mm. in diameter; spikelets usually with more than 10 flowers.

Some of stems arching and rooting at tip; tubercle continuous with body of achene ....3. *E. rostellata.*
Stems erect, not rooting at tip; tubercle sharply differentiated from body of achene.

Stems mostly 0.5 to 1.5 mm. in diameter; style 3-cleft; achene plano-convex, keeled on the convex side.

Spikelet slender, acute ....4. *E. Parishii.*
Spikelet ovoid, obtuse ....5. *E. montevidensis.*
Stems mostly 1.5 to 2 mm. in diameter; style 2-cleft; achene biconvex
6. *E. palustris.*

1. **E. acicularis** (L.) R. & S. Cambria, in moist swale under pines (*7707*). Plants immature, and identity therefore uncertain. Plants growing at the Laguna near San Luis Obispo also had the appearance of this species, but no specimens were collected.

2. **E. parvula** (R. & S.) Link. "Salt marshes, San Luis Obispo and Humboldt counties," according to Mason in "A Flora of the Marshes of California." No specimens seen.

3. **E. rostellata** Torr. WALKING SEDGE. Locally abundant in boggy spot, upper Arroyo Grande above County Park (*7908*).

4. **E. Parishii** Britt. In wet places, either permanent or temporary, frequent toward the interior in watersheds of Salinas and Cuyama rivers. Not yet found near the coast.

5. **E. montevidensis** Kunth. Rare in wet places in eastern part: Ortega Spring, Palo Prieto Canyon (*Twisselmann 2924*).

6. **E. palustris** (L.) R. & S. COMMON SPIKE-RUSH. Common in permanently or temporarily wet soil in coastal lowlands and eastward to upper Salinas Valley; sometimes associated with *E. Parishii*. According to many authors, the correct name of this species is *E. macrostachya* Britt.

3. **Scirpus** L. BULRUSH

Inflorescence surrounded by 2 to several long leafy bracts.
    Spikelets 3 to 6 mm. long, in a large compound umbel ....1. *S. microcarpus*.
    Spikelets 10 to 25 mm. long, in a usually congested umbel of heads
                                                            2. *S. robustus*.
Inflorescence with a single bract which looks more or less like a continuation of the stem, sometimes also with one or two shorter sharp-pointed scale-like bracts.
    Small tufted annuals (or perennials?) with a cluster of fibrous roots; spikelet solitary (rarely 2 or 3).
        Bract averaging about as long as spikelet, rarely twice as long, or sometimes obsolete; scales mostly brown, obtuse or mucronulate ....3. *S. cernuus*.
        Bract more than twice as long as spikelet; scales green, acute or acuminate.
                                            4. *S. koilolepis*.
    Perennials with rhizomes, 3 dm. tall or more; spikelets (rarely 1 or) 2 to many.
        Stems sharply triangular, usually much less than 2 m. tall; spikelets in a single dense head.
            Some of leaves bearing long blades; stem under 3 mm. in diameter, firm (with small intercellular spaces), its sides flat or nearly so
                                          5. *S. americanus*.
            Leaves reduced to sheaths, or one of them bearing a short blade; stem over 3 mm. in diameter, soft (with large intercellular spaces), its sides concave ....6. *S. Olneyi*.
        Stems round or bluntly triangular, usually 2 m. tall or more; spikelets in a compound umbel or umbellately arranged heads.
            Stem with 3 rounded angles, bright green ....7. *S. californicus*.
            Stem terete, dull or grayish green ....8. *S. acutus*.

1. **S. microcarpus** Presl. Common around springs and along streams, often in shade, near coast and in Santa Lucia Mts.

2. **S. robustus** Pursh. Common in low wet places in coastal area. Less widespread in interior: reported in Temblor Range by Twisselmann (as *S. paludosus* Nelson). Plants on the coastline have long dark brown spikelets, grading into shorter and paler ones inland.

3. **S. cernuus** Vahl var. **californicus** (Torr.) Beetle. Common in moist ground near coast; for example, seepage spots on sea-cliffs, brackish marshes, and poorly drained clay depressions.

4. **S. koilolepis** (Steud.) Gleason. Rare near coast: Cambria, in moist spots in pine woods (*7854*), the only locality known in the county and apparently the southern-most in California.

5. **S. americanus** Pers. THREE-SQUARE. Common in wet places, most often in saline soil, especially in hollows among coastal sand-dunes. In the interior notably plenti-ful along San Juan River.

6. **S. Olneyi** Gray. Occasional in marshes near coast, and rare in interior: Ortega Spring, Palo Prieto Canyon, according to Twisselmann.

7. **S. californicus** (C. A. Mey.) Steud. Widespread in coastal marshes, sometimes locally abundant: around Morro Bay; between Pismo Beach and Oceano, and south-ward. Often grows with *S. acutus*.

8. **S. acutus** Muhl. TULE. Common in low wet places near coast; occasional and perhaps temporary in wet spots in Salinas Valley. Formerly plentiful over consider-able areas, but in recent years extensive stands have been destroyed, as near Oceano. Specimens from San Luis Obispo Co. have been identified by A. A. Beetle as *S. ru-biginosus* Beetle, but it seems highly improbable that the local plants belong to more than one species. Also, the differences between these plants and *S. validus* Vahl hardly seem great enough to be significant.

4. Cladium R. Br.

1. **C. californicum** (Wats.) O'Neill. SAW-GRASS. Forming a dense stand on border of bog in Black Lake Canyon (*7341, 7353*).

5. Carex L. SEDGE

Spikelets on a plant all essentially alike, usually containing both pistillate and stam-inate flowers (sometimes the plants dioecious).
    Spikelets with staminate flowers uppermost, or plants dioecious.
        Spikelets in a simple spike, which may be short and head-like.
            Plants dioecious; stem smooth; perigynia pale; beak of perigynium nearly as long as the body . . . . .1. *C. Douglasii.*
            Plants usually monoecious; stem scabrous on the angles; perigynia brown, with beak about ⅓ to ½ as long as the body.
                Usually with long-creeping rhizomes, forming a sod, sometimes tufted; mature inflorescence over 5 mm. wide; bracts not usually tapering into a long filiform point.
                    Stems mostly less than 25 cm. tall; mature inflorescence mostly less than twice as long as broad; scales dark brown
                                    2. *C. pansa.*
                    Stems mostly over 25 cm. tall; mature inflorescence mostly more than twice as long as broad; scales light brown.
                              3. *C. praegracilis.*
                With short thick rhizomes, forming tufts; inflorescence less than 5 mm. wide; lower bracts tapering into a long filiform point
                            4. *C. tumulicola.*
        Spikelets in a somewhat branched spike, which may be very dense.
            Plants not forming large tussocks; perigynia broadest above the base, not subcordate.
                  Pistillate scales gradually narrowed into a very short awn
                              5. *C. alma.*
                  Pistillate scales rounded above, abruptly awned . . . .6. *C. Dudleyi.*

      Plants forming large tussocks; perigynia broadest at the subcordate
      base . . . . .7. *C. Cusickii.*
    Spikelets with staminate flowers at base.
      Perigynia not thin-margined . . . . .8. *C. leptopoda.*
      Perigynia thin-margined
        Perigynia 3 to 3.5 mm. long . . . . .9. *C. subfusca.*
        Perigynia 3.5 to 4.5 mm. long .10.. *C. Harfordii.*
Spikelets of two distinctly different kinds, the terminal staminate (sometimes with
a few pistillate flowers) and the lateral pistillate (sometimes with a few staminate
flowers).
    Perigynia pubescent (hairs visible at magnification of 10 diameters).
      Pistillate spikelets few-flowered (rarely with as many as 10 flowers), the
      lower ones on long pedicels from the basal leaf-axils.
        Body of perigynium globose, abruptly narrowed into the beak.
          Perigynia several-ribbed longitudinally, in addition to the 2 lat-
          eral keels . . . . .11. *C. globosa.*
          Perigynia, aside from the 2 lateral keels, not evidently nerved
                            12. *C. brevicaulis.*
        Body of perigynium ovoid, gradually narrowed into the beak
                            13. *C. Brainerdii.*
      Pistillate spikelets many-flowered (mostly with more than 10 flowers, rarely
      as few as 5); basal leaves without axillary spikelets.
        Lowest bract long-sheathing; lower spikelets on long pedicels; scales
        ovate-acute; perigynia 3-angled .14. *C. triquetra.*
        Lowest bract with sheath not over 5 mm. long; spikelets sessile or the
        lowest short-pedicelled; scales lanceolate-acuminate; perigynia not 3-
        angled . . . . .15. *C. lanuginosa.*
    Perigynia glabrous (or in *C. obispoensis* "appressed-puberulent" but the hairs
    invisible under a hand lens).
      Lowest bracts with sheath over 2 cm. long; lowest spikelets with pedicels
      longer than sheaths.
        Leaf-tufts not surrounded by fibers at base.
          Leaves 1.5 to 3 mm. wide; pistillate spikes 2 or 3; perigynia 3.5
          mm. long . . . . .16. *C. mendocinensis.*
          Leaves 5 to 10 mm. wide; pistillate spikes 3 to 10; perigynia 6 to
          8 mm. long . . . . .17. *C. obispoensis.*
        Leaf-tufts surrounded by coarse fibers at base, formed by disintegra-
        tion of older leaves . . . . .18. *C. luzulina.*
    Lowest bracts consisting of blade only, or with sheath very short (in *C.
    spissa* sometimes to 2 cm. long); lowest spikelets with short pedicels or ses-
    sile (pedicel to 3 cm. long in *C. spissa*).
      Styles 3.
        Plants 10 to 20 dm. tall; leaves 7 to 14 mm. wide; pistillate spike-
        lets 60 to 140 mm. long . . . . .19. *C. spissa.*
        Plants mostly 4 to 7 dm. tall; leaves 1 to 3 mm. wide; pistillate
        spikelets 6 to 23 mm. long . . . . .20.. *C. serratodens.*
      Styles 2.
        Beak of perigynium bidentate and hispidulous . .21. *C. Barbarae.*
        Beak of perigynium entire, smooth.
          Forming tussocks in rocky stream-beds; leaves 1 to 4 mm.
          wide; pistillate spikelets 1 to 3 cm. long; achenes not con-
          stricted at the middle . . . . .22. *C. senta.*
          Forming large masses in coastal marshes; leaves 4 to 8 mm.

wide; pistillate spikelets 4 to 10 cm. long; achenes constricted
at the middle ....23. *C. obnupta.*

1. **C. Douglasii** Boott. Caliente Mt., in deep dry sand on ridge about 2 miles
west of highest peak (*8328*); locally plentiful but rarely flowering.

2. **C. pansa** Bailey. Occasional in sand near the sea: just north of Piedras Blancas
Point (*7659,* immature but apparently belonging to this species); White Lake south
of Arroyo Grande. The fact that plants showing the key-characters of *C. praegracilis*
are also found on coastal dunes indicates that a genetic difference exists. On the
other hand, plants identified by *Carex* specialists as *C. praegracilis* show so much
variation among themselves that the validity of *C. pansa* as a species must be judged
as doubtful.

3. **C. praegracilis** W. Boott. Common near the coast, especially in hollows among
sand-dunes, and occasional in moist places in the interior.

4. **C. tumulicola** Mkze. Woods and north-facing slopes in northwestern corner of
county: mouth of Arroyo de la Cruz (*7944*); San Carpoforo Creek just above Windy
Point (*8989*). These two collections are distinguishable from local plants of *C. praegracilis* by the features mentioned in the key, but it is uncertain to what extent these
differences hold true in other regions.

5. **C. alma** Bailey. Sycamore Canyon, west slope of La Panza Range, in rocky
stream-bed (*6994*). Known in the Santa Lucia Mts. of Monterey Co., but not yet
found in the San Luis Obispo Co. portion of that range.

6. **C. Dudleyi** Mkze. Rare in moist places in Santa Lucia Mts. toward the north:
Pine Mt. (*8016*); Waterdog Creek, *Hardham,* acc. to Leafl. West. Bot. 10: 80.

7. **C. Cusickii** Mkze. Forming tussocks in wet soil or even in shallow water: dune
lakes southwest of Arroyo Grande southward to Oso Flaco Lake (*7909*); boggy spots
in valley of upper Arroyo Grande (*7906*).

8. **C. leptopoda** Mkze. San Carpoforo Creek just above Windy Point, in wet places
in shade of boulders (*8995*).

9. **C. subfusca** W. Boott. *C. montereyensis* Mkze. Frequent in moist places near
coast, perhaps always in sandy soil, inland to Huasna district (*Eastwood & Howell
4143*). Perigynia with prominent ventral nerves (the distinguishing mark of *C. montereyensis*) are often found on the same plant with nerveless perigynia. Not being
willing to accept two species on one plant, I can find no reason for placing the
plants under two different names.

10. **C. Harfordii** Mkze. *C. sub-bracteata* Mkze. *C. gracilior* Mkze. Frequent in
moist places (or near Cambria in rather dry pine woods): hills near coast and in
Santa Lucia Mts. (mostly west slope); notably absent from areas of serpentine. Careful study of specimens has shown that the three names listed have been applied
arbitrarily to plants which constitute a unit geographically, ecologically, and in
aspect. The features alleged to separate the three "species" are often combined in a
single plant. As in *C. subfusca,* virtually nerveless perigynia and prominently nerved
ones may be found in the same spikelet. No useful purpose is served by attempting
to maintain a classification having such an esoteric basis.

11. **C. globosa** Boott. Common near coast, mostly on brushy or openly wooded
sandy hills, sometimes on open hilltops or bluffs near the sea, where the plants become dwarfed. Extending into Santa Lucia Mts., but it is uncertain how extensively
"true" *C. globosa* occurs in that range. See note under *C. Brainerdii.*

12. **C. brevicaulis** Mkze. Specimens from two northern coastal localities show the

key-character which is used to distinguish *C. brevicaulis* (perigynia nerveless or nearly so): Cambria, in pine woods (*6635*); San Carpoforo Creek (*Eastwood & Howell 5990*). These plants are entirely similar to some forms of *C. globosa* in aspect and show parallel variation in size. There is no geographic or ecologic barrier between these plants and typical *C. globosa*; whether or not a genetic barrier exists can not now be determined. It seems probable that *C. brevicaulis* will eventually prove to be a minor variant of *C. globosa* determined by a slight genetic difference. It is significant in this connection that a collection from between Rocky Butte and Pine Mt. in the Santa Lucia Mts. (*7896*) has mostly perigynia corresponding to *C. globosa*, but a few perigynia in the very same spikelet correspond to *C. brevicaulis*. The same mixed condition has been found in several collections from other parts of California.

13. **C. Brainerdii** Mkze. This species, like *C. brevicaulis,* is doubtfully distinct from *C. globosa*. Certain plants in the Santa Lucia Mts. correspond exactly with published descriptions of *C. Brainerdii* and closely match specimens from the Sierra Nevada and northern California. Because all such plants in the South Coast Ranges have until now been assumed to be *C. globosa,* it is certain that many occurrences have been overlooked. As in other parts of its range, *C. Brainerdii* here is typically an associate of yellow pine: Pine Top Mt. (*8887*). A collection from between Rocky Butte and Pine Mt. (*7896*) includes one plant referable to *C. Brainerdii* mixed with *C. globosa*. This may suggest that the two sometimes form mixed colonies without interbreeding, or alternatively that they represent parts of a single interbreeding population. Further field studies as well as garden studies of these plants are greatly needed. Plants of this section of *Carex* are often abundant in the Santa Lucia Mts., as in upper Lopez Canyon, but it is unknown whether *C. Brainerdii* or typical *C. globosa* is the more plentiful in this mountain range, because the plants have been passed over in collecting, on the assumption (perhaps erroneous) that they belong to the common and widespread species of coastal California.

14. **C. triquetra** Boott. "Clay soil of flats and rocky slopes": Huasna district, *Eastwood & Howell 4153* in 1937; not found in the county since.

15. **C. lanuginosa** Michx. Local in boggy places: Upper Arroyo Grande (*7907*); Nipomo Mesa (*Eastwood & Howell 3871*).

16. **C. mendocinensis** Olney. Wet ground in cypress grove, headwaters of Chris Flood Creek (*Hardham 6160*).

17. **C. obispoensis** Stacey. Forming large clumps in dry to moist soil: serpentine areas in Santa Lucia Range from Rinconada Mine (*6115*) northward; Perfumo Canyon in San Luis Range (*Eastwood & Howell 5931*); outside of serpentine area near mouth of Arroyo de la Cruz (*6684, 7951*). This sedge is not known outside of San Luis Obispo Co., but there is every reason to expect its occurrence in southern Monterey Co. It reaches its greatest abundance along the ridge northwest of Cuesta Pass, where it grows luxuriantly under cypresses.

18. **C. luzulina** Olney. Edges of streams, north slope of Cypress Mt. (*Hardham 4770* and *5660*). In appearance these plants resemble specimens identified as *C. luzulina* from north coastal California less than they do *C. albida* Bailey, which is localized in Sonoma Co. The Cypress Mt. plants may represent an unnamed endemic species coordinate with *C. albida* or, more credibly, both may be local races of the variable species *C. luzulina*.

19. **C. spissa** Bailey. Sawgrass Sedge. Forming large masses around springs and

along streamlets: on serpentine near San Luis Obispo and northward; also outside the area of serpentine rock north of Arroyo de la Cruz. One large colony even grew on the ocean bluff.

20. **C. serratodens** W. Boott. Occasional along streamlets or in stream-beds in Santa Lucia Mts., usually in rocky places, often on serpentine. Also "between the Laguna and San Luis Obispo" (*Eastwood & Howell 5937*) and in American Canyon, La Panza Range (*9835*).

21. **C. Barbarae** Dewey. Mostly in rather dry sandy soil near coast, sometimes in moist places, often in shade, extending inland to upper Salinas Valley (7 miles west of Pozo, *Eastwood & Howell 2298*). This species and *C. obnupta* are so similar superficially that, in the absence of a large series of mature specimens having achenes, it is nearly impossible to work out the exact local distribution of the two species. An additional source of difficulty is that plants of this type are often sterile. This fact would seem to indicate that interspecific hybridization may occur.

22. **C. senta** Boott. Forming dense tussocks in rocky stream-beds: common throughout Santa Lucia Mts.; Perfumo Canyon in San Luis Range; inland to upper Salinas River. According to J. T. Howell (oral communication), most plants of this region which have been called *C. nudata* are actually *C. senta*. (It should be mentioned in passing that this represents a reversal of an opinion earlier expressed by Mr. Howell in "The Vascular Plants of Monterey County," p. 162). In fact, not a single specimen of unmistakable *C. nudata* from San Luis Obispo Co. has been seen.

23. **C. obnupta** Bailey. Wet places near coast: south end of Morro Bay (*7867*, the plants immature and identification therefore tentative); plants near Piedras Blancas Point are certainly this species, showing the distinctive achenes of the section to which it belongs.

Palmae. PALM FAMILY

1. Phoenix L. DATE PALM

1. **P. canariensis** Hort. ex Chabaud. CANARY ISLAND PALM. Around planted trees, seedlings come up in great profusion. Usually these are not allowed to grow, but when left undisturbed they may develop to mature size, as has happened along the bed of at least one streamlet at San Luis Obispo.

Araceae. ARUM FAMILY

1. Zantedeschia Spreng.

1. **Z. aethiopica** (L.) Spreng. CALLA LILY. Sparingly escaping or persisting from abandoned plantings in moist places, as near Arroyo Grande and Cambria.

Lemnaceae. DUCKWEED FAMILY

Fronds[1] flat, variously shaped, more than 1 mm. long.
    Fronds not linear or narrowly oblong, usually with roots.
        Frond with more than 1 root . . . . .1. *Spirodela.*
        Each frond with 1 root only, sometimes none . . . . .2. *Lemna.*
    Fronds linear or narrowly oblong, rootless . . . . .3. *Wolffiella.*
Fronds globular or flattened on top, not over 1.5 mm. in diameter . . . . .4. *Wolffia.*

---

1. "Frond" is the convention term for the reduced vegetative body of the *Lemnaceae.*

1. **Spirodela** Schleiden

1. **S. polyrrhiza** (L.) Schleiden. Seen only once in the county, in Pismo Creek at Pismo Beach.

2. **Lemna** L. Duckweed

Frond less than 5 mm. long, round or oval, without a long "stalk."
    Frond symmetrical, not narrowed at one end.
        Frond when fully developed 2 to 4 mm. long.
            Frond flat on both surfaces . . . . 1. *L. minor.*
            Frond strongly convex on lower surface . . . . 2. *L. gibba.*
            Frond 1 to 2 mm. long . . . . 3. *L. minima.*
        Frond asymmetrical, slightly narrowed at one end . . . . 4. *L. valdiviana.*
Broad portion of frond 6 to 10 mm. long, narrowed into a long "stalk," budding at right angles so as to appear 3-lobed . . . . 5. *L. trisulca.*

1. **L. minor** L. Apparently very common on the surface of pools and sluggish streams. Possibly, however, some of the plants taken for this species on casual observation are really *L. gibba* or *L. minima*.

2. **L. gibba** L. Marshes in the sand-dunes at Oceano (*Nobs & Smith 810,* "fronds gibbous to flat").

3. **L. minima** Phil. Floating on pools or slow streams: Ortega Spring, Palo Prieto Canyon (*Twisselmann 3112*); abundant on Salinas River one mile southwest of Pozo (*Twisselmann 2370*); upper Arroyo Grande (*R. D. Page 212*).

4. **L. valdiviana** Phil. Plants, identified at the time as this species, were found growing mixed with *Wolffiella* in Black Lake Canyon; no specimens at present available.

5. **L. trisulca** L. Ivy-Leaved Duckweed. Abundant in Oso Flaco Lake (*Piehl 63,777*).

3. **Wolffiella** Hegelm.

1. **W. lingulata** Hegelm. Submerged in fresh water, Black Lake Canyon to Oso Flaco Lake.

4. **Wolffia** Horkel

Frond not flattened on top, 0.5 to 1 mm. in diameter . . . . 1. *W. columbiana.*
Frond flattened on top, 0.7 to 1.5 mm. in diameter . . . . 2. *W. arrhiza.*

1. **W. columbiana** Karst. "Dune Lake," according to Mason, "A Flora of the Marshes of California," p. 341. The chain of lakes south of Arroyo Grande are often called the dune lakes, but no one of them, as far as I can learn, is customarily designated specifically as "Dune Lake."

2. **W. arrhiza** (L.) Wimm. Oso Flaco Lake, according to Mason.

Juncaceae. Rush Family

Leaves firm, usually stiff, in several species reduced to bladeless sheaths, never with cobwebby hairs; sheaths open; capsule with many small seeds . . . . 1. *Juncus.*
Leaves thin and soft, flat, with sparse cobwebby hairs when young; sheaths closed; capsule 3-seeded . . . . 2. *Luzula.*

1. **Juncus** L. Rush

Inflorescence apparently lateral, the involucral bract appearing exactly like a continuation of the stem.

Flowers in small heads which are arranged in a panicle; involucral bract usually about equalling inflorescence, rarely up to twice as long . . . . .1. *J. acutus.*
Flowers in a panicle, the panicle often dense but the flowers not in heads; involucral bract about equalling inflorescence or usually much longer.
 Rhizome with long branches, not forming dense round tufts but sometimes rather dense irregular patches; perianth 4 to 6.5 mm. long.
  Perianth dark brown or with dark brown stripes, 5 to 6.5 mm. long; growing on dunes or in salt-marshes . . . . .2. *J. Leseurii.*
  Perianth pale, sometimes with darker stripes, 4 to 5 mm. long; not on dunes or in salt-marshes.
   Stem 1 to 2 m. tall, 3 to 5 mm. thick near base, with groups of subepidermal fibers and vascular bundles in 3 or 4 rows
                    3. *J. textilis.*
   Stem rarely as much as 1 m. tall, 1 to 2 mm. thick near base, without groups of fibers and with vascular bundles in 1 or 2 rows.
    None of basal sheaths bearing blades. . . . . .4. *J. balticus.*
    Some of basal sheaths bearing a long slender stem-like blade
                5. *J. mexicanus.*
 Rhizome with short branches, forming a dense round tuft; perianth 2 to 3 mm. long.
  Perianth-segments spreading in fruit; capsule subglobose; stamens 6
                6. *J. patens.*
  Perianth-segments ascending; capsule obovoid; stamens 3 7. *J. effusus.*
   Inflorescence loose; capsule green or pale brown . . . .var. *pacificus.*
   Inflorescence compact; capsule medium or dark brown
                var. *brunneus.*
Inflorescence terminal, the involucral bract, if any, not appearing exactly like a continuation of the stem.
 Annuals with a small cluster of fibrous roots.
  Stems branching above the base.
   Perianth 4 to 8 mm. long; capsule 3 to 4 mm. long, oblong, somewhat pointed above . . . . .8. *J. bufonius.*
    Flowers mostly solitary, well spaced . . . . .var. *bufonius.*
    Flowers crowded in small clusters at ends of some of branches
                var. *congestus.*
   Perianth 3 to 4 mm. long; capsule 2 to 3 mm. long, broadly oval, rounded above . . . . .9. *J. sphaerocarpus.*
  Stems branching only at base, the flowers solitary (sometimes 2 or 3) on scape-like peduncles.
   Seeds faintly reticulate . . . . .10. *J. uncialis.*
   Seeds longitudinally ridged.
    Seeds with longitudinal ridges and fine transverse lines
              11. *J. brachystylus.*
    Seeds with longitudinal ridges not more prominent than the transverse ones . . . . .12. *J. capillaris.*
 Perennials, usually with rhizomes (*J. tenuis* tufted).
  Blades not at all septate, with flat surface facing the stem.
   Leaves all basal, less than 1.5 mm. wide; flowers often crowded but not in true heads, each with 2 bractlets .13. *J. tenuis.*
   Leaves not all basal, 1.5 to 4 mm. wide; flowers in heads which are subtended by bracts, but the individual flowers without bractlets.
    Heads 8 or fewer; perianth dark brown . . . . .14. *J. falcatus.*
    Heads numerous; perianth green or slightly tinged with brown
              15. *J. macrophyllus.*

Blades more or less septate (the septa visible externally as transverse ridges), either terete or, if flat, facing the stem edgewise.
Blades terete or nearly so, with complete septa.
Capsule dark brown, definitely longer than perianth; blades firm, scarcely 1 mm. in diameter .... 16. *J. articulatus.*
Capsule pale brown, equalling perianth or barely exceeding it; blades soft and spongy, over 1 mm. in diameter ..... 17. *J. dubius.*
Blades flat, with incomplete septa.
Heads numerous (nearly always more than 15); perianth pale, 3 to 4 mm. long .... 18. *J. xiphioides.*
Heads 15 or fewer; perianth dark brown, 4 to 5 mm. long
19. *J. phaeocephalus.*

1. **J. acutus** L. var. **sphaerocarpus** Engelm. Forming large clumps often 2 meters tall, on southern shores of Morro Bay; Avila; Pismo Beach and southward, in salt-marshes and in moist hollows bordering dunes.

2. **J. Leseurii** Bol. Extensively spreading by rhizomes on low dunes bordering beaches, in hollows among the dunes, and in coastal salt or brackish marshes.

3. **J. textilis** Buch. Basket Rush. Known at only two localities in the county, where it may have been introduced from farther south: Steiner Creek ("Serrano Canyon") near San Luis Obispo; Oak Park district (*7405*). At both these places the plants grow in dry soil; this is apparently not true of the species elsewhere. This rush was used by California Indians in weaving baskets.

4. **J. balticus** Willd. Common toward the interior in dry to moist soils; apparently absent from the coastal area where *J. Leseurii* occurs; extending to Temblor Range according to Twisselmann.

5. **J. mexicanus** Willd. Dry to moist places, often in shade, in Santa Lucia Mts. and hills eastward: Franklin Creek; summit between Santa Margarita and Creston; 2 miles west of Pozo; to be expected elsewhere. Seems to merge into *J. balticus.*

6. **J. patens** E. Meyer. Common in moist or drying spots from coast eastward through Santa Lucia Mts. to upper Salinas River.

7. **J. effusus** L. var. **pacificus** Fern. & Wieg. Moist places in Santa Lucia Mts.: Chris Flood Creek (*8999*) and almost certainly elsewhere.

Var. **brunneus** Engelm. Frequent in wet spots near the coast.

8. **J. bufonius** L. var. **bufonius.** Toad Rush. Common throughout in places which are wet in winter.

Var. **congestus** Wahl. Often associated with the typical form, and almost as common: Cambria; Santa Rita Creek; upper Cammatta Creek; Calf Canyon, where locally far more plentiful than var. *bufonius.*

9. **J. sphaerocarpus** Nees. Occasional in temporarily moist places away from the coast: between Bee Rock and the Nacimiento (*Hardham 3223*); Oak Flat west of Paso Robles (*Hardham 3159*); Calf Canyon (*9607*); 18 miles east of Creston on La Panza Road (*8471*).

10. **J. uncialis** Greene. Near Estrella (*Eastwood & Howell 4197,* identified by F. J. Hermann).

11. **J. brachystylus** (Engelm.) Piper. In winter-moist spots, uncommon or overlooked: between Bee Rock and the Nacimiento (*Hardham 3223c*); near Estrella (*Eastwood & Howell 4198*); 18 miles east of Creston (*8472*). Our plants differ from typical *J. brachystylus* as found elsewhere in having the peduncles nearly always one-flowered. In appearance they closely match *J. uncialis,* from which they are distinguishable by their different seeds. F. J. Hermann and subsequent authors have

used the name *J. Kelloggii* Engelm. for this species, but neither the original description of *J. Kelloggii* nor a photograph of the type specimen agrees with *J. brachystylus*. The photograph suggests a very young, and perhaps somewhat abnormal, plant of *J. bufonius*.

12. **J. capillaris** Hermann. In loose sand where flooded during rains, Wilson Canyon creek about 1 mile south of its confluence with Middle Branch of Huerhuero (*9907*).

13. **J. tenuis** Willd. var. **congestus** Engelm. Occasional near coast in moist, usually sandy, soil; extending into Santa Lucia Mts.

14. **J. falcatus** E. Meyer. Occasional in moist to dry sandy soil near coast: Cambria (*7223*); Price Canyon (*9172*); Oak Park district (*6699*). The plants are somewhat variable in inflorescence. Most of those found locally correspond to *J. falcatus* var. *paniculatus* Engelm. from the Mendocino Co. coast, or vary toward *J. orthophyllus* Cov. of the interior mountains. Individual stems in each locality show all gradations from var. *paniculatus* with several flower-heads to the one-headed plants which are commonly regarded as typical *J. falcatus*.

15. **J. macrophyllus** Cov. Mostly in stream-beds, frequent from Santa Lucia Mts. eastward through La Panza Range.

16. **J. articulatus** L. In wet places at San Luis Obispo. Although generally regarded as native over a large part of North America as well as Eurasia, this species is doubtless introduced here.

17. **J. dubius** Engelm. Locally common, usually in stream-beds, in upper Salinas watershed from Atascadero to La Panza Range; Stoney Creek in Huasna River watershed. The forma *rugulosus* (Engelm.) Hoover has not yet been found here but probably will be, as it is known in Monterey Co. and from Santa Barbara Co. southward.

18. **J. xiphioides** E. Meyer var. **xiphioides.** *J. phaeocephalus* Engelm. var. *paniculatus* Engelm. as often identified by botanists, but perhaps not as to type specimen. Common along streams and around springs in both eastern and western parts of county. In addition to *J. xiphioides*, Twisselmann reported *J. oxymeris* Engelm. as being common in the Temblor Range, but no specimens from there have been seen. Herbarium specimens identified as *J. oxymeris* and as *J. xiphioides* resemble one another so closely, down to the most trivial detail, as to cast serious doubt upon the reliability of the features which have been used by authors to segregate the plants into two species.

19. **J. phaeocephalus** Engelm. *J. phaeocephalus* var. *glomeratus* Engelm. *J. xiphioides* var. *glomeratus* (Engelm.) Hoover. Wet (or formerly wet) places near the coast, where commoner than *J. xiphioides*; rarely inland to Salinas Valley: Atascadero (*7118*). This species seems to merge into *J. xiphioides*; but, until an exhaustive study of the genetics of *Juncus* has been made, it is perhaps premature to classify both as elements of the same species. Such a study would probably result in a drastic reduction in the number of recognized species in the genus. Several of the recognized species are segregated on the basis of characters seemingly as inconstant as those which differentiate *J. phaeocephalus* from *J. xiphioides*.

2. Luzula DC. Wood-Rush

1. **L. campestris** (L.) DC. Common in woods in western part. The correct name for these plants may be *L. multiflora* (Retz.) Lejeune, *L. comosa* E. Meyer, or *L. subsessilis* (Wats.) Buch. All these names have been used by different authors for coastal

Californian plants. There is considerable variation but scarcely any basis for separating the plants into clearly distinguishable groups. A collection from Price Canyon (*6741*) has stems barely exceeding the leaves, congested inflorescence, and dark brown flowers only 2.5 mm. long. One from near the Rocky Butte Fire Lookout (*9073*) has taller stems, more open inflorescence, and lighter brown flowers 4 to 5 mm. long. Examination of a large number of specimens from various parts of California reveals no correlation of the differences mentioned. European material identified by European botanists as *L. campestris* and as *L. multiflora* shows a similar range of variation.

Liliaceae. Lily Family

Plants with bulbs; flower-color and inflorescence various.
  Sepals resembling the petals in length and color, often differing slightly in shape.
    Leaves basal (except for reduced bract-like ones), arising directly from the bulb; flowers erect or spreading.
      Pistil with 3 distinct styles; perianth-segments with green or yellow gland near base . . . . 1. *Zigadenus.*
      Pistil with 1 style; perianth without conspicuous nectaries.
                              2. *Chlorogalum.*
    Leaves all borne above base of stem; flowers (in our species) pendent.
      Flowers not orange; anthers basifixed; style 3-branched . . 3. *Fritillaria.*
      Flowers (in our species) large, bright orange with dark spots; anthers attached near middle; style not branched . . . . .4. *Lilium.*
  Sepals unlike the petals in shape and color, and usually either shorter or longer.
    Bulbs with membranous coats; sepals shorter than petals or about as long; petals not brown-tipped, the margin with short hairs or (usually) none.
      Leaves neither channeled nor keeled; style slender, short; stigma-lobes slender; pedicels recurved at least in fruit . . . . .5. *Calochortus.*
      Leaves longitudinally channeled above, often keeled on the back; style none (often the top of the ovary not developing seeds and therefore simulating a stout style); stigma-lobes stout; pedicels erect in flower and fruit . . . . .6. *Mariposa.*
    Bulbs with fibrous coats; sepals (in our species) longer than petals; petals with reddish-brown tips, long-fringed . . . . .7. *Cyclobothra.*
Plants without bulbs, the stems rooting at base; flowers white, in a panicle
                              8. *Asphodelus.*

1. Zigadenus Michx. Zygadene

Perianth-segments 4 to 5 mm. long, barely equalling the stamens or shorter
                              1. *Z. venenosus.*
Perianth-segments 7 to 15 mm. long, distinctly longer than stamens.
  Perianth-segments 9 to 15 mm. long; top of bulb 10 cm. deep or less
                            2. *Z. Fremontii.*
    Plants 25 to 70 cm. tall; flowers in a panicle . . . . .var. *Fremontii.*
    Plants 10 to 25 cm. tall; flowers in a simple raceme . . . . .var. *minor.*
  Perianth-segments 7 to 9 mm. long; top of bulb buried 15 cm. deep or more
                          3. *Z. brevibracteatus.*

    1. **Z. venenosus** Wats. Death Camass. Hog's Potato. Widespread along watercourses in the Santa Lucia and La Panza ranges and the hills between; extending south to Huasna district (*Eastwood & Howell 4157*). Occasional plants having the small flowers with relatively long stamens of *Z. venenosus* and growing, like it, in

wet soil along streams, but with branched rather than simple inflorescence, may possibly represent *Z. fontanus* Eastw. A collection from upper Navajo Creek (*7877*) shows proportionately few plants with branched inflorescence, but one from Stoney Creek (*8047*) consists entirely of branched plants.

2. **Z. Fremontii** (Torr.) Wats. var. **Fremontii.** STAR ZYGADENE. Common in various soils and in both sunny and shaded positions from coast eastward to La Panza Range, but absent from the more level lands of the Salinas Valley. Very vigorous plants which come up after fires in woods or chaparral are probably var. *Inezianus* Jepson; apparently it is an environmental form.

Var. **minor** (H. & A.) Jepson. Grasslands near coast from near San Luis Obispo northward. Bulbs transplanted to the garden produce taller stems with more numerous flowers, thus approaching typical *Z. Fremontii.*

3. **Z. brevibracteatus** (Jones) Hall. Around La Panza Range, especially on the east side. The identification of these plants is tentative. They seem not exactly typical of *Z. brevibracteatus* but resemble it more closely than any other known species.

2. **Chlorogalum** Kunth

Flowers white to purple, open in late afternoon; perianth-segments strongly recurving, 15 to 23 mm. long; bulb-coats coarsely fibrous . . . . 1. *C. pomeridianum.*
    Stem erect, branching only above the base.
        Stem 6 to 25 dm. tall . . . . var. *pomeridianum.*
        Stem 3 to 6 dm. tall . . . . var. *minus.*
    Stems with widely divaricate branches from the base, the plant wider than tall
                                                  var. *divaricatum.*
Flowers blue, open all day; perianth-segments spreading, 5 to 7 mm. long; bulb-coats membranous . . . . .2. *C. purpureum.*

1. **C. pomeridianum** (DC.) Kunth var. **pomeridianum.** SOAP PLANT. Common in the hills, except along the immediate coast, eastward through La Panza Range.

Var. **minus** Hoover. Areas of serpentine or chemically similar rock: ridge northwest of Cuesta Pass and near San Luis Obispo to hills east of Morro Bay. Bulbs planted in the garden produce taller stems than in the wild. This fact suggests that "var. *minus*" may be merely an environmental form.

Var. **divaricatum** (Lindl.) Hoover. Common along the coast from Cambria northward; in some places very plentiful. Also on Pt. Sal Ridge, Santa Barbara Co. In cultivation, the plants retain their low spreading manner of growth.

2. **C. purpureum** Bdg. var. **reductum** Hoover. Hard-packed gravelly and pebbly soil, 18 miles east of Creston on La Panza road; upper Cammatta Canyon, *Twisselmann 2039*; abundant in the very small area where it occurs. Plants growing around oaks, where the soil is looser and contains some humus, become somewhat taller. Typical *C. purpureum* is a similarly localized endemic around Jolon, Monterey Co.

3. **Fritillaria** L.

Largest leaves on lower part of stem, near the surface of the ground; bulb without bulblets resembling rice-grains.
    Herbage deep green; bracts broadly lanceolate or elliptical; flowers faintly or not noticeably odorous . . . . .1. *F. biflora.*
    Herbage glaucous; bracts narrowly lanceolate or linear-lanceolate; flowers strongly odorous . . . . .2. *F. agrestis.*

Lower half (approximately) of stem leafless, the lowest leaves well above the base; base of bulb bearing many easily detached bulblets resembling rice-grains.
  Perianth-segments 25 to 30 mm. long ....3. *F. lanceolata.*
  Perianth-segments 15 to 20 mm. long ....4. *F. viridea.*

1. **F. biflora** Lindl. CHOCOLATE LILY. MISSION BELLS. Grasslands, open woods, and sparse chaparral from coast inland to near Creston. Usually in clay soils, but the most vigorous plants were found growing in silt loam near San Simeon. Large stands of both this species and the following show much variation in the proportions of brown-purple and of green in the flowers.

2. **F. agrestis** Greene. Eastern parts of county, in calcareous clay, from Highland district (about 10 miles east of Creston) to La Panza district, Cottonwood Pass, and higher parts of Temblor and Caliente ranges. Often found in large numbers, and not destroyed by ordinary plowing of grain-fields. Possibly *F. succulenta* Elmer is the correct name for the plants here called *F. agrestis.*

3. **F. lanceolata** Pursh. CHECKER LILY. In shady woods, widespread through Santa Lucia Mts. and San Luis Range. The broad solitary leaves which grow from non-flowering bulbs are frequently seen, but in our area the bulbs rarely produce flowering stems. Our plants are all of the narrow-leaved phase which has been called var. *floribunda* Benth. or *F. mutica* Lindl.

4. **F. viridea** Kell. Ridge northwest of Cuesta Pass, growing in leaf-mold in *Quercus durata* thicket (*8734*); only one plant found in flower.

4. **Lilium** L.

1. **L. pardalinum** Kell. LEOPARD LILY. Occasional around springs and along brooks: San Simeon Creek (*8028*); Tassajera Creek; Lopez Canyon; Perfumo Canyon (*Eastwood & Howell 5918*).

5. **Calochortus** Pursh

Flowers pendent; petals arched so that their tips touch or come close together, hairy all over inner surface ....1. *C. albus.*
Flowers erect, bowl-shaped; petals hairy only toward base ....2..*C. uniflorus.*

1. **C. albus** Dougl. FAIRY LANTERN. WHITE GLOBE-TULIP. Common in shaded places, or along the north coast in open exposures, from Salinas River westward; rare, if not absent, east of that river. Plants with white flowers, sometimes tinged with pink or green, are commonest. A form with pink to deep rose-red flowers occurs in the Santa Lucia Range from Old Creek and Santa Rita Creek northward. This color variant has been identified with var. *rubellus* Greene, which was based on a collection from Pacific Grove, Monterey Co. The only plants recently seen at the type locality of var. *rubellus* have flowers which are at most pink-veined and slightly pink-tinged.

2. **C. uniflorus** H. & A. PINK STAR-TULIP. Clay soil of coastal meadow, 1.8 miles north of Arroyo de la Cruz (*7447* in 1948). This southernmost stand of the species, the only one known in the county, seems to have been completely destroyed in a "range-improvement" program.

6. **Mariposa** (Wats.) Hoover. MARIPOSA LILY

Gland on the flat surface of the petal, not excavated.
  Gland round, quadrate, or longitudinally oval, not longer transversely than lengthwise of petal.

Petals white to lilac (rarely pale yellow), the color markings, if any, restricted to the lowest 2/5 or less.
  Capsule oblong, obtuse; petals with red area around gland, otherwise unmarked ....1. *M. catalinae.*
  Capsule linear in outline, narrowed toward the top.
    Petals 3 to 5 cm. long; old leaves forming persistent sheathing bases; plants of dry places.
      Petals white (purplish on back) or rarely pale yellow, with red area around gland and usually a small red spot immediately above it; stem bulblet-bearing underground
2. *M. simulans.*
      Petals uniformly lilac, sometimes with a small purple spot at the very base; stem without bulblets. ....3. *M. splendens.*
    Petals about 25 mm. long, white (sometimes pale purple outside our range), often with a brown spot surrounding the gland; old leaf-bases delicate, early disappearing; plants of moist places; stem bulblet-bearing ....4. *M. Palmeri.*
  Petals white to purple or red, with a dark spot near center and usually a fainter brownish or reddish spot near apex, sometimes entirely rose-red or deep red ....5. *M. venusta.*
Gland lunate to transversely oblong.
  Bulblets several; gland transversely oblong; petals white, purplish on outside ....6. *M. argillosa.*
  Bulblet usually one; gland lunate; petals (in our plants) bright yellow, rarely cream-color ....7. *M. lutea.*
Gland deeply excavated, appearing on back of petal as a bulge ......8. *M. clavata.*
  Petals deep yellow; anthers deep purple; hairs on petal markedly clavate; glandular hairs much branched.
    Plants 1 to 4 dm. tall, with internodes well over 2 cm. long; leaves ascending or curved outward ....var. *clavata.*
    Plants up to about 1 dm. tall, with internodes 2 cm. long or less; leaves curved downward ....var. *recurvifolia.*
  Petals light yellow; anthers yellow to light or medium purple; hairs on petal gradually and slightly enlarged upward; glandular hairs less extensively branched ....var. *pallida.*

1. **M. catalinae** (Wats.) Hoover. Pismo Beach, *Jones* in 1934, cited by Ownbey (Ann. Mo. Bot. Gard. 27: 454. 1940). There has been no confirmation of this record; but, as the species is found near Gaviota Pass in Santa Barbara Co., its occurrence in southwestern San Luis Obispo Co. is certainly not impossible.

2. **M. simulans** Hoover. Widespread through central part of county, where it is endemic: Tassajera Creek and near Atascadero to a few miles north of Creston, eastward to La Panza district and southward to Trout Creek (a tributary of Huasna River). In some years and in some places, it has flowered in great abundance. Occasionally plants with light yellow flowers, instead of the usual white, are found.

3. **M. splendens** (Dougl.) Hoover. LILAC MARIPOSA. Usually in calcareous soils, either sandy or clay, back from the coast: hills west of Paso Robles to Caliente Range, upper Arroyo Grande, and Cuyama Canyon.

4. **M. Palmeri** (Wats.) Hoover. Along streamlets where soil is wet during growing season but drying in summer, in interior parts of county: Mariana Creek (*7244*) and American Canyon in La Panza Range; Stoney Creek.

5. **M. venusta** (Dougl.) Hoover. Grasslands and open woods, frequent from sum-

mits and east slope of Santa Lucia Mts. eastward to Temblor Range; abundant on Caliente Mt. Largely absent from the area occupied by *M. simulans,* but growing with that species in the La Panza district. Highly variable in the basic color of the petals and in the color, size, and shape of the markings. Plants having white flowers are the most widely distributed, and in many places no others are found. In a few localities may be found flowers ranging through pink to rose-red and to purple. A form having petals almost uniformly deep red is found on the Mt. Abel road, Kern Co., and probably grows on Caliente Mt. and in the Temblor Range. A form with light yellow flowers is known but has not been seen in our area. There is also variation in vegetative features. In the Santa Lucia Mts. the stems grow straight up from the bulb and bear usually one bulblet, sometimes two, in the lowest leaf-axils. Plants growing in clay soils in the eastern part of the county generally have several bulblets. Often the bulbs of this form produce, instead of a flowering stem, a more or less horizontal subterranean stem which disperses the bulblets through the soil. In the most arid localities, few flowers or none are produced except in exceptionally rainy years, and the plants reproduce more by bulblets than by seeds. In all of its variations, this mariposa impresses one with its beauty.

6. **M. argillosa** Hoover. Stiff clay "adobe" soils: San Luis Valley and surrounding hills, extending to estuary of Osos Creek on Morro Bay. This species is defined as including three distinguishable and geographically separated populations (subspecies), of which our plants constitute one. The second occurs around Pt. Sal, Santa Barbara Co. The third, which includes the type of the species, is found in the hills bordering the Santa Clara Valley, with an outlying occurrence near Somersville, Contra Costa Co.

7. **M. lutea** (Dougl.) Hoover. YELLOW MARIPOSA. Clay or sandy loam soils near coast: Laguna near San Luis Obispo; near Cambria and northward. Also an isolated occurrence in Salinas Valley 2 miles east of Templeton (*Wiggins 2066*). A very few plants north of Arroyo de la Cruz have cream-colored instead of deep yellow flowers, but I do not believe these plants to be the same genetically as the cream-colored to white forms of this species which are so abundant in the Sierra Nevada foothills and the North Coast Ranges. Plants growing on ocean bluffs are of short stature, and this feature is retained in cultivation.

8. **M. clavata** (Wats.) Hoover var. **clavata.** Usually in clay soils, often derived from serpentine, or in rocky places: widespread in Santa Lucia Mts.; Perfumo Canyon and summits above in San Luis Range. San Luis Obispo is the type locality.

Var. **recurvifolia** Hoover. A very local dwarfed variant found 1.8 miles north of Arroyo de la Cruz, in clay soil of a coastal field. The plants when cultivated retain their vegetative peculiarities indefinitely.

Var. **pallida** Hoover. Sandy or rocky soils of the interior, in areas of sandstone or sometimes granite: La Panza Range to Temblor and Caliente Ranges.

7. **Cyclobothra** D. Don

1. **C. obispoensis** (Lemmon) Hoover. Hills around San Luis Valley, from ridge northwest of Cuesta Pass to divide between Perfumo Canyon and See Canyon in San Luis Range, Reservoir Canyon, and southward nearly to Arroyo Grande. Usually on serpentine, but toward Arroyo Grande growing on sandstone. Found nowhere but in this area.

8. Asphodelus L.

1. **A. fistulosus** L. Well established and spreading from a trash-dump 1 mile west of Nipomo, among brush in sandy soil (*8365* in 1956).

Convallariaceae. LILY-OF-THE-VALLEY FAMILY

Leaves well developed, parallel-veined.
 Flowers rotate, in an erect raceme or panicle .... 1. *Smilacina.*
 Flowers campanulate, drooping, solitary or in a few-flowered umbel.
                   2. *Disporum.*
Leaves reduced to thin dry scales, those on the branches subtending clusters of short filiform green branchlets .... 3. *Asparagus.*

1. Smilacina Desf. FALSE SOLOMON'S-SEAL

Flowers in a panicle; filaments broad, more conspicuous than perianth-segments
                1. *S. racemosa.*
Flowers in a raceme; filaments narrow, less conspicuous than perianth-segments
                2. *S. stellata.*

1. **S. racemosa** (L.) Desf. Occasional in moist shady woods: San Luis Range (See Canyon, *7510, 8866*; observed in Diablo Canyon); Santa Lucia Mts. (Santa Rita Creek and probably elsewhere). Our plants are referred to the rather poorly defined var. *amplexicaulis* (Nutt.) Wats., having leaves widest near base rather than near middle.

2. **S. stellata** (L.) Desf. var. **sessilifolia** (Baker) Hend. More or less shaded and moist places: common near coast (Black Lake Canyon, most canyons in San Luis Range, mouth of Osos Creek, and northward) and frequent in canyons of the Santa Lucia Mts. Locally much commoner than the preceding species, which sometimes grows with it. The typical form of *S. stellata,* found from the Sierra Nevada to the Atlantic Coast, has narrower leaves than this variety.

2. Disporum Salisb.

1. **D. Hookeri** (Torr.) Nich. FAIRY BELLS. Moist shady woods in canyon of Santa Rita Creek west of Templeton (*8465*), the southernmost occurrence of the species and the only one known in the county.

3. Asparagus L.

1. **A. officinalis** L. ASPARAGUS. Occasionally escaped from cultivation in moist, often saline soils: near shore of Morro Bay at mouth of Osos Creek; Twisselmann Ranch south of Cholame.

Amaryllidaceae. AMARYLLIS FAMILY

Stamens 6, with anthers all alike.
 Involucre of 2 to 4 distinct or partly united bracts; plants with odor and taste
  of garlic or onion .... 1. *Allium.*
 Involucre of several distinct bracts; plants lacking garlic or onion odor.
   Perianth-segments distinct, or joined at base to form a tube not more than
    1 mm. long.

Leaves several, very narrow (less than 2 mm. wide), neither keeled nor channeled; flowers greenish white (in our species); filaments without cup-shaped appendage ....2. *Muilla.*
Leaves 1 or rarely 2, keeled below and doubly channeled above; flowers yellow; filaments forming either at base or part way up a cup-like pocket from which the upper portion of the filament arises
3. *Bloomeria.*
Perianth-segments united for at least 2 mm. from the base ...4. *Triteleia.*
Stamens 6, the alternate 3 with smaller anthers, or 3 and alternating with staminodia.
Larger anthers sessile and bearing 2 wing-like appendages; smaller anthers with filaments broadened toward base ....5. *Dichelostemma.*
Stamens with anthers 3, alternating with staminodia ....6. *Brodiaea.*

1. Allium L.

Leaves 2 to several, more or less flattened even if very narrow; ovary without crests or merely with low rounded projections on top.
Perianth bowl-shaped to rotate, the segments spreading from the base.
Bulbs elongate, with reddish coats, the roots borne on an oblique base
1. *A. haematochiton.*
Bulbs round or ovoid, not red.
Stems and roots arising from short rhizomes rather than directly from the bulb; scape more than 15 cm. tall; bracts acuminate, 15 mm. long or more; perianth-segments 10 to 14 mm. long ....2. *A. unifolium.*
Stems and roots arising directly from the bulb; scape less than 15 cm. tall; bracts broadly acute or obtuse and mucronate, 10 mm. long or less; perianth-segments about 5 mm. long ....3. *A. Hickmanii.*
Perianth-segments ascending from the base, forming a cup-shaped perianth, or curved outward above.
Middle bulb-coats with rectangular reticulations; mid-vein of perianth-segments usually conspicuous, purple or sometimes green 4. *A. lacunosum.*
Perianth-segments 5 to 9 mm. long ....var. *lacunosum.*
Perianth-segments 4 to 6 mm. long ....var. *micranthum.*
Middle bulb-coats with strongly undulate ("serrate" or V-shaped) reticulations; mid-vein of perianth-segments not usually conspicuous.
Bulb reddish under the coats; umbel compact, with pedicels 6 to 15 mm. long; perianth-segments white or pink, all shaped nearly alike
5. *A. amplectens.*
Bulb not reddish; umbel loose, with pedicels 15 to 30 mm. long; perianth-segments rose-red or deep purple-red, the petals narrower than the sepals ....6. *A. peninsulare.*
Petals not white-margined or conspicuously undulate
var. *peninsulare.*
Petals with usually white, closely undulate margins var. *crispum.*
Leaf solitary, the blade terete; ovary bearing 6 prominent appendages ("crests") on top.
Stamens shorter than perianth ....7. *A. fimbriatum.*
Stamens longer than perianth ....8. *A. Howellii.*

1. **A. haematochiton** Wats. Locally plentiful in areas of serpentine rock in the hills around San Luis Obispo, the type locality and the northernmost known for the species.
2. **A. unifolium** Kell. Along streamlets where soil is wet in winter, Santa Lucia

Mts.: Santa Rita Creek; west side of York Mt. (*7995*); east slope of Pine Mt. Also along coast: Arroyo del Oso and northward. Greene described a variety *lacteum* from San Luis Obispo Co. as having white flowers, but the plants I have seen are all pink-flowered.

3. **A. Hickmanii** Greene. Open grassland along the northern coast: San Simeon (*Rose 36,080*); mesa summit south of Arroyo de la Cruz (*8164*). The plants are small and largely hidden by the surrounding grass.

4. **A. lacunosum** Wats. var. **lacunosum.** Rare in typical form, growing on serpentine: west base of Mt. Bishop near San Luis Obispo (*7163*); Perfumo Canyon (*Eastwood & Howell 5933*); east of Morro Bay and northwest of Cuesta Pass. The bulbs tend to form large dense clusters. In the interior are found plants with stems averaging taller and pedicels averaging longer, but otherwise they seem inseparable from typical *A. lacunosum*. Such plants have been called *A. Davisiae* Jones and range eastward to the mountains of Kern Co. and the Mojave Desert. Representative collections: Black Mt., La Panza Range, *Hardham 686*; 8.7 miles south of Simmler, *Hardham 3192*; Soda Lake Road at Werling Ranch, *Twisselmann 4389*. This variant is worthy of notice, but it does not seem practical to separate it even as a variety, because it is not clearly enough defined, geographically or otherwise. It seems to be exactly analogous to the *"Muilla serotina"* variant of *Muilla maritima*.

Var. **micranthum** Eastw. Hard rocky soil east of Paso Robles, *Chester Dudley in 1936*; La Panza Range (Pine Mt., *7238*) and bordering mesas to the northeast (18 miles east of Creston on La Panza road, *7692*). Locally abundant in the latter locality and showing considerable variation in flower size. Not on serpentine, at least ordinarily. The bulbs are generally solitary or divide to form only a few in a cluster. *Hoover 7692* shows small-flowered plants representative of var. *micranthum* mixed with larger-flowered plants which approach the *"A. Davisiae"* form.

5. **A. amplectens** Torr. Northeastern base of Cypress Mt., *Bacigalupi 7401*.

6. **A. peninsulare** Lemmon var. **peninsulare.** In shady woods, trail from Stoney Creek to Colwell Mesa (*7967*); summits of Temblor Range, acc. to Twisselmann. Much less common than the following variety, which does not seem sufficiently distinct to rank as a species. Dried specimens of the two are sometimes virtually indistinguishable.

Var. **crispum** (Greene) Jepson. Common over most of the interior in open woods, open brush, or grassland: summits of Santa Lucia Mts. (Pine Mt.) to Cottonwood Pass, Polonio Pass, Temblor Range, and Cuyama Valley. Paso Robles is the type locality.

7. **A. fimbriatum** Wats. On serpentine in southern parts of Santa Lucia Range: Rinconada Mine (*8521*); near Hi Mt. (*8774*).

8. **A. Howellii** Eastw. Mostly in friable gypseous clay soils, sometimes near serpentine, common through eastern part of county from Cottonwood Pass to Cuyama Valley. Often in patches of clay where little else grows.

2. **Muilla** Wats.

1. **M. maritima** (Torr.) Wats. Brushy places, open woods, and treeless plains, generally in sandy calcareous or somewhat saline soils: Occasional near coast (head of Carpenter Canyon north of Arroyo Grande, *6688*; upper Arroyo Grande; Nipomo Mesa, *Eastwood & Howell 3886A*); commoner in interior from Atascadero eastward to Carrizo Plain. Plants from near Creston eastward, often growing among sage-

brush (*Artemisia californica*) and buckwheat (*Eriogonum fasciculatum* var. *polifolium*), have taller stems and larger corms. This form has been called *M. serotina* Greene. Since the soil where such plants grows is neither moist nor fertile, their larger size doubtless has a genetic basis, but the flowers seem quite identical with those of smaller plants.

<h3 style="text-align:center">3. Bloomeria Kell.</h3>

Plant 15 cm. tall or usually more; perianth-segments barely joined at base; filaments slender, inserted in a cup-like structure at the base .... 1. *B. crocea.*
    Cup at base of filament without hair-like appendages.
        Filament-base with a blunt point on either side ....var. *crocea.*
        Filament-base with an acute point on either side ....var. *aurea.*
    Cup-like filament-base bearing 2 long hair-like appendages ......var. *montana.*
Plant less than 15 cm. tall; perianth-segments joined at base to form a tube 1 mm. long; lower part of filament broad, 3 to 5 mm. long, ending in a cup from which the slender upper part arises ....2. *B. humilis.*

    1. **B. crocea** (Torr.) Cov. var. **crocea.** Clay soils, often on serpentine, around San Luis Obispo, intergrading thoroughly with var. *aurea.* Plants around Cayucos and Morro Bay have not been studied in sufficient detail, so it is impossible to state exactly where is the northern limit of this typical phase of the species.

    Var. **aurea** (Kell.) Ingram. Golden Stars. Common over most of the interior, in sunny or shaded positions and in soils ranging from sandy loam to stiff clay: Santa Lucia Mts. eastward to Temblor Range and west to coast at San Simeon. Flowering in great profusion in some spots and in favorable years.

    Var. **montana** (Greene) Ingram. Intergrades with var. *aurea* have been observed at Rinconada Mine (*8361*) and eastward through La Panza Range. The variety *montana* occurs unmixed with the other varieties on the south (Santa Barbara Co.) side of Cuyama Canyon. Some form of the species is plentiful on Caliente Mt. but has not been seen in bloom; probably it will prove to be var. *montana.* This variety should also be sought in the southern part of the Temblor Range.

    2. **B. humilis** Hoover. Highly localized, but locally plentiful, in coastal grassland between San Carpoforo Creek and Arroyo de la Cruz, and on the mesa south of Arroyo de la Cruz. The differentiation of this species from *B. crocea* is discussed in Plant Life 11: 21–22. 1955, and, more importantly, in Plant Life 12: 37–38. 1956.

<h3 style="text-align:center">4. Triteleia Dougl. ex Lindl.</h3>

Perianth bowl-shaped, not narrowed at base, the segments joined for 2 to 4 mm. at base; stamens of equal length and attached at same level ....1. *T. hyacinthina.*
Perianth with funnelform tube 4 mm. long or more; stamens either alternately long and short or attached alternately at 2 levels.
    Perianth yellow or white; stamens with dilated 2-forked filaments of two lengths
                                    2. *T. ixioides.*
        Perianth yellow, the segments rotate or slightly ascending; pedicels less than 3 times as long as flowers ....var. *ixioides.*
        Perianth white, the segments rotate to reflexed; pedicels often more than 3 times as long as flowers ....var. *Cookii.*
    Perianth normally blue; stamens with slender filaments, of equal length but attached at two levels ....3. *T. laxa.*

    1. **T. hyacinthina** (Lindl.) Greene. Apparently very rare along our northern coast: 1 mile north of Arroyo de la Cruz (*9423*, only two plants found).

    2. **T. ixioides** (Ait. f.) Greene var. **ixioides.** Golden Brodiaea. Moist places in

sunny exposures or in open woods: Santa Lucia Mts. from Santa Rita Creek northward, intergrading with the localized variety following.

Var. **Cookii** Hoover. Known only in a small area of the Santa Lucia Range from near Cypress Mt. to Pine Mt.

3. **T. laxa** Benth. Churchbells Locally found in open oak woods, the plants very few: 2 miles northeast of Santa Margarita (*8516*); Atascadero Creek south of Atascadero. The plants are not of the coastal form found farther north but have smaller anthers and thus resemble plants from the San Joaquin Valley and southern Sierra Nevada. It is barely possible that they represent a casual, or even an intentional, introduction.

5. **Dichelostemma** Kunth

1. **D. pulchellum** (Salisb.) Heller. Schoolbells. Very common in all sections of the county in a wide range of habitats.

6. **Brodiaea** Smith

Staminodia 6 to 11 mm. long; filaments 2 to 4 mm. long, flat on back; anthers 4 to
7 mm. long . . . . .1. *B. coronaria.*
    Scape above ground distinctly longer than pedicels . . . . .var. *coronaria.*
    Scape above ground from much shorter than pedicels to about equalling them.
        Staminodia violet, the same color as the perianth; scape above ground usually nearly as long as pedicels . . . . .var. *kernensis.*
        Staminodia normally lighter in color than perianth (rarely both white); scape very short above ground . . . . .var. *macropoda.*
Staminodia 4 to 6 mm. long; filaments 1 mm. long, channeled on back; anthers 3 to
5 mm. long . . . . .2. *B. jolonensis.*

1. **B. coronaria** (Salisb.) Engler var. **coronaria.** Harvest Brodiaea. Plants with relatively tall scapes have been found in our area only at the summit between upper Salinas River and Stoney Creek, in an open field (*7958*), and on the northeast spur of Black Mt., La Panza Range (*Twisselmann 2933*). These localities are widely isolated from other known occurrences of typical *B. coronaria,* and the plants may possibly have more in common genetically with var. *kernensis.*

Var. **kernensis** Hoover. In rather moist spots from Santa Margarita to Creston and eastward to north end of La Panza Range (*7865*). These plants are referred to var. *kernensis* chiefly because of having staminodia of the same violet color as the perianth, not lighter or white. Also, the scapes tend to be longer than in var. *macropoda,* but this difference does not always hold. Another possibly distinctive feature of var. *kernensis* is that the staminodia are abruptly incurved at the very tip.

Var. **macropoda** (Torr.) Hoover. Represented by two separate populations. One, in the interior, has corms without offsets and is locally abundant about 18 miles east of Creston on the La Panza road. The other, found in the coastal area from Cambria northward, produces abundant offsets. The latter has smaller flowers and approaches *B. jolonensis.*

2. **B. jolonensis** Eastw. A local race occurring in heavy clay soils around San Luis Obispo is vegetatively very similar to the north-coastal form of *B. coronaria* var. *macropoda* mentioned above, but it shows the flower characters which serve as recognition marks for *B. jolonensis.* The plants are dwarfed as compared with the forms of *B. jolonensis* found elsewhere. Reported by Twisselmann from Palo Prieto Creek and Carrizo Plain, but no specimens seen.. The occurrence of the typical form of the species, as represented in Monterey Co., is to be expected.

Agavaceae. Century-Plant Family

Leaves fleshy only at base, not spiny-toothed; flowers white; ovary superior   1. *Yucca.*
Leaves fleshy throughout, spiny-toothed on margins; flowers yellow; ovary inferior
                                                                          2. *Agave.*

1. Yucca L.

1. **Y. Whipplei** Torr. Often very plentiful in Santa Lucia Mts. southward to Cuyama Canyon; summits of San Luis Range, east to La Panza Range, Caliente Range, and southern part of Temblor Range. Probably all our plants are referable to subsp. *percursa* Haines. In our coastal region, this species is restricted to serpentine rock, but toward the interior it grows on sandstone and granite.

2. Agave L.

1. **A. americana** L. Century Plant. Often planted. A large clump has persisted for years, blooming each year and gradually increasing, near the east end of San Luis Mt. at San Luis Obispo. How long this clump has been there is unknown, but certainly it seems established to the point where none but human agency could cause its disappearance. Another large clump was noticed near a long-abandoned homesite in Diablo Canyon.

Trilliaceae. Trillium Family

1. Trillium L.

1. **T. sessile** L. var. **angustipetalum** Torr. Occasional in moist deeply shaded places in San Luis Range (See Canyon; Coon Creek; Sycamore Canyon) and Santa Lucia Mts. It is remarkable that our plants have narrow petals like those of the remote Sierra Nevada, instead of being the broad-petaled form found northward in the Coast Ranges.

Iridaceae. Iris Family

Plants without corms; flowers not in spikes.
    Sepals spreading or recurved, markedly larger than the petals; style with 3 petaloid branches . . . . .1.*Iris.*
    Perianth-segments all alike; style unbranched, or the branches not petaloid
                                                                          2. *Sisyrinchium.*
Plants with corms; flowers in spikes.
    Axis of spike bent horizontally at base; bracts firm, green . . . . .3. *Freesia.*
    Axis of spike straight or flexuous; bracts thin-membranous, brownish 4. *Sparaxis.*

1. Iris L.

1. **I. Douglasiana** Herbert. Brushy slopes, moist ravines, and open bluffs along coast: Hazard Canyon, where perhaps it has been exterminated; near mouth of Arroyo de la Cruz and northward. All of our plants have violet flowers, but farther north the color is highly variable.

2. Sisyrinchium L.

Flowers blue (rarely varying to white); stems more or less branched, at least above
                                                                          1. *S. bellum.*
Flowers yellow; stems simple, scapose . . . . .2. *S. californicum.*

1. **S. bellum** Wats. Blue-Eyed Grass. Common on open hills and coastal bluffs, on borders of salt-marshes, and in wooded areas; less plentiful eastward to La Panza Range. Usually summer-dormant, but in moist places sometimes blooming throughout the year. Plants along the coast from Morro Bay northward are low, with relatively broad leaves and large intense blue flowers. At the other extreme, the plants toward the interior, particularly east of the Salinas River, are taller, with narrow leaves and smaller paler flowers. A varietal distinction would be useful, but the difficulty is in deciding what kind of plant should be regarded as the typical *S. bellum*.

2. **S. californicum** (Ker) Dryand. Golden-Eyed Grass. In wet spots along coast from Piedras Blancas Point (*6965*) northward.

3. Freesia Klatt

1. **F. refracta** Klatt. Common garden flower, spreading aggressively in cultivation; naturalized at Cambria according to J. T. Howell.

4. Sparaxis Ker

1. **S. grandiflora** Ker. Also naturalized at Cambria according to J. T. Howell.

Orchidaceae. Orchid Family

Flowers sessile, small, in the axils of reduced bracts.
    Perianth with a spur . . . . .1. *Habenaria.*
    Perianth without a spur . . . . .2. *Spiranthes.*
Flowers short-pedicelled, rather large (25 to 35 mm. in diameter), in the axils of leafy bracts. . . . . .3. *Epipactis.*

1. Habenaria Willd. Rein-Orchis

Well developed leaves few, all near base of stem, those above the base reduced and scale-like . . . . .1. *H. unalascensis.*
    Flowers green.
        Spike less than 15 mm. thick, with flowers not much crowded; spur typically short but variable . . . . .var. *unalascensis.*
        Spike more than 15 mm. thick, with flowers usually much overlapping; spur at least twice as long as perianth-segments . . . . .var. *elata.*
    Flowers white . . . . .var. *maritima.*
Well developed leaves several, gradually becoming smaller from base upward
                                    2. *H. dilatata.*

1. **H. unalascensis** (Spreng.) Wats. var. **unalascensis.** Frequent in dry shady woods in Santa Lucia Range, usually rooted in decaying forest litter. Specimens from the vicinity of Pine Mt. and Rocky Butte are called var. *unalascensis* because of their slender and rather loose spikes, but the flowers have long spurs as in var. *elata.*

Var. **elata** (Jepson) Correll. Wooded areas and dense chaparral in western part, probably the commoner variety locally. Plants are rather plentiful, but only a small proportion bloom in most years, because of our climate. *Habenaria Michaelii* Greene, which Correll and others regard as essentially identical with this variety, was based on a specimen from "San Luis Obispo County."

Var. **maritima** (Greene) Correll. Coastal hill south of Arroyo de la Cruz, only one plant in poor condition seen in flower. This variety is distinguished by its stout short stems and white fragrant flowers, but these features are not always consistently combined. Within the city limits of Monterey, plants have been seen with white flowers but with tall, rather slender stems and without the faintest fragrance.

2. **H. dilatata** (Pursh) Hook. var. **leucostachys** (Lindl.) Ames. WHITE REIN-ORCHIS. Bog in Black Lake Canyon (*7338, 7352*).

2. Spiranthes Rich. LADIES'-TRESSES

Lip abruptly narrowed below the broad apex, without papillae at base
                                                                    1. *S. Romanzoffiana.*
Lip not narrowed below the tip, with a pair of papillae at base .....2. *S. porrifolia.*

1. **S. Romanzoffiana** C. & S. Near mouth of Arroyo de la Cruz, on north-facing slope (*8264*).

2. **S. porrifolia** Lindl. Moist places in pine woods at Cambria. The plants are locally rather plentiful, but they seldom bloom at that locality. It is doubtful whether this is really a species distinct from *S. Romanzoffiana*. Aside from the unsatisfactory nature of herbarium specimens, flowering material is too scarce to form a basis for proper judgment on the matter.

3. Epipactis Rich.

1. **E. gigantea** Dougl. STREAM ORCHIS. Widely scattered in wet places in western part: San Simeon Creek; south fork of Old Creek; Morro Creek; Lopez Canyon; Stoney Creek; Black Lake Canyon.

Saururaceae. LIZARD-TAIL FAMILY

1. Anemopsis Hook.

1. **A. californica** Hook. YERBA MANSA. Permanently moist places in valleys of the interior: Palo Prieto Canyon; Shandon; Creston. Common along the San Juan River, according to Twisselmann.

Salicaceae. WILLOW FAMILY

Leaves more than twice as long as wide; scales persistent after flowers open; flowers without a disk . . . . .1. *Salix.*
Leaves less than twice as long as wide; scales falling as flowers open; flowers with a disk . . . . .2. *Populus.*

1. Salix L. WILLOW

Lower leaf-surface lighter in color than the upper, tomentose to puberulent or glabrous.
    Lower leaf-surfaces densely soft-tomentose . . . . .1. *S. Coulteri.*
    Lower leaf-surfaces glabrous to densely pubescent but not thick-tomentose.
        Scales dark; stamens 2; leaves remotely serrulate to entire, pubescent on lower surface or in age glabrate . . . . .2. *S. lasiolepis.*
            Leaves narrowly oblong to oblanceolate or obovate, mostly over 15 mm. wide.
                Leaves mainly oblong to oblanceolate . . . . .var. *lasiolepis.*
                Leaves obovate to broadly oblanceolate or elliptic . .var. *Bigelovii.*
                Leaves linear to linear-lanceolate or narrowly oblanceolate, 14 mm. wide or less . . . . .var. *Braceliniae.*
        Scales yellow; stamens 3 to 10; leaves closely serrulate to entire, glabrous on lower surface (except sometimes near base).

Petioles without glands ....3. *S. laevigata.*
Branchlets glabrous ....f. *laevigata.*
Branchlets pubescent ....f. *araquipa.*
Petioles with wart-like glands near base of blade ......4. *S. lasiandra.*
Leaf-surfaces both essentially alike in color, usually appressed-silky, at least when young.
Leaves green, tending to be early glabrate ....5. *S. melanopsis.*
Leaves silvery or gray-silky.
Stigmas 1 mm. long; style present though very short; capsule hairy or glabrate ....6. *S. Hindsiana.*
Leaves 3 to 10 mm. wide; style 0.5 mm. long ....var. *Hindsiana.*
Leaves 2 to 3.5 mm. wide; style 0.1 to 0.2 mm. long ......var. *Parishiana.*
Stigmas 0.5 mm. long; style none; capsule glabrous ....7. *S. exigua.*

1. **S. Coulteri** And. Moist rocky places in Santa Lucia Range: Steiner Creek, Chorro Creek, Tassajera Creek, Morro Creek and coast near Monterey Co. line. The localities near San Luis Obispo, where the species apparently reaches its southern limit, are on serpentine, but northward it grows on other kinds of rock.

2. **S. lasiolepis** Benth. var. **lasiolepis.** CALIFORNIA PUSSY-WILLOW. ARROYO WILLOW. Our commonest willow, found everywhere along stream-courses in western part. While considerable water is needed for the establishment of seedlings, full-grown individuals may survive where there is no surface indication of moisture. One of the most notable occurrences is on the dunes from Oceano southward, where the tips of the branches continue to grow indefinitely, even though the original base of the tree has been deeply buried in shifting sand. Largely absent from the hot arid eastern part, even along streams, but occasional in Temblor Range, according to Twisselmann.

Var. **Bigelovii** (Torr.) Bebb. Black Lake (*Eastwood 14,349*, identified by C. R. Ball). Perhaps most of the willows on the sand-dunes belong to this variety.

Var. **Braceliniae** Ball. From Salinas River eastward; perhaps the prevalent form in that area.

3. **S. laevigata** Bebb. f. **laevigata.** RED WILLOW. Common along streams and in moist valley lands in both coastal and interior regions. Varies from a shrub to a fair-sized tree. Probably a minor genetic variant of *S. lasiandra.*

Forma **araquipa** Jepson. Oak Park district north of Arroyo Grande, and very probably elsewhere.

4. **S. lasiandra** Benth. Cambria (*Eastwood 13,679*), a specimen without catkins identified by C. R. Ball. This species differs from *S. laevigata* only in having glands near the summit of the petiole. When *S. lasiandra* and *S. laevigata* are studied genetically, they probably will be regarded as constituting one species.

5. **S. melanopsis** Nutt. var. **Bolanderiana** (Rowlee) Schneider. Among rocks in stream-beds in Santa Lucia Mts.: San Carpoforo Creek (*8996*) and to be expected elsewhere.

6. **S. Hindsiana** Benth. var. **Hindsiana.** SANDBAR WILLOW. In sand where subject to flooding along Salinas River and its tributaries. Similar shrubs in the bed of the Santa Maria River may also be this species.

Var. **Parishiana** (Rowlee) Ball. "San Luis Obispo Co.," according to McMinn in "Illustrated Manual of California Shrubs." It is unknown what proportion of our plants may be referable to this variety. If descriptions can be taken at face value, the only basis for separating *S. Hindsiana* var. *Parishiana* from *S. exigua* seems to

be the highly subjective distinction between a style which is vanishingly short and
no style at all.

7. **S. exigua** Nutt. Along streams in interior: Shandon (*7585*); Palo Prieto Creek
(*Twisselmann 914*); abundant along Cuyama River in Cuyama Valley. From the
literature it would appear that staminate shrubs of this and of *S. Hindsiana* are indistinguishable. The identification of the two collections reported as this species is
accordingly based on pistillate material.

2. Populus L. POPLAR

Leaves longer than wide, minutely crenate or crenate-serrate, pale green or white on
back . . . . 1. *P. trichocarpa.*
Leaves at least as wide as long, conspicuously crenate, the same shade of green on
both surfaces . . . . 2. *P. Fremontii.*

1. **P. trichocarpa** T. & G. BLACK COTTONWOOD. Of general occurrence along
streams near the coast, extending inland to Salinas River.

2. **P. Fremontii** Wats. FREMONT COTTONWOOD. Very common in the interior, following the Salinas River, the Cuyama River, and their tributaries, occasionally extending into canyons in the hills.

Myricaceae. BAYBERRY FAMILY

1. Myrica L.

1. **M. californica** C. & S. CALIFORNIA WAX-MYRTLE. Frequent in moist ground
near the coast, along streams, around springs and marshes, and in hollows among
sand-dunes.

Juglandaceae. WALNUT FAMILY

1. Juglans L. WALNUT

1. **J. Hindsii** Jepson. CALIFORNIA BLACK WALNUT. This native of central California has been extensively planted, mostly as grafting stock for English walnut but
often for its own sake. Beside the trees which have persisted from abandoned plantings in the hills, walnuts have to some extent spread spontaneously along certain
streams, notably San Luis Creek and Atascadero Creek. A few planted trees of *J.
californica,* the native walnut of southern California, are in Morro Bay State Park,
but all the walnuts I have seen growing wild here are clearly referable to *J. Hindsii.*

Garryaceae. SILK-TASSEL FAMILY

1. Garrya Dougl.

Lower leaf-surfaces covered with tangled curly hairs.
    Leaf-blades rounded at base, broadly rounded toward apex and obtuse, abruptly
    acute, or mucronate, the width usually more than half the length . . 1. *G. elliptica.*
    Leaf-blades mostly cuneate at base, narrowed toward the acute or acuminate
    apex, the width mostly half the length or less . . . . 2. *G. Veatchii.*
Lower leaf-surfaces with straight appressed hairs . . . . 3. *G. flavescens.*

1. **G. elliptica** Dougl. SILK-TASSEL BUSH. Occasional on north-facing slopes near
coast: Coon Creek in San Luis Range; Price Canyon; upper Arroyo Grande. This
far south, individuals are few and widely spaced.

2. **G. Veatchii** Kell. Common along summits and on east slope of Santa Lucia

Range, extending almost to Salinas River near Santa Margarita and near Pozo. *Garrya Veatchii* and *G. elliptica* are well separated ecologically, but reliable characters for distinguishing them are hard to find.

3. **G. flavescens** Wats. Represented by only a few shrubs on high ridges of La Panza Range and Caliente Range. One shrub was also found in a canyon bottom on Garcia Mountain south of Pozo (*9558*). Authors have called Californian shrubs of this species var. *pallida* (Eastw.) Bacigalupi, but the type specimen of *G. pallida* has the lower leaf-surfaces and young fruits only sparsely hairy and thus is perhaps more closely related to *G. Fremontii* than to *G. flavescens*.

Betulaceae. Birch Family

1. Alnus Hill. Alder

Leaf-margins not revolute . . . . 1. *A. rhombifolia.*
Leaf-margins narrowly revolute . . . . 2. *A. oregona.*

1. **A. rhombifolia** Nutt. White Alder. Frequent along coastal streams, extending into Santa Lucia Mts.: San Simeon Creek, Santa Rosa Creek, Arroyo Grande, and many other streams.

2. **A. oregona** Nutt. *A. rubra* Bong., not Marsh. Red Alder. Mouth of San Carpoforo Creek. Possibly overlooked elsewhere for *A. rhombifolia,* some variants of which resemble it closely.

Fagaceae. Beech Family

Leaves mainly acuminate, entire, covered on the back with golden-brown fuzz; nuts enclosed in a spiny involucre (bur) . . . . 1. *Castanopsis.*
Leaves not acuminate, entire to toothed or lobed, without golden fuzz on back (except young leaves of *Quercus chrysolepis*); nuts with base enclosed in an acorn-cup.
    Catkins erect; leaves with many lateral veins from midrib (usually 10 or more on either side), closely and evenly toothed . . . . 2. *Lithocarpus.*
    Catkins drooping; leaves with few lateral veins (fewer than 10, usually about 5 on either side of midrib), the margins various . . . . 3. *Quercus.*

1. Castanopsis Spach. Chinquapin

1. **C. chrysophylla** (Dougl.) A. DC. var. **minor** (Benth.) A. DC. Golden Chinquapin. Locally plentiful in western part of San Luis Range, particularly between Coon Creek and Islay Creek, growing on shale. Vigorously crown-sprouting after fires.

2. Lithocarpus Bl.

Medium to large trees; leaves mostly 7 to 12 cm. long   *L. densiflorus* var. *densiflorus.*
Shrubs to small clump-forming trees; leaves mostly 3 to 7 cm. long . . . . . var. *parvus.*

1. **L. densiflorus** (H. & A.) Rehder var. **densiflorus.** Tanbark Oak. Santa Lucia Mts., mostly at higher altitudes or in deep canyons, not continuously distributed but sometimes locally abundant. The best developed tanbark oak forest known in the county is on the ridge east of Cuesta Pass.

Var. **parvus** Hoover. In the vicinity of Cuesta Pass, a shrubby form of the species, with leaves noticeably smaller than those of the tree form, is common. The shrubs with the smallest leaves, however, occur near the coast in the San Luis Range, where the species is rare.

3. **Quercus** L. Oak

Leaves lobed, the lobes often few-toothed or lobulate.

  Lobes, or their teeth or lobules, tipped with bristles; acorn-cups with overlapping scales.

    Leaves deeply lobed, the larger lobes with a few teeth or lobules

                                      1. *Q. Kelloggii.*

    Leaves shallowly lobed, the lobes not or rarely toothed or lobulate

                                        2. *Q. morehus.*

  Lobes not tipped with bristles; acorn-cups warty.

    Leaves 2 to 4 cm. long, the lobes or their teeth spine-tipped.

      Leaves ovate or elliptic in outline . . . . 10. *Q. dumosa* (hybrid?).

      Leaves oblong, about twice as long as wide . . . . 3. *Q. MacDonaldii.*

    Leaves 4 to 11 cm. long, the lobes obtuse, less commonly acute or mucronate, but never spine-tipped.

      Leaves 1½ to 2 times as long as wide, the lower surface with dense stellate hairs which overlap to form a thin close felt; acorn-cups deep; acorns elongate . . . . .4. *Q. lobata.*

      Leaves 1 to 1½ times as long as wide, rarely much longer, the lower surface with more widely spaced hairs which do not overlap; acorn-cups shallow; acorns short and stubby . . . . .5. *Q. Garryana.*

Leaves toothed to entire (when coarsely toothed, as in *Q. Douglasii*, the leaves may perhaps be described as shallowly lobed).

  Lower leaf-surfaces bright green like the upper, either glabrous or with tufts of hair in the forks of the principal veins; acorn-cups with overlapping scales.

    Leaves coarsely toothed, the teeth bristle-tipped . . . . .2. *Q. morehus.*

    Leaves either entire or with teeth spinose, not bristle-tipped.

      Leaves about twice as long as wide, the margins flat or sometimes curved downward, the lower surface entirely glabrous 6. *Q. Wislizeni.*

        Tree with well developed trunk . . . . .var. *Wislizeni.*

        Shrub with several stems from base . . . . .var. *frutescens.*

      Leaves little longer than wide, except as the broadly revolute edges may conceal the true shape, often almost orbicular; lower surface usually with tufts of hair in forks of veins . . . . .7. *Q. agrifolia.*

        Leaves, except for tufts of hair in forks of veins, glabrous

                                   var. *agrifolia.*

        Leaves on both surfaces somewhat stellate-pubescent, in addition to tufts of hair in forks of veins . . . . .var. *oxyadenia.*

  Lower leaf-surfaces pale glaucous-green to white, uniformly pubescent or puberulent, or early glabrate; acorn-cups usually either warty or tomentose but sometimes visibly scaly.

    Lower leaf-surfaces persistently pubescent or puberulent.

      Leaves (except on juvenile shoots) with margins curved downward, tomentose on lower surface . . . . .8. *Q. durata*

      Leaves with margins not curved downward, puberulent on lower surface.

        Deciduous (most leaves falling in December or January); leaves coarsely few-toothed (2 to 4 teeth on each side), the teeth pointing forward, or sometimes entire . . . . .9. *Q. Douglasii.*

        Evergreen (leaves persistent until new growth develops in spring); leaves entire to spinose-dentate (4 or more teeth on each side) or rarely lobed and with several spine-tipped teeth . .10. *Q. dumosa.*

Upper leaf-surfaces nearly glabrous (puberulent only along midrib) ....var. *dumosa*.
Upper leaf-surfaces puberulent ....var. *turbinella*.
Leaves when fully developed glabrous and whitish on lower surface, when young with thin grayish or golden-yellow tomentum.
Leaves (except on juvenile shoots) mostly entire; acorn-cups thick, the rim not flaring ....11. *Q. chrysolepis*.
Leaves prominently spinose-dentate; acorn-cups thin, with flaring rim
12. *Q. Dunnii*.

1. **Q. Kelloggii** Newb. Black Oak. Santa Lucia Mts., mostly on higher summits or on lower north-facing slopes, locally abundant west of Templeton, and between Rocky Butte and Pine Mt.

2. **Q. morehus** Kell. Oracle Oak. Rare or overlooked in southern part of Santa Lucia Range: 2 miles northwest of Cuesta (*Bolt 710*); 3¼ miles southwest of Dove (*Axelrod 559*); 4.1 miles southwest of Pozo (*Gifford 732*); Oak Park district north of Arroyo Grande. Generally assumed to be a hybrid between *Q. Kelloggii* and *Q. Wislizeni*. No trees of *Q. Kelloggii* are present in the San Luis Obispo Co. localities where *Q. morehus* is known. Conversely, *Q. morehus* has not been collected or reported in those parts of the Santa Lucia Range where *Q. Kelloggii* grows. The two trees at Oak Park are deciduous, as they were virtually leafless in late January, 1968. The lower surfaces of their leaves are completely glabrous. The shrubby variety of *Q. Wislizeni* grows close by, but there is nothing else present with which it might hybridize to produce trees showing the distinctive features of *Q. morehus*.

3. **Q. MacDonaldii** Greene. A specimen from San Carpoforo Creek (*Chester Dudley in 1927*) resembles the type collection of *Q. MacDonaldii* from Santa Cruz Island. The drawing representing *Q. MacDonaldii* in Abrams's "Illustrated Flora of the Pacific States" differs in showing obtuse-lobed leaves and may have been made from a specimen of *Q. lobata*. Both the type and the Dudley collection may be of hybrid origin, as suggested by the rarity and sporadic occurrence of such individuals, but it is not easy to identify possible parents. Another possible interpretation is that these specimens are vigorous crown-sprouts of *Q. Wislizeni* var. *frutescens*. The leaves on such shoots are always sharply toothed and sometimes may be lobed as well.

4. **Q. lobata** Neé. Valley Oak. Very common from summits of Santa Lucia Range eastward to near Cholame and San Juan River; valley of upper Arroyo Grande. Found on hillslopes in areas of relatively high rainfall, but restricted to low valleys toward the arid interior. Paso Robles (originally El Paso de Robles) was named for this tree, which in general appearance resembles the European oak *Q. Robur*, called "roble" in Spain. According to Jepson in "The Trees of California," an exceptionally large tree grew on the Old Kentucky Ranch west of Paso Robles. Individuals of this species are the largest oaks in America, and probably the largest in the world.

5. **Q. Garryana** Dougl. Oregon Oak. Ridges and north slopes in higher parts of Santa Lucia Mts. toward the north. Only one collection showing the distinctive acorns of this species has been made: between Rocky Butte and Pine Mt. (*8405*). Trees in other places have identical leaves and therefore are believed to be *Q. Garryana* rather than *Q. lobata*: Pine Top Mt. (*8889*); trail to Cypress Mt. from Cam-

bria-Adelaida road (*8390*). At the latter locality *Q. lobata* was also found (*8391*) and seemed readily distinguishable on the basis of the leaf characters. I am restricting the name *Q. lobata* to those trees in which the stellate hairs on the lower leaf-surfaces form a close dense felt, as in trees growing between Salinas and Monterey, the presumed type locality of *Q. lobata*. Trees which were present in lower Lopez Canyon, before construction of the dam, produced the short stout acorns of *Q. Garryana* and probably belonged to this species.

6. **Q. Wislizeni** A. DC. var. **Wislizeni.** INTERIOR LIVE OAK. The tree form of this species is seen occasionally along the summits of the Santa Lucia and La Panza ranges. Some notably fine large trees occur on the south (Santa Barbara Co.) side of Cuyama Canyon near Clear Creek. Although there has been no opportunity to explore the adjacent section of San Luis Obispo Co., comparable trees undoubtedly exist there.

Var. **frutescens** Engelm. One of the "scrub oaks," a common component of chaparral from near the coast inland to La Panza Range.

7. **Q. agrifolia** Neé var. **agrifolia.** COAST LIVE OAK. Our commonest native tree near the coast, growing in ravines, on north-facing slopes, and on stabilized sandy plains; common also eastward through La Panza Range. Where exposed to strong and frequent winds from the sea, individuals do not reach tree size. The shrubby form has been named var. *frutescens* Engelm. but is probably not genetically significant. A remarkable variant found near the coast, especially among the pines around Cambria, differs from most trees of the species in having much larger leaves, their margins flat rather than curved downward, and the tufts of hair in the forks of the veins on the lower surface (normally the most reliable mark of recognition for *Q. agrifolia*) may be lacking. These large-leaved trees should surely be distinguished by name from the prevailing form, but they just possibly may represent the type to which the name *Q. agrifolia* was originally applied.

Var. **oxyadenia** (Torr.) J. T. Howell. Near Santa Margarita and eastward. The trees have not been studied enough to warrant a definite statement about distribution, but it is likely that a large proportion of those growing in the interior will be found to be of this variety, or intermediate toward it.

8. **Q. durata** Jepson. LEATHER OAK. Rocky places in Santa Lucia Range and San Luis Range, particularly near San Luis Obispo. Notably abundant on serpentine, but sometimes growing on other types of rock. On the Atascadero-Morro Bay road east of the summit, some individuals develop a single trunk and grow as small trees, in contrast to the usual low shrubby form of the species.

9. **Q. Douglasii** H. & A. *Q. Alvordiana* Eastw., as to type. BLUE OAK. Very common on interior hills from Santa Lucia Mts. eastward to Cottonwood Pass, west edge of Carrizo Plain, lower Cuyama Valley, and apparently on summits of Temblor and Caliente Ranges. The last mentioned trees have not been sufficiently studied and may not be typical. The type specimen of *Q. Alvordiana,* from Kern Co., seems from its appearance to be merely a specimen of *Q. Douglasii* with rather small leaves for the species. The name, however, has been used also for specimens which to me seem quite typical *Q. dumosa,* and for various apparent hybrids. Vast numbers of blue oaks east of the Salinas River have been destroyed, first to provide land for raising grain, and more recently evidently from an aversion against trees in general. This species reaches the coast in southern Monterey Co., but in San Luis Obispo Co. is not found on slopes directly facing the ocean.

On the south slope of Pine Mt. above San Simeon is a grove of trees which have all the appearance of *Q. Douglasii* except that the leaves are smaller than average. The few acorns that were found were remarkably small and slender for the species. This form should probably be named as a local variety.

10. **Q. dumosa** Nutt. var. **dumosa**. Scrub Oak. Common in Santa Lucia Range, and eastward at least to La Panza Range. Notably abundant west of Paso Robles and in the hill country east of Santa Margarita. Often a component of chaparral, or as frequently mixed among larger trees such as *Q. agrifolia* and *Q. Douglasii*. The leaves, acorns, and acorn-cups are so variable that it is difficult to find reliable means of distinguishing *Q. dumosa* from other species. The shrubby or thicket-forming habit (with several trunks from the base) is fairly constant, but sometimes small trees with a single trunk bear acorns and leaves exactly like those found on shrubby individuals of *Q. dumosa*. Occasional aberrant individuals are probably hybrids between *Q. dumosa* and *Q. Douglasii*. Lobed leaves, which are rare in *Q. dumosa*, could result from hybridization with *Q. lobata*, but may rather be simply one more manifestation of the great variability of this species.

Var. **turbinella** (Greene) Jepson. Mostly on north-facing slopes, often mixed with junipers, from hills near north end of La Panza Range eastward to Temblor and Caliente Ranges. *Quercus turbinella* subsp. *californica* Tucker (type locality, Caliente Mt.) is identical with *Q. dumosa* in every detail except one: the presence of minute pubescence on the upper leaf-surfaces. (Hairs are present along the midrib even in typical *Q. dumosa*). Exactly parallel variations in the size and shape of the leaves, the acorns, and the acorn-cups are found in both. Any one of these variations might be taken as the basis for a subspecies or a variety as reasonably as the pubescence of the leaves. Such a trivial difference, probably having a very minor genetic basis, warrants no more than varietal rank in classification at the most. It may be remarked incidentally that *Q. agrifolia* has a comparable variant, var. *oxyadenia* (except that the difference involves both surfaces of the leaves), and that contemporary authorities on the oaks are not for that reason classifying var. *oxyadenia* as a distinct species. It is logically faulty to give this sort of difference greater weight in *Q. dumosa* than in *Q. agrifolia*. Twisselmann reported *Q. Douglasii, Q. dumosa,* and *Q. turbinella* from the Temblor Range but did not state how he was able to distinguish them.

11. **Q. chrysolepis** Liebm. Goldcup Oak, Maul Oak. Occasional on high rocky ridges and in deep canyons of Santa Lucia Range; occurring sparingly on highest summits of La Panza and San Luis ranges. Reaching its best local development in Lopez Canyon, where not only are the trees very large, but some of them produce the largest acorns seen in any oak.

12. **Q. Dunnii** Kell. *Q. Palmeri* Engelm., a later name according to J. M. Tucker. Holly-Leaf Oak. Highly localized about 4 miles west of Paso Robles on the Peachy Canyon Road, where only a few clumps grow in disintegrated shale along the flat bottom of a narrow valley, associated with *Q. agrifolia* and *Q. dumosa*. The species apparently is not otherwise known farther west or north than Deep Creek at the north base of the San Bernardino Mts. (according to Munz). Characteristically growing in the form of small thickets, with several trunks which probably have developed from one seedling, and sprouting from the roots. Some individuals of *Q. dumosa* imitate *Q. Dunnii* rather closely in vegetative features, but the two are readily distinguishable in the field once they have been identified. The spar-

ingly produced acorns, and especially the acorn-cups (broad, shallow, with flaring rim, and velvety inside), are the parts which are outstandingly distinctive.

Urticaceae. Nettle Family

Leaves opposite; plants with stinging hairs.
    Pistillate calyx of distinct sepals; stem firm ....1. *Urtica.*
    Pistillate calyx saccate, minutely toothed at orifice; stem slender and weak
                                               2. *Hesperocnide.*
Leaves alternate; plants without stinging hairs ....3. *Parietaria.*

1. Urtica L. Nettle

Tall perennial; leaves gray-hairy on lower surface ....1. *U. holosericea.*
Annual not over 5 dm. tall; leaves glabrous except for the stinging hairs 2. *U. urens.*

1. **U. holosericea** Nutt. Common in western part along streams, in marshes, and around springs; also on shifting, apparently dry sand-dunes. Less common in the interior, but reported from Temblor Range by Twisselmann, and also seen in Cuyama Valley.

2. **U. urens** L. Occasional weed in sandy soils, generally in cultivated ground: Morro Bay; near Edna; upper Arroyo Grande.

2. Hesperocnide Torr.

1. **H. tenella** Torr. Frequent in sandy soils in the hills, in positions sheltered from wind, usually in shade except when growing after fire.

3. Parietaria L.

1. **P. floridana** Nutt. Widely scattered, not common, where sheltered by cliffs or overhanging rocks: west edge of Carrizo Plain (*6173*); coast just south of Monterey Co. line (*6668*); Price Canyon.

Loranthaceae. Mistletoe Family

Plants with green leaves ....1. *Phoradendron.*
Plants yellow, with leaves reduced to scales ....2..*Arceuthobium.*

1. Phoradendron Nutt.

Leaves broadly oblanceolate to elliptic or orbicular; parasitic on dicotyledons.
    Leaves broadly elliptic or oval to orbicular, the width from about ¾ the length
    to equalling it ....1. *P. flavescens.*
    Leaves broadly oblanceolate or obovate to elliptic, the width about ½ to ⅔
    the length ....2. *P. villosum.*
Leaves oblanceolate to narrowly oblong; parasitic on juniper and cypress
                                            3. *P. Bolleanum.*

1. **P. flavescens** (Pursh) Nutt. var. **macrophyllum** Engelm. Occasional in Salinas River watershed, parasitic on various trees, most commonly on cottonwoods: San Miguel (*6465*); on *Robinia* (locust) in San Juan River Valley (*Twisselmann 1766*).

2. **P. villosum** Nutt. Oak Mistletoe. Common in interior on various species of oaks, particularly *Quercus agrifolia* and *Q. Douglasii*; sometimes on other kinds of trees. *Quercus agrifolia* near the coast is usually free from this parasite, but some of the trees southeast of San Luis Obispo near the Arroyo Grande creek are infested.

3. **P. Bolleanum** (Seem.) Eichler var. **densum** (Torr.) Fosb. Juniper Mistletoe. Occasional on California juniper in eastern part. Not yet detected on *Cupressus Sargentii* in San Luis Obispo Co., but reported as growing on that species elsewhere.

2. Arceuthobium Bieb.

1. **A. campylopodum** Engelm. Pine Mistletoe. On *Pinus radiata* at Cambria, and on *P. Sabiniana* in the interior. Diligent search will probably reveal its presence locally on other species of pines.

Polygonaceae. Buckwheat Family

Flowers not in involucres (the calyx in *Lastarriaea* tubular and involucre-like).
  Leaves with membranous stipular sheaths (ocreae).
    Calyx white to rose-red or green, with 5 equal petal-like lobes
                                                   1. *Polygonum.*
    Calyx green to brownish-red, of 3 very small outer sepals and 3 larger inner ones . . . . .2. *Rumex.*
  Leaves without membranous sheaths.
    Leaves oblanceolate or broader; bracts spineless or tipped with straight spines.
      Leaves opposite, fan-shaped or broader than long, sparsely pilose-ciliate; plants spineless . . . . .3. *Pterostegia.*
      Leaves alternate, oblanceolate to ovate, gray-woolly; upper leaves and bracts tipped with spines . . . . .4. *Hollisteria.*
      Leaves linear; bracts and calyx lobes tipped with hooked spines
                                                5. *Lastarriaea.*
Flowers borne in umbels or singly in tubular to campanulate or bowl-shaped involucres.
  Teeth of involucre spine-tipped (in *Centrostegia Thurberi* sometimes merely apiculate).
    Bracts opposite or whorled, not 3-lobed; involucre containing a single flower . . . . .6. *Chorizanthe.*
    Bracts alternate, 3-lobed or 3-parted; involucre bearing usually 2 to several flowers, rarely only 1.
      Bracts much reduced, 2 to 4 mm. long; involucral teeth approximately equal . . . . .7. *Centrostegia.*
      Bracts represented by well developed 3-lobed leaves (often reduced toward tips of branches); involucral teeth unequal . . . . .8. *Mucronea.*
  Teeth or lobes of involucre not spine-tipped . . . . .9. *Eriogonum.*

1. Polygonum L.

Flowers in narrow raceme-like panicles with reduced bracts much shorter than the flowers.
  Perennial with erect stems from a thick base; leaves mostly over 4 cm. wide; calyx rose-red . . . . .1. *P. coccineum.*
  Annuals (probably) with stems decumbent at base and rooting at the nodes; leaves not over 3 cm. wide; calyx white, pink, or green.
    Sheaths not ciliate; calyx white or pale pink . . . . .2. *P. lapathifolium.*
    Sheaths long-ciliate; calyx rose-pink or green.
      Panicle short, densely flowered; calyx pink . . . . .3. *P. Persicaria.*
      Panicle elongate, sparsely flowered; calyx green with white-margined lobes . . . . .4. *P. punctatum.*

Flowers in small clusters in axils of upper leaves.
    Perennial with woody stems ....5. *P. Paronychia.*
    Annuals, the stems tough but not woody.
        Leaves toward tips of branches little reduced, surpassing the flowers
                                            6. *P. arenastrum.*
        Leaves toward tips of branches much reduced and barely equalling the
        flowers or even shorter ....7. *P. aviculare.*

1. **P. coccineum** Muhl. Wet places near coast from Laguna near San Luis Obispo to Oso Flaco Lake.

2. **P. lapathifolium** L. Wet places: Paso Robles (*Chester Dudley* in 1926), Santa Margarita Creek (*10,028*), and probably elsewhere.

3. **P. Persicaria** L. Lady's Thumb. Occasional in fresh water or wet soil in coastal area.

4. **P. punctatum** Ell. var. **leptostachyum** (Meisn.) Small. Dotted Smartweed. Common in low wet places near coast, extending inland at least to upper Arroyo Grande.

5. **P. Paronychia** C. & S. Local on coastal sand-dunes in vicinity of Hazard Canyon (*7187, 9042*).

6. **P. arenastrum** Jord. ex Bor. Knotweed. Common weed in cultivated ground and by roadsides, in moist to dry soils, not usually in sand. This species has usually been included in *P. aviculare* L.; use of the name *P. arenastrum* is based on the identification by Mertens and Raven as that species of *Twisselmann 1366* from Palo Prieto Pass. Most specimens showing the well-developed upper leaves of *P. arenastrum* come from the coastal area.

7. **P. aviculare** L. Knotweed. Mainly in Salinas Valley: San Marcos Creek, between San Miguel and Paso Robles (*Chester Dudley* in 1935, reported as *P. argyrocoleon* Steud.); Atascadero; Santa Margarita; Pozo. Immature plants which seem to be the same were collected at Choice Valley near our eastern border (*E. McMillan* 153). These plants are distinguished by having greatly reduced leaves in the inflorescence, exactly as the preceding species might be expected to produce late in the season. The question of the distinctness or identity of the two can therefore not be considered settled until the plants are carefully compared at all stages of growth.

2. **Rumex L.**

Plants with a fleshy tap-root, with perfect flowers; leaves not hastate.
    Inner sepals entire or minutely denticulate.
        Stems erect, with largest leaves at base.
            Inner sepals without callous grains, in fruit becoming red or pink and
            8 to 14 mm. long ....1. *R. hymenosepalus.*
            Inner sepals with callous grains, in fruit green or eventually turning
            brown, 2 to 6 mm. long.
                Inner sepals in fruit 4 to 6 mm. long, with broad margins bordering the callous grain ....2. *R. crispus.*
                Inner sepals in fruit 2 to 3 mm. long, the callous grain almost as wide as the sepal ....3. *R. conglomeratus.*
        Stems decumbent to ascending, without basal leaves.
            Leaves more than 3 times as long as wide; inner fruiting sepals 2.5 to 4 mm. long ....4. *R. salicifolius.*
            One or more of inner sepals with a callous grain.
                One inner sepal with a callous grain ....f. *salicifolius.*
                All three inner sepals with callous grains ......f. *transitorius.*

Inner sepals without callous grains .... f. *ecallosus.*
Leaves 2 to 3 times as long as wide; inner fruiting sepals 4 to 5 mm.
long .... 5. *R. crassus.*
Inner sepals bordered with long slender teeth.
Perennial; branches divaricate at wide angles .... 6. *R. pulcher.*
Annual; branches ascending .... 7. *R. fueginus.*
Plants with slender rhizomes, dioecious; lower leaves hastate .... 8. *R. angiocarpus.*

1. **R. hymenosepalus** Torr. Canaigre. Sandy soils, usually where subject to periodic flooding, Cuyama Valley, Santa Maria Valley, southern Temblor Range, and apparently spreading northward in the interior: Red Hills southeast of Shandon (*Twisselmann 1825*).

2. **R. crispus** L. Curly Dock. Common in low places, in moist to rather dry soil, and in irrigated ground everywhere. The dried reddish brown fruiting panicles are used in flower arrangements.

3. **R. conglomeratus** Murr. Along streams and in marshy places, common near the coast, occasional in wet or irrigated spots eastward.

4. **R. salicifolius** Weinm. f. **salicifolius.** Wet ground or stream-beds, apparently uncommon in typical form but widely scattered: Jack Lake; Gillis Canyon near Shandon (*Howell 24,380*).

F. **transitorius** (Rech. f.) J. T. Howell. Occasional along the coast: San Simeon Bay (*Eastwood 15,036*); Los Osos (*8977*). The latter was exceptional in growing in dry sand, not in moist soil as is usual.

F. **ecallosus** J. T. Howell. Drying rocky bed of Navajo Creek, east side of La Panza Range (*7250*) and probably elsewhere.

5. **R. crassus** Rech. f. Wet places along the coast: south of Hazard Canyon (*7188*); common from Cambria northward.

6. **R. pulcher** L. Roadsides, pastures, abandoned fields, and borders of marshes, usually in clay soils, occasional in Santa Lucia Mts. and frequent near coast: Adelaida; Cambria; San Luis Obispo; Arroyo Grande.

7. **R. fueginus** Phil. Golden Dock. Frequent in fresh-water or brackish coastal marshes, at least from Morro Bay southward: mouth of Osos Creek; Oceano; Jack Lake; Oso Flaco Lake.

8. **R. angiocarpus** Murb. Sheep Sorrel. Common in sandy coastal fields and occasional elsewhere. A collection from south of Hazard Canyon (*9036*) shows the inner sepals firmly adherent to the achene and thus is *R. angiocarpus* rather than *R. Acetosella* L., which seems identical in all other respects. It is rather unlikely that there would be two different sheep sorrels introduced here, so it is tentatively assumed that all our plants are the same. In Great Britain *R. Acetosella* is regarded as an indicator of acid soil. The soils where sheep sorrel grows here are probably not markedly acid: another reason for suspecting that our plants may not be genuine *R. Acetosella.*

3. **Pterostegia** F. & M.

1. **P. drymarioides** F. & M. Common, particularly around rocks, in hills from the coast to the Temblor Range, in all except the hottest and driest localities.

4. **Hollisteria** Wats.

1. **H. lanata** Wats. Common in clay or sandy, but usually hard-packed, soils in eastern part from Cottonwood Pass and Cholame Valley through Carrizo Plain to Cuyama Valley.

5. **Lastarriaea** Remy

1. **L. coriacea** (Goodman) Hoover. Often abundant in sandy soils, particularly near the coast, but also scattered through the interior.

6. **Chorizanthe R. Br. ex Benth.**

Involucral teeth with white, pink, or rose-red margins.
  Involucral teeth connected by a membrane which forms a flaring rim from the mouth of the tube.
    Lower cauline leaves alternate; involucres in condensed cymes or heads at top of stem or in upper leaf-axils .1. *C. membranacea.*
    Cauline leaves whorled or opposite at the lowest node; involucres in a branching cyme, mostly congested at ends of branches . . . .2. *C. Douglasii.*
  Involucral teeth distinct down to the mouth of the tube.
    Plant hairy but appearing bright green; margins of involucral teeth white; calyx-lobes entire . . . . .3. *C. diffusa.*
    Plant gray-villous; margins of involucral teeth pink; calyx-lobes erose.
      Pink margin of involucral teeth broader than midrib . . .4. *C. pungens.*
      Pink margin of involucral teeth very narrow or obsolete
                                                            5. *C. angustifolia.*
Involucral teeth uniform in color, without white or pink margins.
  Involucral teeth alternately longer and shorter, the 3 longer ones approximately equal.
    Stems from prostrate to (in *C. Breweri*) widely spreading from a short suberect base.
      Involucre 3-angled . . . . .6. *C. polygonoides.*
      Involucre 6-ribbed.
        Plant not yellowish; calyx white or pink.
          Leaves many; plant gray-villous; involucres congested in head-like clusters . . . . .5. *C. angustifolia.*
          Leaves few; plant strigose, brownish or purplish; involucres less crowded . . . . .7. *C. Breweri.*
        Plant yellowish; calyx yellow . . . . .8. *C. procumbens.*
    Stems erect or ascending, cymosely branched.
      Leaves all basal . . . . .9. *C. staticoides.*
      Lower nodes with leaves . . . . .10. *C. Xanti.*
  Longer 3 involucral teeth unequal.
    Involucral teeth 6, the longest not dilated and leaf-like.
      Longest involucral tooth shorter than the tube.
        Involucral tube, if saccate, not ventricose on one side.
          Plants when vigorous prostrate, erect only if dwarfed; pubescence of involucre appressed or mainly so; calyx white, rarely varying to rose-pink . . . . .11. *C. obovata.*
          Plants typically erect, rarely spreading; pubescence of involucre ascending or spreading; calyx purple.
            Involucres·spreading-villous on lower part; outer calyx-lobes typically bilobed but sometimes entire 12. *C. biloba.*
            Involucres with inconspicuous ascending hairs; outer calyx-lobes entire . . . . .13. *C. Palmeri.*
        Involucral tube saccate below, ventricose on one side
                                                            14. *C. ventricosa.*
      Longest involucral tooth longer than the tube.
        Longest involucral tooth straight.

Entire plant light green, the leaves rather sparsely hairy; involucral teeth yellowish green, sometimes slightly brown-tinged; calyx minute, green, barely exserted at the mouth of the involucre at anthesis . . . . .15. *C. uniaristata.*
Plant gray with appressed hairs, except the sometimes yellowish involucres; involucral teeth dark purple or brown; calyx evident, white, its lobes well exserted at anthesis
16. *C. rectispina.*
Longest involucral tooth hooked like the others
17. *C. Clevelandii.*
Involucral teeth 5, the longest dilated, more or less leaf-like
18. *C. Watsonii.*

1. **C. membranacea** Benth. Occasional in hard-packed gravelly or rocky soil in hills of the interior: Paso Robles eastward to Palo Prieto Pass and east side of La Panza Range. Goodman separated this species from *Chorizanthe* as the genus *Eriogonella,* but its close relationship to *C. Douglasii* seems to me clear enough to require keeping both in the same genus.

2. **C. Douglasii** Benth. Sandy soils of the interior from Nacimiento River to west edge of Carrizo Plain, usually in areas of sandstone or granite, forming scattered dense patches where little else grows. In May, 1963, between Creston and Shandon, these bright rose-red patches, interspersed among the soft blue of thistle-sage and assorted other colors, produced a strikingly beautiful effect on the open low hills or among scattered oaks.

3. **C. diffusa** Benth. var. **diffusa.** In sand near the coast: south side of Morro Bay; Nipomo Mesa. The following variety is locally more common, or at least more often noticed.

Var. **nivea** (Curran) Hoover. Characterized by broad showy white margins on the involucral teeth, this variety is prevalent from hills on the south side of San Luis Valley southward over Nipomo Mesa to western Santa Barbara Co.

4. **C. pungens** Benth. Coastal sand, near Hazard Canyon (*7185, 9035*); perhaps also on San Simeon Point (Goodman, Ann. Mo. Bot. Gard. 21: 36). Otherwise known only on the coast of northern Monterey Co.

5. **C. angustifolia** Nutt. var. **angustifolia.** Near Santa Maria; perhaps restricted to western Santa Barbara Co., though some of the plants on Nipomo Mesa may belong to this typical variety with only three stamens. Sometimes the involucral teeth may show a very narrow pink margin, thus varying toward *C. pungens.*

Var. **Eastwoodiae** Goodman. Locally abundant in sandy country around the south side of Morro Bay, and on Nipomo Mesa and flats among the sand-dunes to the west. This variety, characterized by more numerous stamens, is far more plentiful than typical *C. angustifolia.*

6. **C. polygonoides** T. & G. Barren hard-packed gravelly or rocky soil, 18 miles east of Creston on La Panza road (*8160*).

7. **C. Breweri** Wats. Areas of serpentine rock, restricted to the southern portion of the Santa Lucia Range from Morro Creek to East Fork of Corral de Piedra Creek. The type locality is San Luis Obispo.

8. **C. procumbens** Nutt. In sand on Nipomo Mesa about 3 miles northwest of Nipomo (*9179*). Not otherwise reported west or north of Los Angeles Co.

9. **C. staticoides** Benth. Frequent in rocky, gravelly, and sandy places from Santa Lucia Mts. eastward to La Panza Range, and in the hills near Arroyo Grande. Par-

ticularly plentiful in chaparral areas which have been recently cleared by fire or other means.

10. **C. Xanti** Wats. Sandy hills, often around sandstone outcrops, from east side of La Panza Range to Cottonwood Pass, Temblor Range, and Caliente Mt. In a year of good rainfall this, like various other species of *Chorizanthe,* forms bright splashes of rosy color on the dry, usually barren hills. The involucres and stems contribute more to the effect than do the flowers.

11. **C. obovata** Goodman. Locally rather common in sandy soils, usually around calcareous sandstone, mostly near coast from 6 miles south of San Luis Obispo near See Canyon (*6142*) to Nipomo Mesa (*7287*), extending inland to Huasna district (*6195*) and in a disjunct area in upper Salinas Valley between Santa Margarita and Pozo. Price Canyon (near Pismo Beach) is the type locality. Resembling *C. Palmeri* in many details and confused with it by authors, but entirely distinct in the field and never found with *C. Palmeri,* even though present in the same general area. While *C. Palmeri* grows in magnesium-rich soils (from ultrabasic intrusive rocks), *C. obovata* grows probably always in calcium-rich soils. Plants of *C. Palmeri* grow erect with only a rare exception and have purple flowers. All but the smallest plants of *C. obovata* are usually prostrate and have nearly always white, or at most slightly pink-tinged, flowers, but in the Huasna district the color varied to a bright rosy red. This showy color form has probably now been completely destroyed by disastrous fires which have been set in that area over the past several years. The distribution of *C. obovata* is centered in San Luis Obispo Co., and it is probably found elsewhere only in western Santa Barbara Co. Specimens from San Benito and Monterey cos. which Goodman cited under this name really belong, I think, to *C. biloba.*

12. **C. biloba** Goodman. Interior hills in clay soils, disintegrating shale, or firm sandy loams, from Paso Robles (the type locality) to Palo Prieto Pass and La Panza district, extending northward into Monterey Co. In favorable years whole hillsides are colored rose-red from the numerous involucres. The flowers themselves are purplish. Although this species and *C. obovata* are entirely distinct ecologically, the problem of differentiating between them calls for much more study than has so far been devoted to it. Evidently not all plants in the territory and in the habitats of *C. biloba* have the bilobed outer perianth-segments which were originally taken as the distinguishing mark of this species. *Chorizanthe biloba* is generally erect, whereas vigorous plants of *C. obovata* are usually prostrate. More constant and more significant differences between them remain to be worked out.

13. **C. Palmeri** Wats. Clay soils in areas of serpentine or partially serpentinized igneous rock: lower Tar Spring Creek on Huasna Road east of Arroyo Grande (*7552*), apparently the southernmost locality; head of Perfumo Canyon in San Luis Range; Rinconada Mine and northward in Santa Lucia Mts. to coast of southern Monterey Co. San Luis Obispo is the type locality. In places abundant enough to color hillslopes bright purple in mid-summer.

14. **C. ventricosa** Goodman. West side of Cottonwood Pass, in clay around a small outcrop of serpentine (*7730*); a rare species found in similar soils northward through southeastern Monterey Co. to San Benito Co.

15. **C. uniaristata** T. & G. Sandy or rocky soils, in areas of sandstones or shale, in eastern part.

16. **C. rectispina** Goodman. Rare, in granite sand or disintegrating shale, La

Panza Range and hills to the west: summit between Creston and Calf Canyon (*7738*); Black Mt. (*Hardham 2279*). The type locality is given as "McGinness, 25 miles northeast of San Luis Obispo"; this must have been in or near the La Panza Range. Plants of this species are never numerous in the few places where it has been found.

17. **C. Clevelandii** Parry. In chaparral burn on sandstone, Franklin Creek west of Adelaida (*Hardham 7136*).

18. **C. Watsonii** T. & G. Sandy soils in eastern part, rare: between San Juan River and Carrizo Plain (*7889*); Temblor Range near north end of Elkhorn Plain (*8133*). These represent outlying western occurrences of what is mainly a Great Basin or desert species.

7. **Centrostegia** Gray

Involucres cylindric, with divergent teeth; flowers several ....1. *C. insignis.*
Involucres saccate or campanulate, with short erect teeth; flowers 2 or 1.
    Involucre without horns at base ....2. *C. Vortriedei.*
    Involucre bearing 3 divergent spine-tipped horns at base ......3. *C. Thurberi.*

1. **C. insignis** (Curran) Heller. *Chorizanthe insignis* Curran. In patches of sterile sand where little else grows, La Panza district to western edge of Carrizo Plain. A rare species, otherwise known only in Indian Valley and near Jolon, Monterey Co., yet in a favorable year the small plants are abundant enough to color the ground pink.

2. **C. Vortriedei** (Bdg.) Goodman. *Chorizanthe Vortriedei* Bdg. Sandy or chalky soils in Santa Lucia Mts., rare: Franklin Creek west of Adelaida (*Hardham 7135*); Nacimiento River (probably the type locality, from label on type collection).

3. **C. Thurberi** Gray. *Chorizanthe Thurberi* (Gray) Wats. Occasional in dry loose sandy soils in eastern part: between San Juan River and Carrizo Plain; summit near head of Carneros Canyon in Temblor Range; Caliente Mt.

8. **Mucronea** Benth.

Bracts clasping; involucres usually not solitary, up to 6 in each axil; calyx-lobes entire ....1. *M. californica.*
    Bracts mostly 15 mm. long or less ....var. *californica.*
    Bracts over 15 mm. long ....var. *Suksdorfii.*
Bracts perfoliate; involucres usually solitary; calyx-lobes fimbriate ..2. *M. perfoliata.*
    Bracts glabrous or sparsely glandular, turning dark red, shining var. *perfoliata.*
    Bracts sparsely but obviously hairy, green or reddish tinged, dull ....var. *opaca.*

1. **M. californica** Benth. var. **californica.** *Chorizanthe californica* (Benth.) Gray. In sand, upper Salinas Valley from Atascadero eastward to north end of La Panza Range and southward to near Pozo; common near coast from around Morro Bay southward. So variable in size of bracts that the extremes look like two quite different plants. (The parallel case of *Polygonum arenastrum* and *P. aviculare* might well be reviewed with this fact in mind). Plants of the Salinas Valley generally have much reduced bracts. This form is sometimes found also near the coast: Price Canyon (*9168*). Many of the coastal plants vary more or less toward var. *Suksdorfii,* as the plants with the largest bracts are called. Throughout the range of the species to the south the same rule seems to hold, size of bracts being strongly correlated with nearness to the ocean. Plants with red bracts covered sandy fields near Nipomo with masses of color in early summer of 1964.

Var. **Suksdorfii** (Macbr.) Goodman. Near the seashore: dunes just south of Hazard Canyon (*9038*). The extreme form is distinctive in appearance, but intermediate plants are far more numerous.

2. **M. perfoliata** (Gray) Heller var. **perfoliata.** *Chorizanthe perfoliata* Gray. Barren sands and loose calcareous clays, from near San Juan River eastward to Temblor Range and in Cuyama Valley.

Var. **opaca** Hoover. Sandy soils in open woodland and chaparral areas: La Panza Range, westward to Santa Margarita and upper Arroyo Grande.

9. **Eriogonum** Michx.

Annuals with slender tap-root.
 Involucres all borne on peduncles.
  Plants woolly (often glabrate above but the lower stems and leaves persistently woolly), with opposite or whorled cauline leaves as well as a basal rosette.
   Flowers well exserted, not covered with wool.
    Outer calyx-lobes incurved above or at least not curved outward, much wider than the erect inner ones.
     Outer calyx-lobes not inflated; inner calyx-lobes narrowly oblong, obtuse; cauline leaves generally linear, sometimes varying to oblong or oblanceolate . . . . 1. *E. angulosum.*
     Outer calyx-lobes more or less obviously inflated; inner calyx-lobes linear-lanceolate, obtuse to acuminate; cauline leaves elliptic or ovate to oblong, rarely lanceolate.
      Calyx creamy or pale yellowish; outer calyx-lobes with purplish or dark green spot, in age becoming inflated toward apex; inner calyx-lobes obtuse to acute
                2. *E. maculatum.*
      Calyx white or pink with deeper rose shadings; outer calyx-lobes inflated from middle upward, in age strongly so; inner calyx-lobes acute to acuminate 3. *E. viridescens.*
    Calyx-lobes all curved outward near tips, the outer little wider than the inner . . . . .4. *E. gracillimum.*
   Flowers very small, almost concealed by a tuft of cottony wool which fills the involucre . . . . .5. *E. gossypinum.*
  Plants glabrous or glabrate above the base, with leaves only in a basal rosette (in *E. Ordii* sometimes a pair of leaves at lowest node).
   Stems not inflated; calyx white to pink.
    Leaves ciliate, otherwise glabrous; calyx with hooked hairs
                      6. *E. inerme.*
    Leaves thinly woolly, tending to be glabrate; calyx appressed-pubescent . . . . .7. *E. Ordii.*
   Stems inflated; calyx yellow . . . . .8. *E. inflatum.*
 Involucres in part sessile along the branches or in the forks of the inflorescence (sometimes nearly all peduncled in the juvenile phase, but always a few sessile).
  Involucres white-woolly.
   Involucres turbinate, hardly longer than wide; leaf-blades rounded at base . . . . .9. *E. Eastwoodianum.*
   Involucres cylindric or somewhat enlarged upward, considerably longer than wide; leaf-blades usually gradually narrowed toward base
                     10. *E. roseum.*

Branches spreading, the ultimate branches bearing fewer than 10
involucres . . . . .var. *roseum.*
Branches erect, some of the ultimate branches much elongate and
bearing more than 10 involucres . . . . .var. *leucocladon.*
Involucres glabrous or glabrate.
Leaf-blades oblong to oblanceolate, gradually narrowed into the petiole.
Stems thinly woolly throughout . . . . .11. *E. gracile.*
Stems glabrous or glabrate except near base . .12..*E. citharaeforme.*
Leaf-blades round or oval, abruptly narrowed into the petiole.
Involucre 2 to 3 mm. long; calyx 2 mm. long.
Ultimate branches bearing several spicately arranged involu-
cres . . . . .13. *E. vimineum.*
Ultimate branches bearing few involucres, often only a termi-
nal one . . . . .14. *E. Covilleanum.*
Involucre 1 to 1.5 mm. long; calyx 1.5 mm. long . .15. *E. Baileyi.*
Perennials, at least the root woody.
Involucres in heads.
Leaves well distributed up the stems.
Leaf-blades broadest near base, ovate or broadly lanceolate
16. *E. parvifolium.*
Leaf-blades tapering toward base, linear to oblong or oblanceolate
17. *E. fasciculatum.*
Leaves green on upper surface.
Upper leaf-surfaces, involucres, and calyces (externally) gla-
brous or nearly so . . . . .var. *fasciculatum.*
Upper leaf-surfaces, involucres, and calyces pubescent
var. *foliolosum.*
Leaves gray on both surfaces .var. *polifolium.*
Leaves all crowded at or near base of stems.
Erect flowering stems simple or branched only at the top, tomentose,
bearing few heads or a single large one . . . . .18. *E. latifolium.*
Stems branching, glabrous, bearing many heads . . . . .19. *E. nudum.*
Involucres not in heads (sometimes 2 in a place).
Involucres in a divaricately branched cyme or "panicle" 20. *E. Heermannii.*
Involucres spicately arranged along the branches.
Stem glabrous, glaucous, inflated; involucres glabrous 21. *E. indictum.*
Stem appressed-woolly, not inflated; involucres appressed-woolly.
Leaves oblong to lanceolate or narrowly ovate, usually well spaced
on lower part of stem . . . . .22..*E. elongatum.*
Leaves orbicular to broadly ovate or obovate, closely crowded
23. *E. saxatile.*

1. **E. angulosum** Benth. Common from Salinas Valley eastward, in friable clays
and firm sandy loams; not usually growing in coarse or loose sand.

2. **E. maculatum** Heller. Occasional in sandy soils in eastern part: San Juan
River; southern Temblor Range; Cuyama Valley.

3. **E. viridescens** Heller. In approximately the same area of the county as *E. angulosum* and growing in the same kinds of soil. Less generally common than *E. angulosum* but quite plentiful in some localities. Like several other interior or "desert"
species, *E. viridescens* extends far down toward the coast along the Cuyama River,
where it was collected before the construction of Twitchell Reservoir.

4. **E. gracillimum** Wats. Frequent in eastern part from La Panza district to Temblor Range and Caliente Mt., extending down Cuyama River like *E. viridescens.* Sometimes growing in light calcareous clay, but generally more characteristic of sandy soils than are *E. angulosum* and *E. viridescens.*

5. **E. gossypinum** Curran. Rare on hot, dry, barren gravelly or sandy slopes, foothills of Temblor Range: Panorama Hills near base of Crocker Grade *(7800)* to northern part of Elkhorn Plain. Otherwise found only in the San Joaquin Valley in Kern Co. and Kings Co.

6. **E. inerme** (Wats.) Jepson. Sandy soil in chaparral area near summit of Black Mt., La Panza Range *(8243).* Abundant and vigorous the year following a fire, but sparingly found at other times.

7. **E. Ordii** Wats. On barren gypseous clay slopes in eastern part: Cholame *(Eastwood 13,897,* type of *E. tenuissimum* Eastw.); frequent in Temblor Range, where locally abundant in spots; hills on north side of Cuyama Valley.

8. **E. inflatum** Torr. & Frem. *E. trichopes* Torr. Usually in friable white calcareous clay, widespread and often abundant in eastern part from Cholame to Cuyama Valley. In our area it is impossible to separate *E. trichopes,* described as having four-lobed involucres, from *E. inflatum,* which authors have described as having five-lobed involucres. Our plants have mostly four-lobed involucres, but some of those on the same plants are five-lobed. The conclusion that *E. trichopes* and *E. inflatum* are the same is reinforced by the facts that both are credited with approximately the same geographic distribution and that they show parallel variation in the degree of inflation of the stems and in the duration of life. Probably all of our plants are annual, but a perennial form of the species is reported elsewhere. South of Soda Lake the plants have been in some years remarkably plentiful, forming a yellow carpet in late May or early June.

9. **E. Eastwoodianum** J. T. Howell. *E. temblorense* Howell & Twisselmann. On shale slopes which are otherwise barren, west side of Cottonwood Pass. The autumnal stage of growth looks very different from the vernal. A collection of the vernal phase from Cottonwood Pass *(7733)* resembles closely the type collection of *E. Eastwoodianum* and has the "roundish" leaf-blades of that plant, not "oblong to elliptic" as described for *E. temblorense* (Leafl. West. Bot. 10: 45). The same seems to be true even of some plants from the type locality of *E. temblorense.* The fact that both *E. temblorense* and *E. Eastwoodianum* are reported to occur in the same portion of Monterey Co. (Howitt and Howell in "Vascular Plants of Monterey County" pp. 57 and 59) supports the suspicion that they are not distinct.

10. **E. roseum** D. & H. var. **roseum.** Sandy or broken shale soils in eastern part: Cottonwood Pass *(Twisselmann 1570);* Palo Prieto Canyon *(Twisselmann 1467);* Cammatti Ranch, where grading into var. *leucocladon;* Mariannas Summit, Temblor Mts. *(Twisselmann 1628);* Caliente Mt.; eastward and northward in San Joaquin Valley.

Var. **leucocladon** (Benth.) Hoover. *E. virgatum* Benth. Sandy soils, more plentiful and widespread than typical *E. roseum,* extending west at least to Salinas River.

11. **E. gracile** Benth. Frequent, sometimes locally abundant, in sandy soils of coastal region south of San Luis Obispo and in Salinas Valley.

12. **E. citharaeforme** Wats. Sandy soils in interior, evidently at only a few scattered localities: "Baron Schroeder's Ranch, Santa Margarita," the type locality

(now called Eagle Ranch); Caliente Mt., abundant in sandy places from peak westward (*8214*). The same entity is found in the mountains of Santa Barbara and Ventura cos. It is possibly a glabrate variant of *E. roseum*.

13. **E. vimineum** Dougl. Rare in Santa Lucia Range: Tassajera Creek, in gravel derived from gray shale (*9678*). The type of subsp. *polygonoides* Stokes ("along the road from San Luis Obispo to the coast," *Stokes 125*) quite closely resembles *E. vimineum* as it occurs elsewhere in California, rather than *E. gracile* to which Munz referred it as a variety. The specimen itself provides no reason for regarding it as anything other than simply *E. vimineum*, although the occurrence is rather unexpected.

14. **E. Covilleanum** Eastw. Occasional on disintegrated shale or gypseous clay banks in eastern part: Cottonwood Pass (*7638*); Sandiego Canyon, Temblor Range (*Eastwood* and *Howell 4111*); head of Morales Canyon, Caliente Range (*8201*).

15. **E. Baileyi** Wats. Very common on sandy flats, especially flood-beds of streams, Salinas Valley to Palo Prieto Canyon and La Panza district, southward to Cuyama River; upper Arroyo Grande. It has been impossible to find any clearly defined difference between our plants and *E. Baileyi* of the intermountain region; hence *E. elegans* Greene is considered a synonym.

16. **E. parvifolium** Smith. Common along coast on rocky hills, sandy flats, and dunes, extending inland only a few miles, to vicinity of San Luis Obispo.

17. **E. fasciculatum** Benth. var. **fasciculatum.** Many of the plants near San Luis Obispo, although varying toward var. *foliolosum*, seem best referred to the typical variety, because the involucres and calyces are nearly glabrous. Plants growing in other localities near the coast have not been closely examined, but ought to be, to determine in which variety they may best be placed.

Var. **foliolosum** (Nutt.) Stokes. Mostly in rocky places, frequent in Santa Lucia Mts. and hills of upper Salinas Valley; particularly common in La Panza Range; Temblor Range, according to Twisselmann, although var. *polifolium* is prevalent in most of that area. Some specimens show a combination of features of this variety and the following.

Var. **polifolium** (Benth.) T. & G. Common in hills of eastern part from Cottonwood Pass to Cuyama Valley, extending westward to hills between Shandon and Creston. Notably plentiful on sandstone in the Caliente Range and in southern Temblor Range.

18. **E. latifolium** Smith. Sea-cliffs and sand-dunes, from Piedras Blancas Point (*6452*) northward.

19. **E. nudum** Dougl. var. **pubiflorum** Benth. Much reduced by agriculture and other human activities, but still common in the Salinas River basin, in rocky or firm sandy soils. All collections now at hand, all of which are from the interior, are referable to var. *pubiflorum*. The species does occur near the coast, however, as at Cayucos. Some of these coastal plants may have glabrous flowers.

20. **E. Heermannii** D. & H. On disintegrating sandstone, Caliente Mt. (*8218, 8269, 8330*).

21. **E. indictum** Jepson. Occasional on shale or clay banks in the interior: east of San Miguel; Cottonwood Pass; Palo Prieto Pass, and southward. Apparently confused by authors with *E. auriculatum* Benth., which seems to be rather an insignificant variant of *E. nudum*, whereas *E. indictum* is markedly distinct. The same opinion was implied by Twisselmann in his Temblor Flora.

22. **E. elongatum** Benth. Common on rocky or firm sandy slopes from near coast (except on hills directly facing the ocean) eastward to Temblor Range.

23. **E. saxatile** Wats. In rocky sites: found by Roman Gankin on Burnett Peak, where the plants were recognizable in early spring by their distinctive vegetative appearance. Not collected in flower.

Chenopodiaceae. Goosefoot Family

Leaves not rigid and sharp-pointed; calyx without scarious wings.
 Branches not fleshy; leaves well developed.
  Leaves slightly fleshy or, if linear, not at all so.
   Flowers perfect, with calyx.
    Calyx without hooked spines.
     Calyx usually 5-lobed; stamens 1 to 5.
      Calyces of flowers in a cluster coherent and indurate in fruit ....1. *Beta.*
      Calyces not coherent or indurate .....2. *Chenopodium.*
     Calyx of 1 sepal; stamen 1 ....3. *Monolepis.*
    Calyx-lobes each bearing a hooked spine ....4. *Bassia.*
   Flowers unisexual: the staminate with calyx, the pistillate enclosed between 2 bracts or in a sac.
    Leaves mealy-pubescent or glabrous; inflorescence not villous.
     Achene enclosed between 2 bracts with margins at least partly free ....5. *Atriplex*
     Achene enclosed in a flattened sac with a very small orifice
               6. *Grayia.*
    Leaves soft-hairy; inflorescence conspicuously woolly-villous
                7. *Eurotia.*
   Leaves fleshy, terete or linear; flowers in axillary clusters: perfect, pistillate, and staminate mixed ....8. *Suaeda.*
  Branches fleshy, becoming woody toward base; leaves reduced to very short fleshy decurrent scales.
   Shrubs; scales alternate ....9. *Allenrolfea.*
   Woody-based herbs; scales opposite ....10. *Salicornia.*
Leaves on branches of mature plants rigid, sharp-pointed; fruiting calyx with broad scarious wings ....11. *Salsola.*

1. Beta L.

1. **B. vulgaris** L. Beet. Commonly escaped from cultivation along the coast, usually in saline soils. Wild plants do not generally have the greatly enlarged root which characterizes many of the cultivated forms of the species.

2. Chenopodium L. Goosefoot

Plant not glandular, more or less mealy.
 Annuals; leaves not truncate or sagittate at base.
  Leaves dark green, not much reduced toward ends of stems (except perhaps in very old plants); terminal panicle hardly larger than the upper axillary ones ....1. *C. murale.*
  Leaves light or grayish green (upper surface bright green in *C. macrospermum*), markedly reduced toward ends of stems; flowers mostly in a terminal panicle which is leafless or with small bracts.

Flowers in round head-like clusters mostly 4 to 6 mm. in diameter

2.. *C. album.*

Flowers not in round clusters, or these less than 4 mm. in diameter.

Leaves, except the reduced upper ones, rhombic-ovate, mostly toothed.

Leaves mealy or glaucous on both sides.

Seeds minutely reticulate . . . . .3. *C. Berlandieri.*

Seeds smooth . . . . .4. *C. strictum.*

Leaves bright green on upper surface, white-mealy on lower.

5.. *C. macrospermum.*

Leaves oblong or lanceolate to linear, entire or the largest with one tooth on each side.

Stems erect; leaves linear to narrowly lanceolate, acute

6. *C. dessicatum.*

Stems spreading; leaves oblong to rhombic-lanceolate, mostly obtuse . . . . .7. *C. carnosulum.*

Perennial with fleshy root; leaves truncate or sagittate at base 8. *C. californicum.*

Plant glandular and aromatic . . . . .9. *C. ambrosioides.*

Stems erect, finely hairy to subglabrous . . . . .var. *ambrosioides.*

Stems widely spreading, villous . . . . .var. *vagans.*

1. **C. murale** L. Pigweed. Very common in cultivated ground in both eastern and western parts of the county. It thrives particularly in the rich soils where livestock have been kept. It also is well established in sandy areas near the coast.

2. **C. album** L. Lamb's Quarters. Common in disturbed ground throughout. A highly variable species. Some of the plants (with greener leaves) seem to be var. *viride* (L.) Moq., while others (with narrower leaves) are probably var. *lanceolatum* (Muhl.) Coss. & Germ. The name *C. album* is here reserved for plants with smooth glossy seeds, but it is improbable that plants with microscopically reticulate seeds represent one or more species really distinct from this.

3. **C. Berlandieri** Moq. Widespread weed, mostly toward the interior. Some of the specimens placed under this name by H. A. Wahl seem hardly distinguishable from *C. murale*; others in outward appearance closely resemble plants identified as *C. album*. There is no obvious geographical or ecological separation from either species.

4. **C. strictum** Roth var. **glaucophyllum** (Aellen) Wahl. In sand just south of Hazard Canyon (*9037*). Identified by H. A. Wahl.

5. **C. macrospermum** Hook. f. var. **farinosum** (Wats.) J. T. Howell. Moist sandy places near the coast, forming dense stands. An unusually robust plant from Morro Bay is labelled by H. A. Wahl var. *halophilum* (Phil.) Standley, but is probably not different genetically.

6. **C. dessicatum** A. Nelson var. **leptophylloides** (J. Murr.) Wahl. A narrow-leaved *Chenopodium* is common in the interior, mostly in sandy soils, particularly in valleys of the Salinas River drainage, the hill country north of Carrizo Plain, and Cuyama Valley. The name here used for this plant is based on identifications by H. A. Wahl.

7. **C. carnosulum** Moq. var. **patagonicum** (Phil.) Wahl. Locally frequent on dry flats among sand-dunes from Oceano southward into western Santa Barbara County; scarce on dunes at south end of Morro Bay (*10,629*).

8. **C. californicum** Wats. Widely distributed, most commonly in sandy soil, in both coastal and interior districts.

9. **C. ambrosioides** L. var. **ambrosioides**. Mexican Tea. Frequent in moist to dry sandy soil, or along stream-beds, mainly near the coast.

Var. **vagans** (Standley) J. T. Howell. Mouth of Morro Creek, on dunes (*9240*), and probably in similar habitats elsewhere.

10. **C. Botrys** L. was found, after completion of the manuscript, in the sandy bed of the Salinas River at Paso Robles. It resembles *C. ambrosioides,* differing most noticeably in its deeply lobed leaves.

3. **Monolepis** Schrad.

1. **M. Nuttalliana** (Schult.) Greene. In bare alkali spots in eastern part, not common: west base of Temblor Range near Simmler-McKittrick road (*8088*); Soda Lake.

4. **Bassia** All.

1. **B. hyssopifolia** (Pall.) Kuntze. Summer weed of irrigated ground in vicinity of Shandon. It may be expected to become common, as it is abundantly established in the San Joaquin Valley and elsewhere.

5. **Atriplex** L. Saltbush

Annuals with slender tap-root.
    Staminate flowers mixed with the pistillate in terminal spikes or small axillary clusters.
        Flowers in terminal spikes.
            Leaves entire to sparingly sinuate-toothed; in coastal salt-marshes
                                              1. *A. patula.*
                Leaves lanceolate to oblong . . . . .var. *patula.*
                Leaves rhombic-ovate, often hastate . . . . .var. *hastata.*
            Leaves prominently several-toothed; interior hills and plains
                                            2. *A. rosea.*
        Flowers in axillary clusters.
            Leaves rhombic-ovate to triangular-hastate or obtuse-triangular.
                Stem erect, branching above base; leaves coarsely toothed
                                          2. *A. rosea.*
                Stem with spreading branches from the base; leaves entire or slightly hastate . . . . .3. *A. expansa.*
            Lower leaves oblong, the upper gradually reduced, crowded, lanceolate to ovate, rounded or cordate at base.
                Plant gray; fruiting bracts 4 to 5 mm. long, the upper margin regularly dentate . . . . .4. *A. coronata.*
                Plant light green; fruiting bracts 2 to 3.5 mm. long, the upper margin with few irregular teeth . . . . .5. *A. vallicola.*
    Staminate flowers in small round clusters in terminal leafless spikes or panicles
                                              6. .*A. Serenana.*
Perennials, shrubby or with a thick root.
    Herbs, often woody at base, less than 3 dm. tall though often widely spreading.
        Leaves oblong or narrowly elliptic to linear.
             Leaves dentate or mostly so; fruiting bracts at maturity becoming reddish and fleshy . . . . .7. *A. semibaccata.*
             Leaves entire; fruiting bracts not becoming red or fleshy.

Root fleshy; staminate flowers mostly mixed with pistillate; fruiting bracts entire ....8...*A. californica.*
Root woody; staminate flowers in bractless terminal spikes; fruiting bracts toothed ....9. A. *fruticulosa.*
Leaves ovate, broadly elliptic, or obovate.
Leaves alternate, thick; plants monoecious ....10. *A. leucophylla.*
Leaves mostly opposite, thin; plants dioecious, or the staminate with a few pistillate flowers ....11. *A. Watsonii.*
Shrubs over 5 dm. tall.
Leaves triangular-ovate to oval or elliptic, at least the lower with well-defined petioles.
Summer-flowering; branches not thorny (in our plants); fruiting bracts winged all around ....12. *A. lentiformis.*
Spring-flowering; branches thorny; fruiting bracts extended into leaf-like blades ....13. *A. spinifera.*
Leaves oblong or oblanceolate to linear, narrowed toward base but without clearly defined petiole.
Fruiting bracts 2 mm. long, without wings ....14. *A. polycarpa.*
Fruiting bracts over 4 mm. long, with 4 conspicuous vertical wings

15. *A. canescens.*

1. **A. patula** L. var. **patula.** Along coastline, rare: 9 miles north of Morro (*J. T. Howell 4485* in 1929).

Var. **hastata** (L.) Gray. Common on borders of beaches and in coastal salt-marshes, especially around Morro Bay.

2. **A. rosea** L. Abundant along eastern border north of Carrizo Plain. A common wheat-field weed according to Twisselmann.

3. **A. expansa** (D. & H.) Wats. Locally plentiful in occasionally rain-flooded alkali soil on Carrizo Plain (*6506*).

4. **A. coronata** Wats. Frequent in alkali soils in eastern part from Cholame Valley to Carrizo Plain. The crowded upper leaves are often short, broad, and cordate as described for *A. cordulata* Jepson. All our plants however, have fruiting bracts shaped as in typical *A. coronata.* The reports by Twisselmann in his "Temblor Flora" of *A. cordulata* and *A. coronata* were both presumably based on plants of the species here referred to.

5. **A. vallicola** Hoover. A species which is quite common in low ground on Carrizo Plain, both north and south of Soda Lake, seems to be essentially like the type of *A. vallicola,* which was collected north of Lost Hills in Kern County, although the latter is overmature and therefore difficult to compare.

6. **A. Serenana** A. Nelson. In friable, not necessarily alkaline, soil: Salinas Valley eastward to Palo Prieto Pass, and Santa Maria Valley. To be expected in vicinity of Carrizo Plain, but not yet collected there. In the Salinas Valley it shows a tendency to spread along roads as a weed.

7. **A. semibaccata** R. Br. Commonly established in firm soils, especially by roadsides and in formerly cultivated areas, mainly near coast but also in eastern part near Grant Lake, according to Twisselmann.

8. **A. californica** Moq. Common in sandy soil along the coast. Not so strictly confined to the coastline as is *A. leucophylla,* it extends inland to flats among the dunes north of Oso Flaco Lake.

9. **A. fruticulosa** Jepson. Frequent in firm alkali soils from Cottonwood Pass and Cholame Valley to Carrizo Plain.

10. **A. leucophylla** (Moq.) D. Dietr. Common on beaches, where it is one of the plants growing closest to high-tide level.

11. **A. Watsonii** A. Nelson. Local in salt marsh on east shore of Morro Bay (*6429*), the most northerly known occurrence of the species.

12. **A. lentiformis** (Torr.) Wats. In moist alkaline places, or in dry soil near streams: Estrella River, Cholame Creek, and Cuyama River. Plants of the Salinas Valley have been referred to *A. Breweri* (or *A. lentiformis* var. *Breweri*), but authors are in marked disagreement as to how this can be distinguished from true *A. lentiformis*. In some cases the name *A. Breweri* is applied to shrubs having larger fruiting bracts, but this form has no clear geographic significance. For example, one collection made along the Estrella River (*8751*) has bracts only 2 or 3 mm. long, whereas another from the same locality has bracts averaging about 4 to 5 mm. long (*Mrs. H. C. Cantelow 1805*). My conclusion is that the shrubs in this area should be called *A. lentiformis* until it can be clearly demonstrated that some difference exists.

13. **A. spinifera** Macbr. Formerly the dominant plant on low level portions of Carrizo Plain. Great areas have been cleared for cultivation, but rather extensive stands of the shrub still may be seen near Soda Lake, and also on steep slopes in strongly gypseous soil at Chalk Mt. in Cuyama Valley. The species is normally dioecious, but rarely staminate and pistillate flowers are found in the same cluster.

14. **A. polycarpa** (Torr.) Wats. Dry hills and plains in eastern part. Large numbers of the shrubs have been destroyed by land-clearing for cultivation and by overgrazing, but the species is even yet a prominent feature of its hot, arid home.

15. **A. canescens** (Pursh) Nutt. Alluvial soils, south end of Carrizo Plain (*8280*, evidently the form called var. *laciniata* Parish) and Cuyama Valley. The species is also spreading from artificial sowings, according to Ian McMillan, along the San Juan River and tributaries. Some of the shrubs in Cuyama Valley have smaller fruiting bracts; these may be var. *linearis* (Wats.) Munz, which has not otherwise been reported north of the Colorado Desert.

6. Grayia H. & A.

1. **G. spinosa** (Hook.) Moq. Hop Sage. Locally plentiful on sandy slopes, low hills north of Elkhorn Plain, west base of Temblor Range (*8139*).

7. Eurotia Adans.

1. **E. lanata** (Pursh) Moq. Frequent in sandy or alkaline soils from Temblor Range and east side of Carrizo Plain to Cuyama Valley.

8. Suaeda Forsk.

Branches erect or ascending; leaves pale; flowers 1 mm. wide . . . . 1. *S. fruticosa.*
Branches trailing; leaves deep green; flowers 2 to 3 mm. wide . . . . . 2. *S. californica.*

1. **S. fruticosa** (L.) Forsk. Abundant in strongly alkaline soils, Carrizo Plain and Cholame Valley.

2. **S. californica** Wats. Beaches around Morro Bay; to be expected elsewhere along our coast. At least some of the plants are very sparsely and inconspicuously hairy, thus varying slightly toward var. *taxifolia* (Standley) Munz.

9. **Allenrolfea** Kuntze

1. **A. occidentalis** (Wats.) Kuntze. In moist soils heavily charged with alkali in Cholame Valley, along San Juan River, at Soda Lake, and in Cuyama Valley.

10. **Salicornia** L.

Flowers borne to tips of branches ....1. *S. virginica.*
Branches continuing beyond flower-bearing axils ....2..*S. subterminalis.*

1. **S. virginica** L. PICKLEWEED. SAMPHIRE. Very common in saline soils along the coast, particularly on the shore of Morro Bay and at Avila.

2. **S. subterminalis** Parish. With *S. virginica* in the salt marsh on the east side of Morro Bay.

11. **Salsola** L.

1. **S. Kali** L. var. **tenuifolia** Tausch. RUSSIAN THISTLE. Common weed in fertile friable soils in the interior, especially in cultivated ground.

Amaranthaceae. AMARANTH FAMILY

1. **Amaranthus** L.

Flowers mostly in leafless terminal panicles or branched spikes.
    Stem erect, simple or branched above base.
        Sepals of pistillate flowers narrowed upward.
            Sepals of pistillate flowers obtuse or abruptly pointed. 1. *A. retroflexus.*
            Sepals of pistillate flowers gradually tapering to a point.
                Sepals of pistillate flowers 3 mm. long; stamens usually 3
                                2. *A. Powellii.*
                Sepals of pistillate flowers 1.5 to 2 mm. long; stamens 5
                                3. *A. hybridus.*
        Sepals of pistillate flowers broadened upward ....4. *A. Palmeri.*
    Stems spreading to prostrate ....5. *A. deflexus.*
Flowers in axillary clusters.
    Central stem erect, the spreading branches curved upward to form a subglobose outline ....6. *A. albus.*
    Stems prostrate ....7. *A. blitoides.*

1. **A. retroflexus** L. PIGWEED. Common weed in summer-irrigated places.

2. **A. Powellii** Wats. Abundant and rank-growing weed in a cultivated field at Edna (*10,094*).

3. **A. hybridus** L. Plants which seem to be this species have often been seen in moist cultivated ground. In the absence of herbarium specimens, their identity is uncertain. Some of them may possibly be *A. Palmeri,* which is similar in superficial appearance.

4. **A. Palmeri** Wats. Apparently a rare weed, scarcely established at Twisselmann Ranch (*Twisselmann 2353* in 1955); 2 miles south of Creston (*10,061* in 1966). Closer attention to these unattractive weeds might reveal that this species occurs more widely.

5. **A. deflexus** L. "Rare garden weed" at Twisselmann Ranch (*Twisselmann 8977* in 1963). To be expected in towns near the coast, but not yet collected.

6. **A. albus** L. TUMBLEWEED. Frequent, mainly toward the interior and most often in sandy soils which have been flooded in warm weather. Formerly called *A. graeci-*

zans L., which recent authors consider to be a different species. *Salsola Kali* (Russian thistle) is a commoner "tumbleweed" in our area than this species.

7. **A. blitoides** Wats. Usually along roadsides, sometimes in low places where water has stood, in the interior. Recent American authors have included this plant in *A. graecizans* L., but Brenan (*Watsonia* 4: 266 and 272. 1961) indicated that the two are different.

Illecebraceae. KNOTWORT FAMILY

Leaves and sepals not spine-tipped . . . . .1. *Herniaria.*
Leaves and sepals spine-tipped . . . . .2. *Cardionema.*

1. Herniaria L.

1. **H. cinerea** DC. Sparingly introduced in sandy soils: Maxey Ranch, east side of Cholame Valley, according to Twisselmann; upper Arroyo Grande.

2. Cardionema DC.

1. **C. ramosissimum** (Weinm.) Nels. & Macbr. SAND MAT. Abundant in sandy places near the coast.

Nyctaginaceae. FOUR-O'CLOCK FAMILY

Involucre 5-lobed; calyx campanulate to tubular-funnelform . . . . .1. *Mirabilis.*
Involucre of distinct bracts; calyx salverform, with slender tube and rotate lobes
                                             2. *Abronia.*

1. Mirabilis L.

Involucre one-flowered.
    Stems herbaceous from a fleshy root; calyx 3 to 5 cm. long . . . . . . .1. *M. Jalapa.*
    Stems brittle, woody below; calyx 10 to 14 mm. long . . . . .2. *M. laevis.*
Involucre several-flowered . . . . .3. *M. Froebelii.*

1. **M. Jalapa.** L. FOUR-O'CLOCK. Escaped from cultivation, near Arroyo Grande according to Twisselmann. Often spreading in gardens and occasionally becoming a pest, but rarely if ever becoming established in wild habitats.

2. **M. laevis** (Benth.) Curran. Occasional on barren rocky (sometimes sandy) slopes: coast near Monterey Co. line; Cottonwood Pass; southern Temblor Range to Cuyama Canyon; north edge of Santa Maria Valley. Evidently absent from central part of county.

3. **M. Froebelii** (Behr) Greene. Sandy soil, barely entering our area from the desert region: east end of Cuyama Valley near Kern Co. line and northward to about the divide at the southeast end of Carrizo Plain.

Abronia Juss. SAND-VERBENA

Perennials with fleshy tap-root; stems prostrate, widely trailing.
    Flowers yellow . . . . .1. *A. latifolia.*
    Flowers not yellow.
        Calyx dark red-purple, 13 to 15 mm. long; leaves fleshy . . . .2. *A. maritima.*
        Calyx bright purple to white, 17 to 22 mm. long; leaves rather thin
                                         3. *A. umbellata.*
Annual with slender tap-root; stems ascending . . . . .4. *A. pogonantha.*

1. **A. latifolia** Esch. Yellow Sand-Verbena. Edge of beaches and adjacent sand-dunes, particularly common near Oso Flaco Lake. An apparent hybrid with *A. maritima* was found near the south end of Morro Bay.

2. **A. maritima** Nutt. Low dunes near the shore, apparently at Piedras Blancas Point (not exactly typical and possibly of hybrid origin); abundant in typical form from just north of Morro Bay southward.

3. **A. umbellata** Lam. Beach Sand-Verbena. Very common in all coastal sandy areas, extending inland to Los Osos, Nipomo Mesa, and vicinity of Santa Maria. Occasional plants from the northern part of our coast seem to indicate hybridization with *A. latifolia* or perhaps rarely with *A. maritima*.

4. **A. pogonantha** Heimerl. Loose sand in the interior, especially in the hill country between Shandon and Creston; rare in southern Temblor Range.

Mesembryanthemaceae. Ice-Plant Family

Epidermal cells greatly swollen with water, the plants appearing as if covered with ice.

Annual; leaves alternate . . . . .1. *Mesembryanthemum.*
Perennial; leaves opposite . . . . .2. *Drosanthemum.*
Epidermis smooth, the cells not swollen.
Flowering stems clustered with basal leaves on the crown of a carrot-like root; fruit dry at maturity . . . . .3. *Conicosia.*
Stems prostrate, extensively creeping; fruit fleshy . . . . .4. *Carpobrotus.*

Mesembryanthemum L.

1. **M. crystallinum** L. Ice-Plant. Occasional in sand along the coast, as at Morro Bay.

2. Drosanthemum Schwantes

1. **D. floribundum** (Haw.) Schwantes. Locally established on sea-cliffs near Cambria. Abundantly cultivated in some coastal localities.

3. Conicosia Schwantes

1. **C. elongata** (Haw.) Schwantes. Well established on dunes near south end of Morro Bay and from Pismo Beach southward.

4. Carpobrotus N. E. Brown

Flowers 6 to 8 cm. wide . . . . .1. *C. edulis.*
Flowers 3 to 5 cm. wide . . . . .2. *C. chilensis.*

1. **C. edulis** (L.) N. E. Brown. Widely planted to control erosion or reduce fire hazard. Escaped along coast, usually in sand, and crowding out *C. chilensis* as well as nearly all other plants associated with it. The flowers may be purple, yellow, or white on plants which otherwise are identical.

2. **C. chilensis** (Mol.) N. E. Brown. Abundant in coastal sands. The flowers are consistently rose-purple (magenta).

Tetragoniaceae. New Zealand Spinach Family

1. Tetragonia L.

1. **T. tetragonioides** (Pall.) Ktze. New Zealand Spinach. Cultivated as a vegetable; commonly escaped on or near beaches.

Portulacaceae. PURSLANE FAMILY

Leaves alternate as well as some basal.
    Ovary half-inferior; petals yellow ....1. *Portulaca.*
    Ovary superior; petals white (often minute) to rose-red.
        Flowers in bracteate racemes; sepals ovate-acute or acuminate, not scarious-margined ....2. *Calandrinia.*
        Flowers in helicoid cymes diverging from main stem-axis; sepals oval or orbicular, scarious-margined ....3. *Calyptridium.*
Leaves opposite (often the pair more or less united) or basal only.
    Petals 5; our species annuals ....4. *Montia.*
    Petals many; dwarf tufted perennial with fleshy terete leaves ....5. *Lewisia.*

1. Portulaca L.

1. **P. oleracea** L. PURSLANE. Occasional weed in gardens where watered in summer and in irrigated fields.

2. Calandrinia H. B. K.

Capsule barely equalling sepals or slightly shorter ....1. *C. ciliata.*
    Petals 4 to 6 mm. long; stamens mostly 3 to 6 ....var. *ciliata.*
    Petals 6 to 10 mm. long; stamens mostly 7 to 14 ....var. *Menziesii.*
Capsule conspicuously longer than sepals ....2. *C. Breweri.*

1. **C. ciliata** (R. & P.) DC. var. **ciliata.** Small-flowered plants growing usually in sandy or rocky soil may represent the typical form of the species. Such plants are apparently much less common, but perhaps only less conspicuous, than the following variety.

Var. **Menziesii** (Hook.) Macbr. RED MAIDS. Very common in cultivated ground, especially in upper Salinas Valley.

2. **C. Breweri** Wats. Common in burned chaparral areas, usually in sandy soil, eastward to La Panza Range. Often abundant the year following a fire, but seldom if ever seen otherwise.

3. Calyptridium Nutt.

Capsule slender, usually curved, more than twice as long as sepals
                            1. *C. monandrum.*
Capsule ovoid, less than twice as long as sepals ....2. *C. Parryi.*

1. **C. monandrum** Nutt. Generally in sandy soils, frequent in both coastal and interior districts.

2. **C. Parryi** Gray. In sand, Tierra Redonda Mt. (*Hardham 1867* in 1957).

4. Montia L.

Cauline leaves in several opposite pairs ....1. *M. fontana.*
Cauline leaves in a single pair, usually more or less united.
    Cauline leaves united to form a disk with usually 2 angles on the margin
                            2. *M. perfoliata.*
    Cauline leaves partly united on one side, or distinct.
        Cauline leaves linear-lanceolate to ovate, united above the base on one side.
        Petals showy, 4 to 7 mm. long ....3. *M. gypsophiloides.*
        Petals inconspicuous, 2.5 to 4 mm. long ....4. *M. spathulata.*

> Cauline leaves linear, wholly distinct or very slightly connate at base on one side ....5. *M. tenuifolia.*

1. **M. fontana** L. Moist swale in pine woods at Cambria; probably also elsewhere but easily overlooked. Various names have been used in recent publications for Californian plants of this relationship, but it has been impossible to decide whether any of these names applies specifically to the Cambria plant, and anyhow the alleged differences seem inconsequential.

2. **M. perfoliata** (Donn) Howell. MINER'S LETTUCE. Very common in both shaded and sunny places, in various soils, in all parts except very dry areas or saline soils. Shows much variation in such features as color of plant, size and shape of leaf-disk, length of petals, etc. None of the variants in this county seems to be geographically significant, but there are undoubtedly genetic differences which might justify the separation of more varieties and forms than are usually recognized. Certain plants on Carrizo Plain may represent f. *parviflora* (Dougl.) J. T. Howell; they are remarkably similar to specimens from the state of Washington.

3. **M. gypsophiloides** (F. & M.) Howell. In disintegrated shale, near summit of Temblor Range. An oddly localized occurrence of a species not otherwise known south of Santa Clara Co. and commonest near the coast north of San Francisco Bay. The form found here is probably the same as the Marin Co. plants which have been called *M. gypsophiloides* var. *exigua* (T. & G.) J. T. Howell, but by some authors the name *M. exigua* has been differently interpreted. (See note under *M. tenuifolia*).

4. **M. spathulata** (Dougl.) Howell. Rocky places, often on serpentine farther north. Reported from the Temblor Range by Twisselmann; but, in the absence of any specimens, it is doubtful whether the plant which I interpret as genuine *M. spathulata* (with short pointed cauline leaves joined on one side) actually occurs in our county.

5. **M. tenuifolia** (T. & G.) Howell. Frequent in eastern part in sand, friable clay, or crumbling shale. Apparently this is *M. exigua* (T. & G.) Jepson, as interpreted by Jepson. Whatever its proper name may be, this plant with long narrow cauline leaves which are completely distinct or very slightly connate on one side, seems clearly separable from what I am here calling *M. spathulata*.

5. Lewisia Pursh

1. **L. rediviva** Pursh. BITTER-ROOT. Barren hard-packed gravelly soil, rare in the county: Paso Robles (*Rowntree* in 1933); Burnett Peak, according to R. Gankin.

Caryophyllaceae. PINK FAMILY

Calyx of distinct sepals.
> Stipules absent.
>> Plant glabrous or sparingly pubescent; capsule ovoid or oblong, dehiscent to base.
>>> Petals 2-lobed, or very short and concealed by calyx, or absent [2]
>>>> 1. *Stellaria.*

---

2. The arbitrary characters which are used as a basis for separating *Stellaria*, *Sagina*, and *Arenaria* are, with respect to some of the species, very obscure, and in any event seem to be of little genetic significance. A classification which combined the three into one genus would have greater utility, and at the same time would probably express the phylogeny as accurately as the currently accepted splitting into three genera.

Petals entire or notched at apex, from slightly shorter than sepals to about twice as long.
  Styles as many as sepals . . . . .2. *Sagina*
  Styles fewer than sepals . . . . .3..*Arenaria.*
Plant pubescent throughout; capsule cylindric, dehiscent at apex
                                      4. *Cerastium.*
Stipules present, usually scarious (setaceous in *Loeflingia*).
  Petals evident, little if at all shorter than sepals.
    Leaves apparently whorled (those in axillary clusters as long as those subtending them); leaves of inflorescence reduced to minute bracts . . . . .5. *Spergula.*
    Leaves opposite, not appearing whorled; leaves of inflorescence hardly reduced . . . . .6. *Spergularia.*
  Petals very small, concealed by the sepals.
    Leaves obovate, oblong, or spatulate . . . . .7. *Polycarpon.*
    Leaves subulate, curved outward . . . . .8. *Loeflingia.*
Calyx tubular, 5-lobed above.
  Plants glandular or pubescent, usually both; styles 3 . . . . .9. *Silene.*
  Plants glabrous, not glandular; styles 2 . . . . .10. *Saponaria.*

<h3 align="center">1. Stellaria L.</h3>

Leaves ovate, not much reduced upward; petals evident . . . . .1..*S. media.*
Leaves narrow (except the lowest), reduced in inflorescence to scarious bracts; petals minute or none . . . . .2. *S. nitens.*

1. **S. media** (L.) Cyr. CHICKWEED. Frequent as a weed, especially in shaded places which are moist in winter; abundant under oaks near the coast.

2. **S. nitens** Nutt. Common in "grasslands" and on partly shaded slopes, associated with other annual herbs.

<h3 align="center">2. Sagina L.</h3>

1. **S. occidentalis** Wats. Apparently rare or more probably overlooked: coast near Cambria; burned-over area just south of Price Canyon, in seepage spots in sandy soil (*6751* in 1947).

<h3 align="center">3. Arenaria L. SANDWORT</h3>

Annuals with slender tap-root, growing in dry sandy or rocky sites.
  Leaves, at least the lower ones, mostly over 5 mm. long.
    Leaves firm; petals exceeding sepals, entire . . . . .1. *A. Douglasii.*
    Leaves flaccid; petals barely equalling sepals, usually notched at apex
                                  2. *A. emarginata.*
  Leaves not over 5 mm. long.
    Petals approximately equalling sepals . . . . .3. *A. pusilla.*
    Petals about twice as long as sepals .4. *A. californica.*
Perennial with slender rhizomes, growing in wet soil . . . . .5..*A. paludicola.*

1. **A. Douglasii** Fenzl. Frequent in rocky or sandy places, commonest in Santa Lucia Range, often on serpentine; extending inland to 5 miles east of Creston.

2. **A. emarginata** (H. K. Sharsmith) Hoover. Sycamore Creek, west slope of La Panza Range, on steep shaded slope (*7061*); Rinconada Mine, on serpentine (*6717*).

3. **A. pusilla** Wats. var. **diffusa** Maguire. On dry sand-flats, upper Salinas basin and coastal region, evidently rare: Yaro Creek north of Pozo (*6990*); Jack Lake.

4. **A. californica** (Gray) Brewer. Sandy soils in Salinas River watershed, probably rather common but overlooked. Some of the plants having the aspect of this species have recently been identified as *A. pusilla* var. *diffusa*. A collection from "Navajo Ridge, Red Hill" (*Eben McMillan 17*) is unquestionably *A. californica*.

5. **A. paludicola** Rob. Coastal fresh-water marshes from lower Arroyo Grande to Oso Flaco Lake; scarce or hidden by larger plants.

4. Cerastium L.

Petals conspicuous, twice as long as sepals .... 1. *C. arvense.*
Petals about as long as sepals or a little shorter.
    Annual .... 2. *C. viscosum.*
    Perennial, forming small mats usually in lawns .... 3. *C. vulgatum.*

1. **C. arvense** L. Upper Arroyo Grande 6 miles above County Park, on dry shaded sandstone ledges (*6855* in 1947).

2. **C. viscosum** L. MOUSE-EAR CHICKWEED. Common in burned areas and in cultivated ground, most frequently in sandy soil.

3. **C. vulgatum** L. Mostly a weed in lawns, probably spreading in grass seed mixtures. Our plants may possibly be *C. caespitosum* Gilib. or *C. holosteoides* Fries; which, however, seem hardly distinguishable from *C. vulgatum*.

5. Spergula L.

1. **S. arvensis** L. SPURREY. Abundant in sandy fields near coast; uncommon if not completely absent in the interior.

6. Spergularia J. & C. Presl. SAND-SPURREY

Perennials with fleshy root; sepals 5 to 10 mm. long .... 1. *S. macrotheca.*
Annuals or perennials, the root only slightly if at all fleshy; sepals 2.5 to 5 mm. long.
    Leaves with clusters of shorter ones (condensed branchlets) in their axils; stipules 3 to 5 mm. long, acuminate .... 2. *S. rubra.*
    Leaves in axillary fascicles few or none; stipules 2 to 4 mm. long, hardly if at all acuminate.
        Seeds brown.
            Leaves 1 to 2 cm. long; "stamens 6 to 10" .... 3. *S. Bocconii.*
            Leaves mostly 2 to 4 cm. long; "stamens 2 to 5" .... 4. *S. marina.*
        Seeds black; "stamens 4 to 8" .... 5. *S. atrosperma.*

1. **S. macrotheca** (Hornem.) Heynh. Common on ocean bluffs and in coastal saltmarshes. Perhaps var. *leucantha* (Greene) Rob. is present in Cholame Valley or on Carrizo Plain, but no specimens are at hand to substantiate this possible occurrence. At least one coastal collection (south end of Morro Bay, *9046*) had white petals and in that respect conforms with var. *leucantha*.

2. **S. rubra** (L.) J. & C. Presl. Edges of beaches, roadsides, and hard-packed soil near the ocean.

3. **S. Bocconii** (Scheele) Foucaud. Weedy roadside, Cambria (*Howell 40,816*); Shell Beach (*9187*).

4. **S. marina** (L.) Griseb. Common in moist saline soils, both near the coast and near eastern border of county.

5. **S. atrosperma** R. P. Rossbach. North side of Soda Lake, Carrizo Plain (*8144*) and probably elsewhere in alkali soils of the interior. It seems rather improbable

that the coal-black rather than brown seeds of this plant are sufficient to characterize a species genuinely distinct from *S. marina*. In fact, several of the alleged species
in this genus are based on minute differences which seem rather inconsequential.

7. Polycarpon L.

Leaves in apparent whorls of 4 or opposite, obovate to broadly oblong
                                                                           1. *P. tetraphyllum.*
Leaves opposite, oblanceolate or spatulate . . . . .2. *P. depressum.*

1. **P. tetraphyllum** (L.) L. Well established on ocean bluffs near Cambria and as
a troublesome garden weed at San Luis Obispo; occasional in moist soil along
streams.

2. **P. depressum** Nutt. Sandy soil, often in burned-over areas, in coastal region:
Baywood Park (*6926*); just south of Price Canyon (*6752*).

8. Loeflingia L.

1. **L. squarrosa** Nutt. Frequent in sand, both coastal and interior regions, but
inconspicuous and seldom collected: Atascadero (*6776*).

9. Silene L.

Annuals.
    Calyx with about 20 ribs . . . . .1. *S. multinervia.*
    Calyx with 10 ribs.
        Flowers in an open cyme; plant mostly glabrous, usually with transverse
        glandular bands on upper internodes . . . . .2. *S. antirrhina.*
        Flowers in racemes; plant hairy throughout, without glandular bands
                                                                           3...*S. gallica.*
Perennials.
    Petals bright red.
        Plants less than 3 dm. tall; leaves ovate to elliptic-oblong or oblanceolate
                                                                           4. *S. californica.*
        Plants mostly over 3 dm. tall; leaves linear or the lower ones oblanceolate
                                                                           5. *S. laciniata.*
    Petals not red.
        Flowers nodding; stamens and style conspicuously exserted.   6. *S. Lemmonii.*
        Flowers erect; stamens and style not exserted . . . . .7. *S. verecunda.*

1. **S. multinervia** Wats. Occasional in sandy or rocky places, locally abundant
the year following a fire but otherwise seldom seen, coastal area to La Panza Range.

2. **S. antirrhina** L. Occasional on dry sandy or rocky slopes.

3. **S. gallica** L. Windmill Pink. Common weed in cultivated ground, most frequently in sandy soils, sometimes spreading into undisturbed areas.

4. **S. californica** Durand. Indian Pink. Known in the county only from a collection made in Lopez Canyon (*Condit* in 1905) but to be expected elsewhere in the
Santa Lucia Range. Immediately to the southeast, the species is plentiful on the
road to Mt. Abel.

5. **S. laciniata** Cav. var. **angustifolia** H. & M. Occasional on rocky slopes in coastal
region, often on serpentine, from San Simeon Creek southward, and common in
coastal sandy areas from Morro Bay southward.

6. **S. Lemmonii** Wats. Rare in Santa Lucia Mts., generally in rocky places shaded
by oaks or pines: vicinity of Rocky Butte; 7 miles west of Templeton.

7. **S. verecunda** Wats. var. **platyota** (Wats.) Jepson. Just south of Price Canyon,

scarce in crevices of sandstone. *Silene luisana* Wats., reportedly collected at San Luis Obispo, is included by Abrams (Ill. Fl. Pac. St.) in the synonymy of *S. verecunda,* but the Price Canyon plants, at least, are identifiable by Abrams's key as *S. platyota* (*S. verecunda* var. *platyota*). The occurrence here of typical *S. verecunda* has not been established.

10. Saponaria L.

1. **S. officinalis** L. SOAPWORT. Well established in sandy bed of Salinas River at mouth of Rocky Canyon.

Nymphaeaceae. WATER-LILY FAMILY

1. Nymphaea L.

1. **N. polysepala** (Engelm.) Greene. YELLOW POND-LILY. Slow-moving streams and shallow margin of lake, local at Black Lake southwest of Arroyo Grande. A remarkable occurrence of a mainly northern species.

Ceratophyllaceae. HORNWORT FAMILY

1. Ceratophyllum L.

1. **C. demersum** L. HORNWORT. Occasional in fresh-water lakes and ponds near coast: Laguna near San Luis Obispo; Black Lake Canyon; Oso Flaco Lake. Also in Salinas River.

Ranunculaceae. BUTTERCUP FAMILY

Herbaceous plants; leaves alternate or basal.
  Flowers regular; none of sepals, or all of them, spurred.
    Pistil more than 1; fruit a follicle or an achene; flowers not in racemes.
      Pistils many, spirally arranged on the receptacle.
        Sepals not spurred; receptacle globose to ovoid; leaves rarely all basal ....1. *Ranunculus.*
        Sepals spurred; receptacle becoming much elongated in fruit; small annuals with inconspicuous flowers and the leaves all basal
                                           2. *Myosurus.*
      Pistils mostly 5 or fewer, sometimes more, in a single whorl.
        Sepals persistent; plants with deep-seated heavy roots 3. *Paeonia.*
        Sepals readily deciduous; plants with a cluster of fibrous roots.
          Flowers perfect, white or bright-colored; fruit a follicle.
            Flowers drooping, red; petals each with a long slender spur ....4. *Aquilegia.*
            Flowers erect, white or pink; petals none (sepals showy)
                                    5. *Isopyrum.*
          Flowers (in our species) unisexual, green; fruit an achene
                                   6. . *Thalictrum.*
    Pistil 1; fruit a berry; flowers in racemes ....7. *Actaea.*
  Flowers irregular; upper sepal with conspicuous spur ....8. *Delphinium.*
Woody vines; leaves opposite ....9. *Clematis.*

1. Ranunculus L. BUTTERCUP

Petals yellow (in *R. hebecarpus* inconspicuous or none); not aquatic.
  Leaves deeply lobed or divided; stems branched, with well-developed cauline as well as basal leaves.

Petals 8 mm. long or more; achenes without spines or hairs
1. *R. californicus.*
Stem erect with widely spreading branches; leaf-divisions acute
var. *californicus.*
Stems spreading from the base; leaf-divisions mostly obtuse
var. *cuneatus.*
Petals 8 mm. long or less; achenes with spines or hooked hairs.
Petals evident, 5; achene with spines on margin . . . . .2. *R. arvensis.*
Petals inconspicuous, 5 or fewer, or none; achenes with hooked hairs
3. *R. hebecarpus.*
Leaves crenate, mostly basal; flowering stems scapose or few-leaved
4. *R. Cymbalaria.*
Petals white; plants aquatic or on wet mud . . . . .5. *R. aquatilis.*

1. **R. californicus** Benth. var. **californicus.** CALIFORNIA BUTTERCUP. Very common in places where the soil remains moist during the growing season. Not observed east of La Panza Range, but probably not entirely absent from that region. The species is variable. Some plants, especially toward the interior, are rather conspicuously silky-hairy and in that respect suggest var. *canus* (Benth.) B. & W. However, the only clearly defined variant from the typical form of the species in the county is the following.

Var. **cuneatus** Greene. Ocean bluffs and open fields along the coast from near Cambria northward.

2. **R. arvensis** L. FIELD BUTTERCUP. Locally introduced at Atascadero, in black clay in a field which was wet in spring (*7114* in 1947).

3. **R. hebecarpus** H. & A. Common in loose, humus-rich soils on shaded hillsides, usually under trees, in all parts of the county.

4. **R. Cymbalaria** Pursh var. **saximontanus** Fernald. In wet soil, Alamo Creek near Cuyama River (*7140* in 1947). This locality is now covered by the waters of Twitchell Reservoir.

5. **R. aquatilis** L. var. **capillaceus** (Thuill.) DC. WATER BUTTERCUP. On mud, or submerged in ponds and slow-moving streams: Piedras Blancas Point (*6966*); Atascadero (*Rodin 6072*); Stoney Creek (*7961*).

2. **Myosurus** L. MOUSETAIL

Fruiting receptacles surpassing the leaves, on slender scapes.
Receptacle with carpels 2 to 3 mm. thick at base . . . .*M. minimus* var. *minimus.*
Receptable with carpels 1 to 1.5 mm. thick at base . . . . .var. *filiformis.*
Fruiting receptacles surpassed by the leaves, on short stout scapes . . . . . . .var. *apus.*

1. **Myosurus minimus** L. var. **minimus.** MOUSETAIL. Places which are wet in winter, usually in beds of rain-pools. The typical variety of the species is known in San Luis Obispo Co. only on Carrizo Plain north of Soda Lake, where it is scarce as compared with var. *apus.*

Var. **filiformis** Greene. Navajo Creek district, east side of La Panza Range, along streamlet in recently burned chaparral area (*8185*).

Var. **apus** Greene. Locally common in vernal pools north of Soda Lake on Carrizo Plain.

3. **Paeonia** L. PEONY

1. **P. californica** Nutt. CALIFORNIA PEONY. Common in hill country from near coast eastward through La Panza Range.

4. **Aquilegia** L. Columbine

Herbage not glandular; orifice of petal-spur truncate .... 1. *A. truncata.*
Herbage glandular-pubescent; orifice of petal-spur oblique .... 2. *A. eximia.*

1. **A. truncata** F. & M. Damp, usually shaded, places, locally common in watershed of San Carpoforo Creek.

2. **A. eximia** Van Houtte. Widely scattered in seepage spots and along rocky stream-beds in Santa Lucia Range, generally (always?) in areas of serpentine rock: San Simeon Creek; Tassajera Creek; Poly Canyon; Perfumo Canyon (type locality of *A. fontinalis* J. T. Howell, which seems identical with *A. eximia*).

5. **Isopyrum** L.

1. **I. occidentale** H. & A. Moist shaded woods in humus-rich soil, rare in Santa Lucia Range: near Adelaida (*Wiggins 5842*).

6. **Thalictrum** L. Meadow-Rue

1. **T. Fendleri** Engelm. Relatively moist places, common on wooded hillsides and along streams from coast inland to La Panza Range. Most of our plants are completely glabrous and thus correspond to var. *polycarpum* Torr., called in recent references *T. polycarpum* (Torr.) Wats. A few collections from this area in which the lower surfaces of the upper leaves are more or less obscurely glandular-puberulent have, however, been identified by Bernard Boivin (who made a special study of the genus) as true *T. Fendleri*. Variation in the meadow-rues of San Luis Obispo County is so slight that it is highly unlikely that there are even two recognizable forms, not to mention two distinct species. Many specimens from the main area of *T. Fendleri* (Sierra Nevada to Rocky Mountains) seem identical in every detail with some of our coastal plants. My conclusion is that no useful purpose can be served by arbitrarily assigning a portion of the coastal plants to a different species or variety.

7. **Actaea** L. Baneberry

1. **A. rubra** (Ait.) Willd. *A. arguta* Nutt. Rare in moist shady woods in western part: Coon Creek in San Luis Range; west of Templeton on road to Cambria. Available references are unanimous in classifying this American species as distinct from the European *A. spicata* L., in which earlier authors included it, but do not state how the two can be distinguished.

8. **Delphinium** L. Larkspur

Stems 7 dm. tall or more (except in drought-dwarfed plants), stout (over 5 mm. in diameter); flowers never violet.
    Flowers red .... 1. *D. cardinale.*
    Flowers white to pale blue or purple-tinged.
        Stem puberulent, not glaucous; sepals villous .... 2. *D. californicum.*
        Stem glabrous, glaucous; sepals microscopically puberulent or apparently glabrous .... 3. *D. gypsophilum.*
Stems less than 7 dm. tall, relatively slender (or in *D. Parishii* often 5 to 8 mm. thick).
    Primary leaf-divisions narrow, with narrow lobules.
        Stems and petioles puberulent or hirsute; flowers normally violet.
            Petioles and lower part of stem spreading-hirsute ...4. *D. variegatum.*
            Petioles and stems puberulent .... 5. *D. Parryi.*
        Stems and petioles essentially glabrous; flowers sky-blue ..... 6. *D. Parishii.*

Leaves, at least the lower ones, with broad lobes, the lobes entire to sparingly toothed or slightly lobulate.
   Flowers usually violet, sometimes paler or whitish.
      Root with long attenuate branches; stem not excessively fragile at base; lowest leaves deeply lobed . . . . .7. *D. umbraculorum.*
      Root with short thick branches or simply tuberous; stem readily breaking off from root; lowest leaves (in our form) shallowly lobed
                                                                            8. *D. patens.*
   Flowers red . . . . .9. *D. nudicaule.*

1. **D. cardinale** Hook. Scarlet Larkspur. In disintegrated shale mixed with humus, among shrubs or small trees, highly localized at east base of Santa Lucia Mts. near Paso Robles and Atascadero.

2. **D. californicum** T. & G. Occasional on north-facing slopes or thickets near coast from See Canyon in San Luis Range northward. At the southernmost known locality, the site of Lopez Dam on Arroyo Grande, the species doubtless has now been exterminated.

3. **D. gypsophilum** Ewan. In friable light-colored clays, less often in disintegrating shale or sandstone, common and widespread in eastern part of county, extending west to San Miguel. Some of the plants found on Caliente Mt. (*8226*) had light blue instead of the usual white flowers. This "blue *gypsophilum*" closely matches certain Mojave Desert plants which botanists generally include in *D. Parishii*. It is remarkable that in the Mojave Desert there is a single variable entity within which are included both typical *D. Parishii* and larger plants resembling *D. gypsophilum* (except that none are white-flowered). By contrast, in the San Joaquin Valley and the region of Carrizo Plain white-flowered *D. gypsophilum* is mostly a plant of hillslopes, sometimes spreading onto alluvial areas of the plain, whereas blue-flowered plants which I consider to be *D. Parishii* are restricted to alkaline plains.

4. **D. variegatum** T. & G. In grasslands and open woodlands, widespread and sometimes locally plentiful in Salinas River watershed west of La Panza Range. The southern distributional limit of the species is in the area between Santa Margarita and Pozo. This species never grows in rocky places and generally favors firmer soil than *D. Parryi,* but occasionally the two are found close together and give rise to obvious hybrids.

5. **D. Parryi** Gray. Common in hills of the interior, usually in more or less sandy soils, sometimes in fertile clay loam; less common toward the coast. One of the collections from Nipomo Mesa (*Twisselmann 2635*), although on geographic grounds it should be var. *Blochmaniae,* has the relatively small flowers of typical *D. Parryi.* Plants which have been called subsp. *seditiosum* (Jepson) Ewan differ from typical *D. Parryi* in no clearly defined visible feature, and no reason can be seen for continuing the use of that name. The type collection of *D. Parryi* subsp. *Eastwoodiae* Ewan is also here regarded as being essentially the same entity. Hybrids are occasionally found between *D. Parryi* and *D. variegatum* as well as other species, but plants of obviously hybrid origin are remarkably few.

Var. **Blochmaniae** (Greene) Jepson. This name is used for plants, growing in sand on Nipomo Mesa, with unusually large flowers for the species. Very similar plants were formerly found in sandy places on the east side of the San Joaquin Valley (Madera Co. to Stanislaus Co.). Some of the plants growing on serpentine hills around San Luis Obispo have flowers which are as conspicuously "bicolored" as those of var. *Blochmaniae* and, if the plants have sufficient moisture, about as large. They differ from the Nipomo Mesa plants in having shorter stems and fewer flowers.

6. **D. Parishii** Gray. Plants growing in alkali soil among salt-bush (*Atriplex spinifera*) on Carrizo Plain seem to me to belong to this species rather than to any other. Various botanists have identified these plants with *D. recurvatum* Greene. Lewis and Epling (Brittonia 8: 19. 1954), apparently without experimental proof, call them "hybrid derivatives of *D. gypsophilum* and *D. recurvatum*." *Delphinium gypsophilum* is present in the same general area but grows mostly on hillslopes. It was noted in only one place on the alkaline plain. At the same place were also found *D. Parryi* and a few plants which are evident hybrids between that species and *D. Parishii*. With these exceptions, the great majority of larkspurs on the plain, still plentiful in 1967, are quite uniform in their characters. I conclude that the plants in question represent a well marked species rather than a "hybrid swarm" and that they fall well within the range of variation of *D. Parishii*, as it is represented in the Mohave Desert.

7. **D. umbraculorum** Lewis & Epling. Shaded or sunny slopes, most frequently in loose soil derived from disintegrating shale, Santa Lucia Range (mainly east slope) to Arroyo Grande watershed; rarely found east of Salinas River. On geographic and ecologic grounds, I consider *D. gypsophilum* subsp. *parviflorum* Lewis & Epling to be a form of this species having pale flowers but otherwise entirely identical. Hybridization with *D. gypsophilum* may possibly occur, but the pale-flowered plants more probably represent a simple genetic variation within the species.

8. **D. patens** Benth. subsp. **hepaticoideum** Ewan. Caliente Mt., near highest peak, in sandy soil, mostly under small oaks. (*8217*).

9. **D. nudicaule** T. & G. In partial or complete shade, usually in rocky places, occasional in Santa Lucia Range: near Rocky Butte; Santa Rosa Creek above Cambria; Santa Rita Creek; Lopez Canyon.

9. **Clematis L.**

Leaflets 5 or 7; flowers in cymes . . . . 1. *C. ligusticifolia.*
Leaflets 3; flowers mostly solitary on axillary peduncles. . . . . .2. *C. lasiantha.*

1. **C. ligusticifolia** Nutt. Common along streams, mainly near the coast but also on Salinas River and in Horseshoe Canyon, Temblor Range (*Twisselmann 1406*).

2. **C. lasiantha** Nutt. Common in wooded areas in Santa Lucia Range, San Luis Range, and La Panza Range.

Berberidaceae. BARBERRY FAMILY

1. **Mahonia Nutt.**

1. **M. pinnata** (Lag.) Fedde. CALIFORNIA BARBERRY. Rare in rocky places near coast: See Canyon (*6618*); Perfumo Canyon; Devil's Gap on upper Morro Creek; between Cayucos and Cambria.

Lauraceae. LAUREL FAMILY

1. **Umbellularia Nutt.**

1. **U. californica** Nutt. CALIFORNIA LAUREL. BAY. Plentiful in canyons and on rocky hills in the more humid areas, extending inland only through the Santa

Lucia Mts. Some trees along streams become notably large for the species. In the vicinity of San Carpoforo Creek, laurel shows some remarkable examples of wind-pruning, the tops being sheared off perfectly flat. The leaves, which have a very pungent fragrance, are used for flavoring in cookery. The wood is hard and of high quality. This is the same species which is in Oregon incorrectly called "Oregon Myrtle."

Papaveraceae. POPPY FAMILY

Leaves entire or minutely denticulate.
    Herbs; leaves opposite, often crowded at base.
        Carpels lightly joined, in fruit separating and eventually breaking into one-seeded joints . . . . 1. *Platystemon.*
        Carpels united into one pistil; fruit a capsule . . . . 2. *Meconella.*
    Shrubs; leaves alternate . . . . 3. *Dendromecon.*
Leaves coarsely toothed, lobed, or compound.
    Flowers perigynous, the receptacle ("torus") funnel-shaped; flower-buds pointed
                                                                                4. *Eschscholzia.*
    Flowers hypogynous; flower-buds broadly rounded.
        Capsule long and slender, 2-celled . . . . 5. *Glaucium.*
        Capsule short and stout, 1-celled.
            Plants spiny and prickly . . . . 6. *Argemone.*
            Plants not spiny.
                Style present, terminating in a round stigma . . . . . . 7. *Stylomecon.*
                Style absent; stigmas radially arranged on top of ovary 8. *Papaver.*

1. Platystemon Benth.

1. **P. californicus** Benth. CREAM CUPS. Widely distributed in sandy or rocky soils throughout the county, except in the more rugged hill regions. The species is variable, but most of the variation seems to have little geographic significance. West of the Santa Lucia Range, the petals are always cream-color. Some plants of the interior have the petals deep yellow, wholly or in part, but there are extensive pure stands of the cream-colored form in the interior also. In scattered localities, either the cream-colored or the yellow form may be more or less purple-tinged. In dry situations, the leaves are sometimes crowded at the base of the stems.

2. Meconella Nutt.

Stem branched only near base, the leaves crowded; stamens many. . . . . . 1. *M. linearis.*
Stem branched above the base, the upper leaves widely spaced; stamens 6
                                                                     2. *M. denticulata.*

1. **M. linearis** (Benth.) Nels. & Macbr. Frequent in sandy soils, usually in partial shade, in both coastal and interior districts. Variation in this species exhibits a close parallel to that in *Platystemon californicus,* except that the leaves are usually crowded at the base, so as to obscure their opposite arrangement.

2. **M. denticulata** Greene. Sandy or gravelly slopes, La Panza Range and Stoney Creek, uncommon or overlooked. Grows most frequently in partial shade where the soil contains some humus, and most vigorously after a fire. The species has usually been called a variety of *M. oregana* Nutt., from which it is well separated geographically and from which it is distinguishable by having much larger anthers.

3. Dendromecon Benth.

1. **D. rigida** Benth. TREE POPPY. Common on rocky hills as a minor component of chaparral, especially plentiful after fires: hills near coast, including San Luis Range, throughout the Santa Lucia Mountains and east to La Panza Range.

4. Eschscholzia Cham.

Receptacle with spreading rim . . . . .1. *E. californica.*
    Perennials.
        Stems ascending to spreading; herbage slightly to moderately glaucous, not
        pubescent; peduncles long . . . . .var. *californica.*
        Stems prostrate or decumbent; herbage strongly glaucous, often pubescent;
        peduncles mostly short, the stems leafy to the tips . . . . .var. *maritima.*
    Annuals.
        Petals deep orange, 2.5 cm. long or more . . . . .var. *ambigua.*
        Petals light or yellowish orange, less than 2.5 cm. long. . .var. *peninsularis.*
Receptacle without spreading rim.
    Herbage glabrous or very nearly so; flower-buds erect.
        Petals 15 mm. long or more . . . . .2. *E. caespitosa.*
        Petals 7 mm. long or less.
            Stem branching and leafy above base; receptacle in fruit 2 to 3 mm.
            long; mature capsules 1.5 mm. thick . . . . .3. *E. minutiflora.*
            Stem branching only at base, the leaves crowded and peduncles sub-
            scapose; receptacle in fruit 5 mm. long; mature capsules 4 mm. thick
                                    4. *E. rhombipetala.*
    Herbage more or less pubescent (rarely glabrous, and then the buds nodding);
    flower-buds usually nodding (rarely erect, and then the calyx densely pubes-
    cent).
        Peduncles mostly half the height of the plant or more, often subscapose;
        fruiting receptacle 5 to 8 mm. long, 3 to 4 mm. wide at mouth
                                      5. *E. Lemmonii.*
        Peduncles (except the earliest ones) less than half the height of the plant;
        fruiting receptacle 3 mm. long, not over 2 mm. wide at mouth
                                      6. *E. hypecoides.*

1. **E. californica** Cham. CALIFORNIA POPPY. Very common in all parts, showing considerable differences from one climatic region to another. The San Luis Obispo County material, however, fits rather well into four varieties, of which two are perennial and two annual.

Var. **californica.** Plants around Morro Bay (town) and Baywood Park are quite like plants growing near the type locality, San Francisco, and so are regarded as typical. I can find no sufficient basis for separating var. *crocea* (Benth.) Jepson, and therefore include it under var. *californica.* This typical element of the species is much less common in the interior than are annual forms, although it is present on hills of the Temblor Range above McKittrick in Kern Co.

Var. **maritima** (Greene) Jepson. Plants of the coastal dunes, referred to this name, have a very stout root, widely spreading or prostrate stems which are evenly leafy to the ends, very pale foliage which is often (though not always) hairy, and flowers on the average smaller than in var. *californica.* The petals are ordinarily yellow with deep orange base.

Var. **ambigua** (Greene) Jepson. The large-flowered annual plants which are abundant in the interior are placed under this name, although perhaps an earlier published one is available. The name *E. ambigua* Greene was based on a plant collected by Lemmon near Cholame. The same variety is the prevailing form of *E. californica* in the San Joaquin Valley. It is generally absent near the coast, unless possibly in the sandy country near Nipomo. This annual variety is the poppy which in some years makes such brilliant displays in the upper Salinas Valley and eastward.

Var. **peninsularis** (Greene) Munz. This name is used for annual plants in which the flowers, apparently for a genetic rather than an environmental reason, are comparatively small. Certain plants in the La Panza Range are quite similar to the type specimen of *E. peninsularis* Greene. Such plants, however, are neither so common nor so widespread in this county as the large-flowered ones I am calling var. *ambigua*. Intergradation in flower size is particularly notable between Atascadero and Santa Margarita.

2. **E. caespitosa** Benth. Gravelly or rocky slopes, commonly on crumbling shale, less frequently on serpentine and other rock types, mainly west of the summits of Santa Lucia Range but extending up Cuyama Canyon almost to Cuyama Valley. Apparently absent from central and eastern San Luis Obispo County. The flowers are bright orange as in some forms of *E. californica* and in some years bloom in considerable profusion, as for example on the rocky hills facing the ocean south of Avila and on the north side of the Santa Maria Valley near Twitchell Reservoir. Such an occurrence leads one to suspect that reports of displays of *E. californica* being in past years visible from far out at sea were actually based on *E. caespitosa*. The coastal forms of *E. californica* do not occur in such colorful abundance.

3. **E. minutiflora** Wats. West side of Cottonwood Pass, confined to a small outcrop of serpentine (*7732*). The plants of this collection match very closely certain specimens of *E. minutiflora* from the Mohave Desert, being identical even in the most minute detail (e.g., *J. T. Howell 4885* from 3 miles south of Palmdale, Los Angeles Co.). A collection from 4 miles east of Adelaida, *Eastwood* and *Howell 2362*, has been identified by W. R. Ernst as *E. hypecoides* but is also very similar to desert specimens of *E. minutiflora*, particularly in the size of its flowers.

4. **E. rhombipetala** Greene. La Panza district, where found only in a single patch of clay soil which is otherwise barren except for scattered plants adapted to the same soil, such as *Erodium macrophyllum* and *Acanthomintha obovata*. This species is not in the least showy but of great interest from the geographic and ecologic points of view. The total number of collections is very small. *E. rhombipetala* is an endemic of the inner Coast Ranges, known otherwise from Stanislaus, San Joaquin, Contra Costa, and Colusa cos. Because of its rarity, the species has been neglected and erroneously included in the quite different species *E. caespitosa*. Living plants of *E. rhombipetala* are remarkable for their fleshy leaf-bases. The flowers are inconspicuous, but the receptacle and capsule are disproportionately large for the tiny plants.

5. **E. Lemmonii** Greene. In calcareous or gypseous soils ranging from sand to clay, common from hills around Cholame Valley southward to Cuyama Valley, extending westward to the hill country between Creston and Shandon. Cholame is the type locality. In unusual years of good rainfall this species on and near Carrizo Plain forms patches of dazzling color equalling the best displays of *E. californica* var. *ambigua*. The color is deep orange, at times so intense as to suggest orange-red.

When rains have been scanty or poorly spaced, only scattered and stunted plants can be found.

6. **E. hypecoides** Benth. Common on hillslopes north and east of La Panza Range, extending to Cottonwood Pass and summits of Temblor Range, most often in crumbling shale. When present in the same general area as *E. Lemmonii*, it is notable as growing in more rocky places than does that species, but in some localities *E. hypecoides* is known in fertile calcareous clay. It is common on the east slope of the Santa Lucia Range in Monterey Co., but, unless the plant mentioned above under *E. minutiflora* is to be included here, is not known so far west in San Luis Obispo Co. The species here discussed as *E. hypecoides* (on the basis of comparison with the type collection) has been confused by authors with *E. Lemmonii* and is distinctly unlike what has been called "*E. caespitosa* var. *hypecoides.*"

5. Glaucium Mill.

1. **G. flavum** Crantz. Horned Poppy. In a field at Choice Valley School near the Kern Co. line, apparently well established locally. The name is as given by Twisselmann in his Temblor Flora, but the specimen does not exactly match any of the Old-World specimens in the herbarium of the California Academy of Sciences. It is barely possible that our plant represents some other species of *Glaucium*.

6. Argemone L.

1. **A. munita** D. & H. Prickly Poppy. Occasional in eastern part of county, usually in sandy soils, the plants never numerous: Red Hills near Shandon southward to Cuyama Valley.

7. Stylomecon G. Taylor

1. **S. heterophylla** (Benth.) G. Taylor var. **micropetala** Hoover. Wind Poppy. Clay or rocky soils, mainly on north-facing slopes, known definitely only in eastern part of county and seemingly uncommon here, but perhaps largely overlooked.

8. Papaver L.

Peduncles with soft spreading hairs.
    Petals about 2 cm. long, orange with light green base . . . . 1. *P. californicum.*
    Petals 2 to 3 cm. long, typically bright red . . . . 2. *P. Rhoeas.*
Peduncles densely covered with stiff appressed hairs . . . . .3. *P. hybridum.*

1. **P. californicum** Gray. Rocky or sandy hills in woods or chaparral, generally uncommon but, in the year following a fire, sometimes locally abundant: northwest of Cuesta Pass; hills near Price Canyon; Calf Canyon; Stoney Creek; La Panza Range. Between fires, the species almost or entirely disappears. *Papaver Lemmonii* Greene, which is not now regarded as distinct from *P. californicum,* was based on a collection from an unspecified locality in San Luis Obispo County.

2. **P. Rhoeas** L. Red Poppy. Garden flower sometimes escaping from cultivation, apparently established at Paso Robles (*Twisselmann 5965* in 1960).

3. **P. hybridum** L. Well established near eastern border of county in vicinity of Choice Valley. It has been present in this region at least since 1935, when it was collected in an abandoned grain-field 2 miles north of Blackwell's Corner in Kern Co. Our plants have been erroneously referred to *P. apulum* Ten. var. *micranthum* (Bor.) Fedde. Their identification as *P. hybridum* is based on Fedde's key to species

in "Das Pflanzenreich," and on comparison with Old-World specimens named by European botanists.

Fumariaceae. BLEEDING-HEART FAMILY

1. Dicentra Bernh.

Flowers bright yellow; petals not purple-tipped . . . . .1. *D. chrysantha.*
Flowers cream-white; inner petals purple-tipped . . . . .2. *D. ochroleuca.*

   1. **D. chrysantha** (H. & A.) Walp. GOLDEN DICENTRA. Occasional in dry gravelly or rocky places: eastern slopes of Santa Lucia Mountains north of Paso Robles, where plentiful near Nacimiento Dam; hill country east of Santa Margarita to La Panza Range. This species should be excellent for cultivation in flower gardens.

   2. **D. ochroleuca** Engelm. "San Luis Obispo Co.," according to Stern (Brittonia 13: 20. 1961), but no collection or definite locality cited.

Capparidaceae. CAPER FAMILY

1. Isomeris Nutt.

   1. **I. arborea** Nutt. var. globosa Cov. BLADDERPOD. Occasional in eastern part on barren hills and alluvial plains: east base of Polonio Pass; Cammatti Creek south of Shandon; Temblor Range to Cuyama Valley. The shrub is ornamental, although some persons find the odor disagreeable.

Cruciferae. MUSTARD FAMILY

Fruit elongate, lanceolate or oblong to narrowly linear in outline.
  Racemes bractless.
      Stamens conspicuously exceeding petals; siliques stipitate . . . . .1. *Stanleya.*
      Stamens shorter than petals or barely equalling them; siliques not stipitate.
         Fruit a silique, 2-celled, dehiscent from base when mature, leaving the median septum attached to receptacle.
          Plants with a tap-root or rhizomes; leaves usually many.
            Petals yellow, orange, or light green.
              Upper leaves simple, entire or sparingly dentate.
                 Lower leaves pinnate or deeply lobed . . .2. *Brassica.*
                 All leaves entire or dentate . . . . .3. *Erysimum.*
              Upper leaves pinnate, or at least deeply pinnately lobed or hastate.
                 Plants pubescent.
                 Leaves pinnate or deeply lobed, or the upper merely hastate, the divisions toothed or entire
                            4. *Sisymbrium.*
                 Leaves finely divided . . . . .5. *Descurainia.*
                Plants glabrous.
                 Siliques 3 cm. long or more . . . . .4. *Sisymbrium.*
                 Siliques less than 3 cm. long.
                   Basal leaves largest; siliques 1.5 to 3 cm. long
                            6. *Barbarea.*
                   Basal leaves generally absent at flowering time; siliques 12 mm. long or less
                            7. *Rorippa.*
             Petals white to rose-color or deep purple.
               Petals with undulate blade narrower than the claw
                           8. *Streptanthus.*

Petals with plane blade wider than the claw.
 Anthers sagittate ....8. *Streptanthus.*
 Anthers not sagittate.
  Leaves entire or slightly dentate.
   Stem branching above the base
         9. *Matthiola.*
   Stems simple or rarely few-branched above,
    solitary or clustered ....10. *Arabis.*
  Leaves pinnate or deeply lobed.
   Annuals with a tap-root.
    Siliques with beak nearly as long as
     body; flowers showy ....11. *Eruca.*
    Siliques without beak; flowers small
       12. *Cardamine.*
   Perennials without tap-root, rooting at the
    lower nodes.
     Stems floating or ascending; seeds in 2
      rows ....7. *Rorippa.*
     Stems erect from a decumbent base;
      seeds in 1 row ....12. *Cardamine.*
Plants with a small tuber; basal leaf one or few, not attached to
 flowering stem; cauline leaves 2 or 3 ....13. *Dentaria.*
Fruit indehiscent, not longitudinally divided into 2 cells, few-seeded.
 Leaves not fleshy; fruit not jointed ....14. *Raphanus.*
 Leaves fleshy; fruit consisting of 2 one-seeded joints. ..15. *Cakile.*
Racemes leafy-bracted (or flowers solitary in the axils) ....16. *Tropidocarpum.*
Fruit round to elliptic or triangular.
 Seeds 2 or more; fruit usually dehiscent (a silicle).
  Plants without rhizomes; leaves narrow or, if rather broad, pinnately lobed.
   Silicles with broad septum, flattened parallel to it.
    Petals showy; silicles 2-seeded, readily dehiscent .....17. *Alyssum.*
    Petals inconspicuous; silicles many-seeded, dehiscent only in age
       18. *Heterodraba.*
   Silicles with narrow septum, flattened at right angles to it.
    Silicles many-seeded.
     Silicles elliptic ....19. *Hutchinsia.*
     Silicles triangular-cuneate ....20. *Capsella.*
    Silicles 2-seeded ....21. *Lepidium.*
  Plants with rhizomes; leaves broad, dentate.
   Weed in valleys and low places; stems erect; leaves not fleshy; fruit
    inflated, shallowly 2-lobed at base only ....22. *Cardaria.*
   Growing on shifting dunes; stems decumbent; leaves fleshy; fruit flat-
    tened, deeply 2-lobed at both ends ....23. *Dithyrea.*
 Seed 1 only; fruit indehiscent, a samara or flattened achene.
  Stem with spreading branches; fruit not wing-margined, with hooked hairs
     24. *Athysanus.*
  Stem simple or with few erect branches; fruit with a narrow to broad thin
   margin (wing), woolly to glabrous ....25. *Thysanocarpus.*

1. Stanleya Nutt.

1. **S. pinnata** (Pursh) Britt. Prince's Plume. Dry hills and washes, in subalkaline
or gypseous soils, occasional in Temblor Range, common in Cuyama Valley, extending down the flood-bed of the Santa Maria River occasionally to Santa Maria.
A very showy plant when in bloom.

2. Brassica L.

Upper leaves with broad clasping base, somewhat glaucous ....1. *B. campestris.*
Leaves narrowed toward base, not clasping, not glaucous.
    Annual, without basal rosette at flowering time; leaves with scattered hairs or subglabrous.
        Petals 6 to 8 mm. long; beak of pod not over ⅕ as long as the body, seedless ....2. *B. nigra.*
        Petals 8 to 12 mm. long; beak of pod about ⅓ as long as the body, usually containing a seed ....3. *B. Kaber.*
    Perennial (often flowering the first year) with basal rosette of leaves; leaves densely hairy ....4. *B. geniculata.*

1. **B. campestris** L. Field Mustard. Very plentiful on open hills and in fields near coast, especially around San Luis Obispo, in various soils; scarce and widely scattered in cultivated areas of the interior.

2. **B. nigra** L. Black Mustard. Growing in great profusion in years of good rainfall on open hillslopes from coast to Santa Lucia Mts., reaching its best development in soils derived from disintegrating shale. Less common in upper Salinas Valley, and apparently absent from drier and more rugged portions of the interior.

3. **B. Kaber** (DC.) Wheeler. Charlock. In clay soils, locally abundant around San Luis Obispo and Nipomo, making a great show of flowering each year as *B. campestris* is declining and before *B. nigra* is at its height. Also occasionally plentiful in the interior: Highland district, Cammatti Ranch, Choice Valley, and (according to Twisselmann) Carrizo Plain.

4. **B. geniculata** (Desf.) Ball. Common in most areas except in woods or chaparral, and apparently not in the most arid portions around Carrizo Plain and Cuyama Valley. Generally in firmly packed soils and especially plentiful along roads.

3. Erysimum L. Wallflower

Biennials; flowering stems solitary or sometimes 2 or 3 from base.
    Petals yellow-orange to russet-orange; siliques not over 1.2 mm. thick
                                              1. *E. capitatum.*
    Petals light yellow; siliques 2 to 3 mm. wide ....2. *E. moniliforme.*
Perennial, shrubby, with many leafy branches below from which the flowering stems arise ....3. *E. insulare.*

1. **E. capitatum** (Dougl.) Greene? Orange Wallflower. For the present I am using this name, perhaps erroneously, for a tall biennial with nearly always a single stem from the root and yellow-orange to russet-orange flowers. This species is frequent on gravelly or sandy slopes from the coastal hills near Avila and Pismo Beach eastward through the Santa Lucia Range and upper Salinas watershed to the east slope of the La Panza Range. The other two kinds of *Erysimum* in the county are completely distinct ecologically and in appearance. Intergrades, if they exist at all, are scarce and hard to find. This orange wallflower is notable for its very slender siliques, hardly 1.2 mm. wide at the most, a feature which seems not to characterize *E. capitatum* over most of its range. In that one respect, our plant suggests *E. teretifolium* Eastw., a local species of Santa Cruz Co. It may therefore be questioned whether *E. capitatum* is the right name for the orange-flowered species occurring here, especially as that name has also been used for the following species.

2. **E. moniliforme** Eastw. On open slopes in crumbling shales and calcium-rich sands and clays, common on hills north and east of La Panza Range, and thence to

our eastern borders. Under favorable conditions this biennial grows tall and has stouter stems than the preceding species, but often it is dwarfed by drought. The flowers are light yellow. These plants also differ from the orange-flowered ones in having stouter siliques (2 or 3 mm. wide when fully developed) which are variable in length and sometimes noticeably constricted between the seeds. In the La Panza district, the only place where the two are known to be contiguous geographically, the orange-flowered species is restricted to the digger pine belt in decomposed granite, whilst the one with light yellow flowers grows in treeless areas or among scattered blue oaks and junipers in calcareous sands and clays or in crumbling shale. The type of *E. moniliforme* is unquestionably a plant of the same sort as is prevalent in eastern San Luis Obispo Co., but an earlier published name may also be applicable to it. Some of these plants of eastern San Luis Obispo Co. have been identified by George B. Rossbach as *E. capitatum*; others as *E. capitatum* var. *Bealianum*. If Rossbach's identification of *E. moniliforme* with *E. capitatum* is correct, then some other appropriate name must be found for our orange-flowered plant with very slender siliques. *Erysimum moniliforme* also requires critical comparison with *E. occidentale* (Wats.) Rob. of interior Washington and Oregon and with *E. asperum* Nutt. of the Rocky Mountain region. Herbarium specimens of both *E. occidentale* and *E. asperum* can be found which differ from individuals of *E. moniliforme* in no outwardly visible way. I suspect that, after adequate studies of the plants in all their habitats and in cultivation, these plants will ultimately be included in *E. asperum*.

3. **E. insulare** Greene. Common in coastal sands around Morro Bay and south of Arroyo Grande. The plants for which I here use this name are perennials, eventually becoming dwarf shrubs with many branches from the base. Their flower color is intense yellow with at times a hint of orange. Plants around Morro Bay have short strictly ascending siliques much stouter than those of the tall orange-flowered biennial which passes as *E. capitatum*. Probably the plants to the south often have longer siliques, although no mature ones from the Nipomo-Oso Flaco area are now at hand for study. Names which have been applied to this species include *E. suffrutescens* (Abrams) Rossb. and *E. suffrutescens* var. *grandifolium* Rossb. To the extent that I can interpret these names, they apply to individual specimens which differ in environment and in stage of growth, not to entities which are genetically and geographically significant. No clearly defined difference can be found between plants from San Miguel and Santa Rosa Islands, representing unquestioned *E. insulare*, and the shrubby wallflower of Morro Bay. *E. suffrutescens* var. *lompocense* Rossb. may include some intergrades between this species and the tall orange-flowered biennial called *E. capitatum*, but at least some of the specimens included in that variety are not readily distinguishable from the latter.

4. **Sisymbrium L.**

Siliques acuminate, erect, appressed to stem . . . . .1. *S. officinale.*
Siliques not acuminate, spreading.
    Leaves markedly dimorphic, the upper ones pinnately divided with very narrow
    segments . . . . .2. *S. altissimum*
    Leaves not dimorphic, gradually reduced upward, the upper mostly pinnately
    lobed.
        Pedicels stout; petals 6 to 8 mm. long . . . . .3. *S. orientale.*
        Pedicels slender; petals 3 to 4 mm. long . . . . .4. *S. Irio.*

1. **S. officinale** (L.) Scop. Hedge Mustard. Common weed in western part, especially in wooded areas.

2. **S. altissimum** L. Tumbling Mustard. Sandy or calcareous soils, common in Cuyama Valley and on Carrizo Plain.

3. **S. orientale** L. In sandy soils, not well established: Gillis Canyon near Shandon, and as scattered plants along Estrella River.

4. **S. Irio** L. Occasional as a weed in eastern part: plentiful in 1967 on Carrizo Plain south of Soda Lake; Elkhorn Plain; Twisselmann Ranch.

5. **Descurainia** Webb & Berthel.

Siliques narrowly linear, 2 to 4 cm. long ....1. *D. Sophia.*
Siliques oblong, 5 to 15 mm. long ....2. *D. pinnata.*

1. **D. Sophia** (L.) Webb. Flix Weed. Grainfields on Carrizo Plain south of Soda Lake, locally abundant in 1967.

2. **D. pinnata** (Walt.) Britt. Common on stabilized dunes and sandy plains near coast and in fine sands or friable clay loams in interior. Our plants do not conform well to the descriptions of the various subspecies named by Detling, but on geographic grounds they all supposedly are subsp. *Menziesii* (DC.) Detling.

6. **Barbarea** R. Br.

1. **B. orthoceras** Ledeb. var. **dolichocarpa** Fernald. Frequent along beds of small streams, or sometimes in places which are moist only while the plants are young: Mt. San Luis; Santa Lucia Mts. to La Panza Range, including the intervening hill country. Our plants are variable but, according to Fernald (Rhodora 11: 140), they apparently all belong to *B. orthoceras* var. *dolichocarpa. Barbarea americana* Rydb., which may be the correct name for our plants, is a synonym of this variety according to Abrams, but other authors equate it with typical *B. orthoceras.*

7. **Rorippa** Scop.

Petals white ....1. *R. Nasturtium-aquaticum.*
Petals yellow.
    Stems spreading; pedicels mostly 1 to 3 mm. long ....2. *R. curvisiliqua.*
    Stems erect; pedicels mostly 4 to 6 mm. long ....3. *R. islandica.*

1. **R. Nasturtium-aquaticum** (L.) Schinz & Thell. Watercress. Common in slow-moving streams and wet ground, mainly in western part but also, according to Twisselmann, in Temblor Range.

2. **R. curvisiliqua** (Hook.) Bessey. Wet places, Salinas River bed at Paso Robles; Arroyo Grande according to Jepson (Fl. Cal. 2: 53).

3. **R. islandica** (Oeder) Borb. var. **occidentalis** (Wats.) Butters & Abbe. In shallow water, Atascadero Lake (*8695*); dune lakes south of Arroyo Grande.

8. **Streptanthus** Nutt.

Blade of petal with plane margins, broader than the claw.
    Leaves dentate or entire, or some of them hastate or shallowly lobed; sepals
    spreading, purple; petals pale lilac-pink ....1. *S. anceps.*
    Leaves mostly deeply to shallowly lobed; sepals erect, green; petals white.
        Stem simple or with narrowly ascending branches; siliques usually deflexed,
        sometimes erect or ascending (if less than 3 cm. long, strongly curved)
                                            2. *S. lasiophyllus.*

Siliques 3 to 6 cm. long, straight or slightly curved.
    Siliques deflexed ....var. *lasiophyllus.*
    Siliques erect or ascending ....var. *inalienus.*
Siliques 15 to 30 mm. long, deflexed, strongly curved outward or even upward ....var. *utahensis.*
Stem with divergent branches; siliques rigidly divaricate-ascending, 2 to 3 cm. long ....3. *S. rigidus.*
Blade of petal with crinkled margins, generally narrower than the claw.
    Stigma not visibly 2-lobed.
        Stems glabrous; upper leaves cordate ....4. *S. tortuosus.*
        Stems hairy, at least sparsely so near base; upper leaves linear to lanceolate.
            Sepals broad, more or less ventricose; siliques ascending or spreading
                                5. *S. glandulosus.*
            Sepals narrow, not ventricose; siliques deflexed ....6. *S. heterophyllus.*
    Stigma 2-lobed.
        Stem green or glaucous, not inflated.
            Upper leaves ovate; siliques 3 to 5 cm. long, flattened, 4 to 5 mm. wide
                                7. *S. californicus.*
            Upper leaves lanceolate; siliques 5 to 11 cm. long, terete, 2 to 3 mm. thick.
                Lower leaves sparsely hirsute or at least pustulate, hardly if at all glaucous; siliques usually deflexed, sometimes spreading or ascending ....8. *S. Coulteri.*
                Lower leaves glabrous, glaucous, not pustulate; siliques erect
                                  9. *S. Parryi.*
        Stem yellow, inflated (except in dwarfed plants) ....10. *S. inflatus.*

1. **S. anceps** (Payson) Hoover. *Thelypodium Lemmonii* Greene. In pale friable clay soils which are probably calcium-rich, rather common in eastern part from hills adjacent to Cholame Valley to La Panza district and Carrizo Plain. Although confined to a special sort of soil, the plants in a good year often form dense stands.

2. **S. lasiophyllus** (H. & A.) Hoover var. **lasiophyllus.** *Thelypodium lasiophyllum* (H. & A.) Greene. Common in friable clays and calcareous sands throughout the interior; less frequent on rocky or sandy hills near the coast.

Var. **inalienus** (Rob.) Hoover. San Luis Obispo (*Eastwood 14,407*). This variety, distinguished only by its erect pods, seems rather inconsequential but is significant geographically. Aside from this one record, it is otherwise known apparently only around San Francisco Bay and in the foothills of the middle Sierra Nevada, whereas var. *lasiophyllus* is widely distributed over the southwestern United States and adjacent Mexico.

Var. **utahensis** (Rydb.) Hoover, n. comb. *Thelypodium utahense* Rydb., Bull. Torr. Club 29: 233. 1902. Chalk Mt., Cuyama Valley (*10,296*).

3. **S. rigidus** (Greene) Hoover. In alkaline clay soil, Cholame Valley (*7655*). The plants are quite different in appearance from *S. lasiophyllus* as it grows in the vicinity.

4. **S. tortuosus** Kell. On a bare hill of white volcanic rock (rhyolite) near Pine Mountain above San Simeon; very few plants found.

5. **S. glandulosus** Hook. Of scattered occurrence on open hillslopes in areas of serpentine rock: San Simeon Creek, upper Santa Rosa Creek, ridge west of Cerro Romauldo, and foothills near San Luis Obispo. Here at the southern end of the range of the species, all the plants have bright purple flowers, not dark purple as

ordinarily in the San Francisco Bay region. Furthermore, the species elsewhere does not generally grow on serpentine, although some of its close relatives do. An occasional specimen from farther north, however, seems to show the lighter flower-color of our plants. This form probably is what Greene named *Euclisia violacea*. The collection mentioned first under that name was made by Palmer at an unknown (according to Greene) locality, but also cited was a collection from San Luis Obispo by M. E. Jones. In the herbarium of the California Academy of Sciences a Palmer specimen bearing "San Luis Obispo" on the label is filed as an isotype of *E. violacea*.

6. **S. heterophyllus** Nutt. Dry gravelly slopes in La Panza Range, abundant and vigorous in a burned area at Navajo Camp in 1952 (*8194*); otherwise seldom seen, or only in greatly dwarfed form. Pale-flowered and dwarfed plants of this species can hardly be distinguished from small plants of *S. lasiophyllus*.

7. **S. californicus** (Wats.) Greene. In coarse gravel to fine sandy soil, southern part of Carrizo Plain, where not seen since 1958 and perhaps exterminated; eastern part of Cuyama Valley.

8. **S. Coulteri** (Wats.) Greene. Rocky or sandy slopes, southern part of Temblor Range; Cuyama Canyon (*7599*).

9. **S. Parryi** Greene. In light-colored friable clays, probably calcium-rich, or in crumbling shale, common in interior from vicinity of Paso Robles eastward to northern Temblor Range and hills on north side of Cuyama Valley. The type locality is "probably near Paso Robles" according to Jepson (Fl. Cal. 2: 27). Cholame is the type locality of *Caulanthus Lemmonii* Wats., which is a synonym.

10. **S. inflatus** (Wats.) Greene. DESERT CANDLE. SQUAW CABBAGE. In gypseous clay soils or crumbling white shale, abundant on lower hills of Temblor Range in years of good rain, sometimes occurring on Carrizo Plain where seeds have been washed down; hills on north side of Cuyama Valley.

9. **Matthiola R. Br.**

Stems slender; silique terminating in 2 horns .... 1. *M. bicornis.*
Stems stout; silique not horned .... 2. *M. incana.*

1. **M. bicornis** (Sibth. & Smith) DC. EVENING-SCENTED STOCK. Garden flower "widely escaped and spontaneous about ranchgrounds" at Twisselmann Ranch.

2. **M. incana** (L.) R. Br. STOCK. Locally, and perhaps temporarily, established on dunes at Oceano Beach in 1967.

10. **Arabis L. ROCK-CRESS**

Herbage glabrous and glaucous above the base .... 1. *A. glabra.*
Herbage pubescent throughout.
    Leaves not auriculate; siliques deflexed .... 2. *A. pulchra.*
    Upper leaves auriculate; siliques spreading or a little descending, typically curved .... 3. *A. sparsiflora.*

1. **A. glabra** (L.) Bernh. TOWER MUSTARD. Occasional on hillslopes, generally under live oaks, at widely scattered places in coastal hills, Santa Lucia Mts., and eastward to La Panza Range. The plants are never numerous in any one locality.

2. **A. pulchra** Jones. Scattered along summit ridge of Caliente Mt. (*8203*), among sandstone rocks and in sandy soil.

3. **A. sparsiflora** Nutt. var. **arcuata** (Nutt.) Rollins. In rock crevices at Devil's Gap, upper Morro Creek. The plants, although apparently closer to this variety

than to any other named entity, differ from other specimens in having shorter siliques which are straight or only slightly arcuate.

11. Eruca Mill.

1. **E. sativa** Mill. GARDEN ROCKET. Well established for several years on west side of Polonio Pass along the old road, in fertile soil derived from disintegrated white shale; also a large dense colony at mouth of Palo Prieto Canyon.

12. Cardamine L.

Perennial, rooting at lower nodes; petals 5 to 8 mm. long .... 1. *C. Gambelii.*
Annuals with slender tap-root; petals 1.5 to 3 mm. long.
    Lower leaves with broad leaflets; siliques 1 mm. wide .... 2. *C. oligosperma.*
    Lower leaves with narrow leaflets; siliques 1.5 to 2 mm. wide ...3..*C. virginica.*

1. **C. Gambelii** Wats. Fresh-water or brackish marshes and borders of lakes, locally common among the dunes from Oceano southward.

2. **C. oligosperma** Nutt. Common in shaded places in western part, mostly in sandy soil, particularly plentiful under oaks on south side of Morro Bay.

3. **C. virginica** L. *Arabis virginica* Trel. *Sibara virginica* Rollins. In moist depression one mile from Creston on road to Shandon (*6792*).

13. Dentaria L. TOOTHWORT

Leaflets of cauline leaves linear to oblanceolate or lanceolate, usually entire
                          D. integrifolia var. *integrifolia.*
Leaflets of cauline leaves ovate to lanceolate, mostly toothed. ......var. *californica.*

1. **D. integrifolia** Nutt. var. **integrifolia.** MILKMAIDS. RAIN BELLS. Most plentiful in moist fields along the coast north of Cambria. Occasionally extending into wooded areas, as near Cambria and San Luis Obispo, where it grades into var. *californica.*

Var. **californica** (Nutt.) Jepson. Very common on wooded hills, especially in the Santa Lucia Range. Probably extending to La Panza Range, although no specimens from so far inland are now at hand. Occasional plants, growing with the usual white-flowered form, have rose-colored petals. *Dentaria integrifolia* and *D. californica* Nutt. were published at the same time. So far as can be learned, Jepson in 1901 first combined the two into one species, using the name *D. integrifolia* (Fl. W. Mid. Cal. 220). The International Code of Nomenclature, article 57, does not permit the currently prevalent use of the name *D. californica* for the combined species.

14. Raphanus L.

Petals white to purple; seeds usually 1 to 3, sometimes as many as 8 ...1. *R. sativus.*
Petals white or pale yellow; seeds usually 4 to 10 ....2. *R. Raphanistrum.*

1. **R. sativus** L. RADISH. Cultivated plant from Europe which has run wild and become very plentiful near the coast, especially around Cambria. Also a weed in cultivated areas of the interior, but not common there.

2. **R. Raphanistrum** L. JOINTED CHARLOCK. Often growing with *R. sativus* near the coast, particularly in formerly cultivated fields. The plants which are assumed to be of this species merge gradually into *R. sativus,* both in flower color and in characters of the fruit. The apparent intergrades may be of hybrid origin, or perhaps genuine *R. Raphanistrum* is not actually represented here.

15. **Cakile Mill.**

Leaves pinnately lobed ....1. *C. maritima.*
Leaves crenate or sinuate-dentate ....2. *C. edentula.*

1. **C. maritima** Scop. Sea Rocket. Very abundant on beaches and dunes all along our coast. This native of Europe has become one of the most plentiful members of what I call the "Beach-Dune" plant community and, of all land plants, one of those which grows closest to the edge of salt water.

2. **C. edentula** (Bigelow) Hook. var. **californica** (Heller) Fernald. In sand, much less plentiful than *C. maritima* but well distributed along the coast: dunes near Piedras Blancas Point; San Simeon Creek Beach; north end of Morro Bay; Pismo Beach.

16. **Tropidocarpum Hook.**

Siliques throughout their length flattened at right angles to the septum
T. gracile var. *gracile.*
Upper part of silique flattened, the septum lacking in lower part ......var. *dubium.*

1. **T. gracile** Hook. var. **gracile.** Common in sandy loam or friable clay soils throughout the interior; westward at least to upper Arroyo Grande.

Var. **dubium** (Dav.) Jepson. Apparently only in region of Carrizo Plain. Dried specimens do not show this to be a well-defined variety, as pods characteristic of it and of var. *gracile* may seemingly be found on the same plant. However, careful comparison of living plants should be made, as well as field studies to determine the exact geographic occurrence of the two varieties.

17. **Alyssum L.**

1. **A. maritimum** (L.) Lam. Sweet Alyssum. Well established in sandy soil along the coast, at least from Morro Bay to Oceano. Occasionally it spreads from gardens elsewhere, but seems unable to thrive outside of cultivation away from the sea.

18. **Heterodraba Greene.**

1. **H. unilateralis** (Jones) Greene. Scattered over eastern part, in friable clay soils, the plants often numerous but the colonies widely separated.

19. **Hutchinsia R. Br.**

1. **H. procumbens** (L.) Desv. Occasional in temporarily moist alkaline soil, Carrizo Plain and Cuyama Valley.

20. **Capsella Medic.**

1. **C. Bursa-pastoris** (L.) Medic. Shepherd's Purse. Common weed in cultivated areas, particularly abundant around San Luis Obispo.

21. **Lepidium L.** Peppergrass

Petals showy, light yellow; silicle not notched at apex; style about 1 mm. long
1. *L. Jaredii.*
Petals inconspicuous, white or greenish, or absent; silicle notched at apex; style nearly or quite obsolete.
    Silicle not extended into wings on either side of apical notch.
        Pedicels not evidently flattened.

> Cauline leaves mostly entire, the lower ones coarsely toothed
> > 2. *L. densiflorum.*
> Leaves mostly deeply pinnately lobed with narrow divisions
> > 3. *L. strictum.*
> Pedicels distinctly flattened.
> > Stems spreading from the base, hairy . . . . .4. *L. lasiocarpum.*
> > Stem erect, simple or with ascending branches above the base, usually
> > glabrous . . . . .5. *L. nitidum.*
> Silicle extended into a thin lobe ("wing") on either side of the apical notch.
> > Apical notch of silicle narrow, the wings parallel.
> > > Stems stout; wings of silicle about half as long as body . .6. *L. latipes.*
> > > Stems relatively slender; wings of silicle about ¼ as long as body
> > > > 7. *L. dictyotum.*
> > Apical notch of silicle broad, the wings pointed and somewhat divergent
> > > 8. *L. acutidens.*

1. **L. Jaredii** Bdg. Growing in otherwise bare alkali spots, in silt or fine sand, Carrizo Plain south of Soda Lake. One of the most highly localized species in the world. In the one place where it is found, the plants are remarkably numerous and, as viewed from a distance, form a conspicuous patch of color in March and early April. The flowers are light yellow. Plants with white flowers but otherwise similar (subsp. *album* Hoover) have been found in western Fresno Co.

2. **L. densiflorum** Schrad. Only an occasional plant is found by roadsides or along railways; presumably introduced from central or eastern North America.

3. **L. strictum** (Wats.) Rattan. Street weed, Paso Robles (*Howell 36,512*). This plant and *L. lasiocarpum* call for careful study to determine whether they should be included under one name. Published descriptions apart, the actual plants show no readily apparent differences.

4. **L. lasiocarpum** Nutt. Santa Maria River bed (*Eastwood 322* in 1906); west side of Soda Lake (*9763*, with glabrous silicles but otherwise conforming to this species); to be expected in sandy soil elsewhere.

5. **L. nitidum** Nutt. Common everywhere on open hills and plains in both interior and coastal districts, except in sterile sandy soils.

6. **L. latipes** Hook. Known from only one record in the county: "2.6 miles northeast of Cayucos," *Belshaw 1767*. The locality seems rather improbable. In the Great Valley this species grows mainly on alkaline plains, and in southern California on the borders of salt-marshes.

7. **L. dictyotum** Gray. Common in eastern part from Cholame Valley to Carrizo Plain, usually on level ground in more or less alkaline soil; upper Arroyo Grande, around saline springs. Specimens from this area which have been called *Lepidium oreganum* Howell do not seem separable from *L. dictyotum* in any precise manner. It has been suggested that *L. oreganum* may be a hybrid between *L. dictyotum* and *L. nitidum* (J. T. Howell in Leafl. West. Bot. 1: 92–94), but none of the San Luis Obispo Co. plants so identified seem to me to have anything of *L. nitidum* in them. Furthermore, such plants always produce a full crop of seeds, which would be a rather unusual occurrence for a true interspecific hybrid. In any case, descriptions in the literature make it seem highly doubtful that our plants conform to the type of *L. oreganum.*

8. **L. acutidens** (Gray) Howell. Frequent on alkali flats from Cholame Valley to Carrizo Plain. In San Luis Obispo Co. no evidence of intergradation with *L. dictyo-*

tum, which would justify the name *L. dictyotum* var. *acutidens,* has been seen. The two when growing together are always clearly distinguishable.

22. **Cardaria** Desv.

1. **C. Draba** (L.) Desv. Hoary Cress. An aggressive pest, which has occupied fertile valley lands near Morro Bay, especially at the mouth of Chorro Creek. Scattered infestations extend to San Luis Valley and may be expected elsewhere.

23. **Dithyrea** Harv. Spectacle-Pod

1. **D. maritima** Davidson. Frequent on low sand-dunes nearest the beach from west side of Morro Bay southward. This coastal perennial with widely spreading rhizomes is markedly different from the desert annual *D. californica* with its slender tap-root.

24. **Athysanus** Greene

1. **A. pusillus** (Hook.) Greene. Widespread over stony or brushy hills and probably common, but very easily overlooked.

25. **Thysanocarpus** Hook.

Cauline leaves auriculate.
    Wing of samara with a circle of conspicuous perforations .... 1. *T. elegans.*
    Wing of samara without perforations .... 2. *T. curvipes.*
        Wing with ribs radiating out from body of fruit .... var. *curvipes.*
        Wing without rays .... var. *eradiatus.*
Cauline leaves not auriculate, narrowed toward base .... 3. *T. laciniatus.*

1. **T. elegans** F. & M. Lace-Pod. In loose soils, locally occurring mainly in upper Salinas Valley, as at Atascadero (*6784*); also in upper Arroyo Grande valley and the hills between San Luis Valley and Arroyo Grande. This species and *T. curvipes* are here readily distinguishable, showing none of the intergrades which are stated to exist elsewhere. However, nothing whatever is known of the genetic relationship between the two, which might conceivably warrant placing *T. elegans* as a variety of *T. curvipes,* as most authors have done.

2. **T. curvipes** Hook. var. **curvipes.** Fringe-Pod. Mainly in sandy or gravelly soils, common throughout the interior (although many of the plants may be var. *eradiatus*). There are no definitely known occurrences west of the Santa Lucia Range. The species is variable.

Var. eradiatus Jepson. Apparently belonging to this variety, which otherwise is known only in the desert region, are plants growing on a sandstone hill 5 miles east of Creston on La Panza road (*6804*), and at western edge of Carrizo Plain (*10,268*). The extent to which such plants occur in this area remains to be studied in detail.

3. **T. laciniatus** Nutt. The commonest species, found plentifully in rocky or loose soils throughout the county. If the varieties *crenatus* (Nutt.) Brewer and *emarginatus* (Greene) Jepson occur here at all, they are too poorly defined in this region to make their separation practical.

Crassulaceae. Stonecrop Family

Leaves opposite; dwarf annuals with very small leaves and minute flowers   1. *Tillaea.*
Leaves alternate, some of them usually forming rosettes.
    Petals rotately spreading; carpels divergent.

Root not tuberous; leaves flat, the rosettes not basal but terminating non-
flowering branches . . . . .2. *Sedum.*
Root tuberous, the crown bearing a rosette of a few terete or clavate leaves
and a flowering stem with reduced leaves . . . . .3. *Hasseanthus.*
Petals erect or the tips curved outward; carpels erect . . . . .4. *Dudleya.*

1. Tillaea L.

Flowers clustered in the axils . . . . .1. *T. erecta.*
Flowers solitary in the axils . . . . .2. *T. aquatica.*

1. **T. erecta** H. & A. In sandy or rocky places, probably in all parts of the county
but easily overlooked and seldom collected. After fires which are severe enough to
destroy all the organic matter in the soil, this species is often noticeable, because
few other plants appear in such places the first year.

2. **T. aquatica** L. In muddy places where water has stood after rains: 4 miles
northeast of Paso Robles (*6870*); with little doubt more widely distributed in the
county.

2. Sedum L. STONECROP

1. **S. spathulifolium** Hook. On rocks at Devil's Gap, upper Morro Creek. A collection by Summers from "San Luis Obispo Co." is the type of *Gormania anomala*
Britt., which is the basis for the currently used name *S. spathulifolium* subsp. *anom-
alum* Clausen, but our plants differ in no clearly evident way from typical *S. spathulifolium.*

3. Hasseanthus Rose

1. **H. Blochmaniae** (Eastw.) Rose. In hard clay soils where other vegetation is
sparse and stunted, often in areas of serpentine rock, frequent in coastal region
from near Cayucos (*Eastwood & Howell 2288*) to San Luis Valley.

4. Dudleya B. & R.

Rosettes over 15 cm. in diameter; flowers spreading or a little drooping; corolla, like
the rest of the plant, heavily covered with white powder . . . . .1. *D. pulverulenta.*
Rosettes seldom over 15 cm. in diameter, usually less; flowers erect; corolla not heavily white-powdered.
    Corolla bright or greenish yellow to orange or red.
        Caudex simple or compactly branched; leaves comparatively thin, not over
        $\frac{1}{3}$ as thick as wide; petals yellow to red.
            Plant 2 dm. tall or less; pedicels slender, those of earliest flowers usu-
            ally much longer than the flowers .2. *D. cymosa.*
            Plant over 2 dm. tall; pedicels stout, mostly short, even those of the
            earliest flowers rarely as long as the flowers . . . . .3. *D. lanceolata.*
        Caudex in age with somewhat elongate branches, the rosettes many; leaves
        thick, usually more than $\frac{1}{3}$ as thick as wide; petals yellow (orange in some
        Monterey Co. forms) . . . . .4. *D. caespitosa.*
    Corolla white to straw-color or cream-color, often with purple markings.
        Rosettes solitary or few in a clump; leaves flat; cyme with usually 4 or more
        elongate branches; petals with purple midrib and purple striations on
        either side . . . . .5. *D. murina.*
        Rosettes many, forming a large clump eventually; leaves terete to semi-
        terete; cyme with usually 1 or 2 main branches, often with a few short

branches below; petals white to cream-color, often with midrib purple-
tinged toward apex but otherwise without markings.

  Branches of caudex just below rosettes 10 to 18 mm. in diameter;
  leaves evergreen, after drying 3 to 7 mm. wide at middle   6. *D. Bettinae.*
  Branches of caudex just below rosettes 4 to 8 mm. in diameter; leaves
  shriveling in summer, in dried specimens less than 2 mm. wide

                                                               7. *D. parva.*

1. **D. pulverulenta** (Nutt.) B. & R. Occasional on rock-faces and in rock-crevices.
Northernmost localities include Black Mt. in La Panza Range, Rocky Canyon near
Atascadero, Lopez Canyon in Santa Lucia Range, Perfumo Canyon and See Can-
yon in San Luis Range. Thence it extends southward at scattered localities to the
Cuyama River and far beyond. The large white rosettes are impressive at all times
of year.

2. **D. cymosa** (Lem.) B. & R. On rocks, widely distributed over California's hill
country but remarkably rare in this county and known only in the mountains of
our southern interior: "road to Pozo" (presumably from Santa Margarita), *Eastwood* in 1928; upper Buckhorn Canyon, La Panza Range, *Twisselman 2169*; Big
Falls Canyon. A form of this species occurs on the south side of Cuyama Canyon
near Clear Creek (Santa Barbara Co.) and will undoubtedly be found also on the
north side within our limits. Another form of *Dudleya,* probably belonging to this
species rather than to *D. lanceolata,* is known on the east slope of the Temblor
Range in Kern Co. but probably does not extend across the San Luis Obispo Co.
line. This was reported by Twisselmann under the name *Echeveria laxa.* Plants in
Big Falls Canyon showing the characters of this species were associated with *D.
lanceolata* and *D. pulverulenta* and varied in such a manner as to suggest possible
hybridization followed by backcrossing.

3. **D. lanceolata** (Nutt.) B. & R. Very common in rocks, sand, and clay of coastal
bluffs and dunes, extending inland in rocky places through Santa Lucia Mts., but
absent from drier areas. Although the plants show very wide variation, there is no
basis for separating them into more than one species. Red or reddish-orange flowers
prevail along the coast from Cambria northward and in most of the places where
this species occurs in the Santa Lucia Range. Plants with greenish-yellow to orange-
yellow flowers are the rule in the coastal area from Cayucos southward. Plants having bright green leaves (*D. lurida* B. & R.) often grow in the same small area as
otherwise identical plants with white-glaucous leaves. Plants which in age form
clumps of many rosettes are most plentiful along the coastline but at times are
found at some distance from the sea. Only by arbitrarily selecting some single feature, such as green versus glaucous leaves, could the plants be grouped into recognizable subspecies or varieties, and such a procedure, I believe, would have no practical value for taxonomic (as distinct from horticultural) purposes. The name *D.
lanceolata* is here considered the best choice for the plants in question. *Cotyledon
Palmeri* Wats. and *Cotyledon lingula* Wats., both based on collections made by
Palmer at San Simeon, are believed to belong in *D. lanceolata.*

4. **D. caespitosa** (Haw.) B. & R. Some of the plants in the vicinity of Morro Bay
are, with some uncertainty, classified as this species. The genus *Dudleya* includes
numerous local races, of which a few are readily enough distinguishable to rank as
species (e.g., *D. parva* and *D. Bettinae*), but the majority are separable by features
which are not reliable enough to serve as a basis for valid species. Whatever the
"typical" form of *D. caespitosa* may be, none of the plants in San Luis Obispo Co.

are identical with forms of the species found from Monterey Co. northward. Yellow-flowered plants on Morro Rock and at the east base of Black Mountain near Morro Bay are for the present referred to *D. caespitosa.* Yellow-flowered plants on White's Point and elsewhere around Morro Bay conform better to *D. lanceolata.* Those on the Nipomo Dunes are regarded as definitely *D. lanceolata,* but those on the dunes near Morro Bay are somewhat doubtful, although I now would include them also in *D. lanceolata.* Red-flowered coastal plants from Cambria northward are variable, and some of them could be interpreted as hybrid derivatives of *D. lanceolata* and *D. caespitosa,* but the preponderance of their features leads me to include them in *D. lanceolata.*

Dudleya farinosa (Lindl.) B. & R. is excluded from the San Luis Obispo Co. flora by intention, not by oversight, although authors have credited it with an extended range along the California coast. As I understand *D. farinosa,* it is restricted to the coast of northern Monterey Co., where it grows always within reach of salt spray from the sea. Old plants develop many long trailing or cliff-hanging caudices with the leaves in a looser spiral than in other species and the rosettes therefore somewhat elongate. The plants are heavily white-powdered throughout, the flowering stems spreading or on cliffs even pendent, and the petals a shade of pale yellow which is not duplicated in any of the variants of *D. caespitosa.*

5. **D. murina** Eastw. Local in hills bordering San Luis Valley, always on serpentine, along base of Santa Lucia Mts. from Chorro Creek (the type locality, incorrectly printed as "Cholla" Creek) to East Fork of Corral de Piedra Creek, and in Perfumo Canyon, San Luis Range.

6. **D. Bettinae** Hoover. Bare rocky places, on serpentine, locally plentiful on first ridge west of Cerro Romauldo near San Luis Obispo; also probably near Whale Rock Dam just south of Cayucos.

7. **D. parva** Rose & Davidson. Very local on serpentine, San Bernardo Creek east of Morro Bay, where plentiful on the steep north-facing slope on south side of creek, and on a small serpentine outcrop on a volcanic hill near the mouth of Chorro Creek. Otherwise known only in a very small area in Ventura Co. The occurrence of three highly localized species in the serpentine areas is remarkable. *Dudleya murina, D. Bettinae,* and *D. parva* are clearly closely related and have similar ecologic requirements; yet each varies but slightly and shows no indication of hybridization with the other two. Even though the total area where the three occur is small, they do not actually grow close together.

Saxifragaceae. SAXIFRAGE FAMILY

Flowers in a cyme or a panicle of cymes.
    Stem bearing well-developed leaves above the base ....1. *Boykinia.*
    Stem leafless above the base, except for reduced bracts.
        Leaves pinnately veined, dentate or crenate ....2. *Saxifraga.*
        Leaves palmately veined and palmately lobed ....3. *Heuchera.*
Flowers in a raceme or solitary.
    Flowers in a raceme; leaves palmately lobed or divided ......4..*Lithophragma.*
    Flowers solitary; leaves entire ....5. *Parnassia.*

1. **Boykinia** Nutt.

1. **B. elata** (Nutt.) Greene. Wet, usually rocky, places in watershed of San Carpoforo Creek and northward along coast. Otherwise unknown in the county, though also found from Santa Barbara Co. to Los Angeles Co.

2. Saxifraga L. Saxifrage

1. **S. californica** Greene. California Saxifrage. Common in the hills, in woods or in open moist places, from coast eastward through La Panza Range.

3. Heuchera L. Alum-Root

Hypanthium with calyx 4 to 5 mm. long at anthesis, villous ...... 1. *H. pilosissima.*
Hypanthium with calyx 2 to 3 mm. long at anthesis (enlarging in fruit), puberulent
2. *H. micrantha.*

1. **H. pilosissima** F. & M. Occasional in fog-dampened, mostly shaded locations near coast, usually on rocks: Morro Rock; Hollister Peak (according to Philip Wells); Sycamore Canyon in San Luis Range; Price Canyon. Plants growing along Islay Creek have not been seen in flower but for ecologic reasons are assumed to be this species rather than *H. micrantha. Heuchera pilosella* Rydb., of which the type locality is "near San Luis Obispo," is regarded as a synonym of *H. pilosissima.*

2. **H. micrantha** Dougl. Locally common in Lopez Canyon and its branches, on shaded rocky banks in humus-rich, often moist, soil.

4. Lithophragma Nutt. Woodland Star

Cauline leaves alternate, widely spaced or solitary; pedicel of lowest flower less than 7 mm. long.
    Pedicel of lowest flower 1 to 2 mm. long; hypanthium truncate or obtusely angled at base ....1. *L. heterophyllum.*
    Pedicel of lowest flower 3 to 5 mm. long; hypanthium obconical, tapering to a pointed base ....2. *L. affine*
Cauline leaves 2, opposite, or sometimes alternate and closely spaced on opposite sides of stem; pedicel of lowest flower usually 7 to 12 mm. long ..3. *L. Cymbalaria.*

1. **L. heterophyllum** (H. & A.) T. & G. Common in loose humus-rich soil in shady woods from the coast to the La Panza Range. Except for the variable feature of the shape of the hypanthium, the San Luis Obispo Co. plants seem to belong to one morphologically, geographically, and ecologically coherent entity. In some other parts of California, however, *L. heterophyllum* and *L. affine,* if correctly identified, are distinct and easily differentiated. Years ago I was able to distinguish the two as they occurred in the lowest foothills of the Sierra Nevada on the basis of their tubers. *Lithophragma affine* had "blackberry-like" tubers made up of red fleshy scales, whereas *L. heterophyllum* had tubers which recalled potatoes in a very small way, having a brown surface and no obviously scaly structure. In that region also, *L. affine* always had three-lobed petals, while in *L. heterophyllum* the petals were entire or sometimes with a short tooth on each side. So far as can be learned, no author has mentioned the difference in the tubers or attempted to give it a wider geographic application, and herbarium specimens are useless for determining the nature of the tubers. It is therefore questionable whether two distinguishable species, other than *L. Cymbalaria,* actually exist in this county.

2. **L. affine** Gray. Woodland Star. If correctly identified, found in the same areas as *L. heterophyllum* but less common, and seemingly showing all degrees of intermediacy. I am not fully convinced, however, that genuine *L. affine* actually extends to this portion of California. Most, if indeed not all, of the specimens represent *L. affine* subsp. *mixtum* Taylor, which evidently is based on a series of intermediates. See note on tubers under *L. heterophyllum.*

3. **L. Cymbalaria** T. & G. Common in wooded areas, often growing with *L. heterophyllum* (or *L. affine*), extending sparingly even to Temblor Range.

5. Parnassia L. Grass of Parnassus

1. **P. palustris** L. var. **californica** Gray. In wet places on serpentine, "Cypress Swamp" on northeast slope of Cypress Mt. (*Twisselmann 3231*).

Grossulariaceae. Gooseberry Family

1. Ribes L.

Stems without spines; flowers in racemes which are usually many-flowered.
    Flowers yellow; leaves not glandular ....1. *R. aureum.*
    Flowers rose-pink to white; leaves glandular.
        Petioles puberulent or apparently glabrous; peduncles puberulent, very minutely glandular if at all so ....2. *R. glutinosum.*
        Petioles and peduncles conspicuously glandular-hairy ....3. *R. malvaceum.*
            Leaves usually minutely white-tomentose beneath; flower-tube 3 to 8 mm. long.
                Flowers pink; flower-tube 5 to 8 mm. long; calyx-lobes oval
                                var. *malvaceum.*
                Flowers white (pink in bud); flower-tube 3 to 5 mm. long; calyx-lobes orbicular ....var. *indecorum.*
            Leaves green and sparsely glandular-hairy beneath; flower-tube 8 to 12 mm. long ....var. *viridifolium.*
Leaves on main branches represented by simple or triple spines, the foliage leaves crowded on short axillary branchlets (spines sometimes absent in *R. quercetorum*); flowers mostly solitary or in loose groups (reduced cymes) of 2 to 4, sometimes in few-flowered loose racemes.
    Ovary glabrous or nearly so; berry without bristles or spines.
        Calyx dark purple or greenish; leaves 2 to 5 cm. wide ...4. *R. divaricatum.*
        Calyx yellow; leaves 1 to 2 cm. wide ....5. *R. quercetorum.*
    Ovary hirsute; berry covered with stiff bristles or slender spines.
        Calyx dark red-purple to purplish or green; flower-parts (except carpels) in 5's.
            Branches with bristles or prickles; leaves pubescent or glandular, at least on back.
                Branches covered with curved or flexuous bristles rather than prickles; filaments fully twice as long as petals ....6. *R. sericeum.*
                Branches with sparse to dense mostly weak prickles; filaments little exceeding petals ....7. *R. Menziesii.*
                    Leaves rather thick; filaments exceeding petals var. *Menziesii.*
                    Leaves thin; filaments barely equalling petals ...var. *Hystrix.*
            Branches without bristles or prickles; leaves glabrous or very nearly so
                                8. *R. californicum.*
        Calyx bright clear red; flower-parts (except carpels) in 4's 9. *R. speciosum.*

1. **R. aureum** Pursh var. **gracillimum** (Cov. & Britt.) Jepson. Golden Currant Usually along streams but sometimes in quite dry places, Salinas River and hills to the west from near Atascadero northward; Middle Branch of Huerhuero Creek.

2. **R. glutinosum** Benth. Pink Flowering Currant. In moist shaded woods and along streams: moist hollows among Nipomo Dunes; Arroyo Grande watershed, including Lopez Canyon; San Luis Range, and northward near the coast.

3. **R. malvaceum** Smith var. **malvaceum.** CHAPARRAL CURRANT. Frequent in hilly areas of western part, widely scattered but seldom plentiful, often a minor component of chaparral. Extending inland to hill country east of Salinas River but in the La Panza Range replaced partly, if not entirely, by var. *viridifolium*.

Var. **indecorum** (Eastw.) Jancz. WHITE CURRANT. Occurring generally in the same areas as var. *malvaceum* but probably not extending so far north: San Luis Range; Poly Canyon at San Luis Obispo; ridge west of upper Lopez Canyon; Calf Canyon and southward. The opened flowers are white, but usually the buds are more or less pink-tinged on the outside. In some places this variety replaces var. *malvaceum* entirely, as along Coon Creek, but more commonly there are pink-flowered shrubs growing near the white-flowered ones. The two are identical except for their flowers. There is some correlation of short flower-tube and nearly orbicular calyx-lobes with white flowers, but intergradation exists in all these features.

Var. **viridifolium** Abrams. So far as known, occurring within our limits only in La Panza Range (east slope, on Pozo-La Panza road, *6721*). Some of the shrubs in the Santa Lucia Range have the lower leaf-surfaces not tomentose, but the relatively short flower-tube excludes them from var. *viridifolium*.

4. **R. divaricatum** Dougl. Moist ground along streams or in hollows among sand-dunes, commonly under willows, of general occurrence along the coast and at scattered localities inland to Salinas River.

5. **R. quercetorum** Greene. YELLOW GOOSEBERRY. Frequent in hilly regions of the interior, mainly in dry gravelly or rocky soil, from eastern foothills of Santa Lucia Mts. eastward to Temblor Range. Paso Robles is the type locality. The flowers are usually greenish yellow, but shrubs found near Atascadero had bright golden-yellow flowers.

6. **R. sericeum** Eastw. SANTA LUCIA GOOSEBERRY. Frequent along streams and in oak woods from Lopez Canyon (*8287*) northward, reaching the coast near the Monterey Co. line. An endemic of the Santa Lucia Range.

7. **R. Menziesii** Pursh var. **Menziesii.** Locally plentiful in canyon bottoms of San Luis Range: Coon Creek; Sycamore Canyon; Diablo Canyon. Forma **Victoris** (Greene) Hoover. This form with pale green rather than deep purple calyx grew with the typical form in Sycamore Canyon (*6609*).

Var. **Hystrix** (Eastw.) Jepson. What seems to be this rather poorly defined variety was collected in Los Osos Valley, under live oaks (*6603*). It is known otherwise only along the coast of Monterey Co. Perhaps these shrubs represent an environmental form growing in deeper shade than usual.

8. **R. californicum** H. & A. Frequent on wooded or brushy hills, east slope of Santa Lucia Mts. from Santa Margarita northward.

9. **R. speciosum** Pursh. FUCHSIA-FLOWERED GOOSEBERRY. Common on hills near the coast and inland through Santa Lucia Mts., in open woods or in comparatively humid spots in the chaparral.

Platanaceae. SYCAMORE FAMILY

1. Platanus L. SYCAMORE

1. **P. racemosa** Nutt. CALIFORNIA SYCAMORE. Common in western portion along stream-courses, extending inland to Navajo Creek east of La Panza Range. Absent from the drier area from Paso Robles to Polonio Pass and southward. Notably fine

trees occur in several places, as along Atascadero Creek and at Sycamore Springs near Avila.

Rosaceae. ROSE FAMILY

Shrubs, trees, or woody vines.
 Leaves either compound or palmately lobed.
  Leaves palmate or palmately lobed.
   Pistils 3 to 5, becoming thin dry few-seeded pods .....1. *Physocarpus.*
   Pistils many, fleshy in fruit ....2. *Rubus.*
  Leaves pinnate ....3..*Rosa.*
 Leaves simple, not palmately lobed (in *Adenostoma* divided into narrow segments on juvenile shoots only).
  Petals absent; receptacle tubular; style in fruit elongate and plumose
                 4. *Cercocarpus.*
  Petals present; receptacle not tubular; style not elongate or plumose in fruit.
   Flowers in pyramidal to oblong panicles.
    Leaves broad, toothed ....5. *Holodiscus.*
    Leaves linear, small, entire ("heather-like") ......6. *Adenostoma.*
   Flowers not in panicles: in cymes, small clusters (reduced cymes or corymbs), racemes, or solitary.
    Ovary superior; flowers perigynous.
     Pistil 1 ....7. *Prunus.*
     Pistils 5 ....8. *Osmaronia.*
    Ovary inferior.
     Leaves firm, evergreen, sharply serrate to base; flowers many in a cyme ....9. *Photinia.*
     Leaves thin, deciduous, toothed only near apex or entire; flowers solitary or few in small corymbs or very short racemes
            10. *Amelanchier.*
 Herbaceous plants.
  Leaves compound; perennials, or rarely fairly large annuals, with evident flowers.
   Petals present; pistils many.
    Leaves (in our species) pinnate, or rarely with 3 leaflets only; receptacle not enlarged or fleshy in fruit ....11. *Potentilla.*
    Leaves palmate, with 3 leaflets; receptacle in fruit becoming enlarged and fleshy ....12. *Fragaria.*
   Petals absent; pistil one.
    Flower-tube bearing barbed prickles ....13. *Acaena.*
    Flower-tube without prickles ....14. *Sanguisorba.*
  Leaves palmately lobed; dwarf annual with minute flowers ....15. *Alchemilla.*

1. **Physocarpus** Maxim. NINEBARK

1. **P. capitatus** (Pursh) Kuntze. In rocky stream-beds and on moist shaded slopes, Santa Rita Creek west of Templeton; upper Las Tablas Creek.

2. **Rubus L.**

Stem erect, branched, without prickles ....1. *R. parviflorus.*
Stems trailing, climbing, or arching, usually with prickles (except *R. ulmifolius*).

    Flowers few in axillary and terminal cymes ....2. *R. ursinus.*
    Flowers many in terminal panicles.
        Prickles absent ....3. *R. ulmifolius.*
        Prickles present ....4. *R. procerus.*

    1. **R. parviflorus** Nutt. var. **velutinus** (H. & A.) Greene. THIMBLEBERRY. Occasional in moist canyons near coast: San Luis Range (See Canyon, *7517*); Santa Lucia Range, especially northward.

    2. **R. ursinus** C. & S. WILD BLACKBERRY. Common along streams and on shaded slopes near coast, often forming impenetrable tangles; occasional eastward at least to Salinas River. Plants with some of the leaves simple have been called *R. vitifolius,* but the plants do not fall into two clearly distinguishable groups. The *"vitifolius"* form occurs only along the northern coast. Brewer and Watson in 1876 combined the two names under one species, using for it the name *R. ursinus.* This far south, the native blackberry seldom produces a good crop of fruit.

    3. **R. ulmifolius** Schott var. **inermis** (Willd.) Focke. Escaped from cultivation on bank of Santa Rosa Creek at Cambria (*Howell 40,792*).

    4. **R. procerus** Muell. HIMALAYA BERRY. In waste ground, Cambria (*Howell 40,765*). This rank-growing blackberry is a pest farther north and may become so here.

3. **Rosa** L. ROSE

Petals 12 to 25 mm. long; calyx-lobes persistent on mature fruit.
    Flowers typically many in a cyme, or under dry conditions often few or solitary; plants over 1 meter tall; prickles usually curved and rather stout; hypanthium without stalked glands ....1. *R. californica.*
    Flowers few in a cyme, often solitary; plants rarely as much as 1 meter tall; prickles straight, slender; hypanthium bearing stalked glands (except in f. *pinetorum*) ....2. *R. spithamaea.*
Petals 8 to 12 mm. long; calyx-lobes early deciduous after flowering
                                    3. *R. gymnocarpa.*

    1. **R. californica** C. & S. CALIFORNIA WILD ROSE. Common along streams, rarely on rocky slopes, extending inland at least to Salinas River.

    2. **R. spithamaea** Wats. f. **spithamaea.** Occasional in woods or among chaparral not far from the coast, in sandy or rocky soil: south edge of San Luis Valley on Carpenter Canyon road; Los Osos; ridge northwest of Cuesta Pass. No basis can be found for recognizing var. *sonomensis* (Greene) Jepson, which is therefore regarded as a synonym. The leaves in San Luis Obispo Co. specimens are not "glaucous" as described for var. *sonomensis* and in a collection from Los Osos (*6348*) are to varying degrees minutely glandular beneath.

    Forma **pinetorum** (Heller) Hoover. Differs strikingly from typical *R. spithamaea* in the absence of stalked glands on the receptacle, but the genetic difference between them must be very slight. At Los Osos the plants constitute a uniform population, identical in all respects except that some have the receptacle entirely devoid of glands while others show sparse to dense glands. Similar variation can be found elsewhere. The "pinetorum" form has also been found at Cambria. The presence of this form in San Luis Obispo Co. makes the geographic distribution of *Rosa pinetorum* Heller coextensive with that of *R. spithamaea.*

    3. **R. gymnocarpa** Nutt. Occasional in Santa Lucia Range: between Rocky Butte and Pine Mt.; South Fork of Old Creek; ridge northwest of Cuesta Pass. This spe-

cies is probably of more general occurrence than it appears to be, because in shaded woods it blooms sparingly or not at all. Only where the ground has been cleared, as on rocky road-banks, are many flowers produced.

4. **Cercocarpus** H. B. K.

Backs of leaves gray-green to white, usually more or less covered with curved or curly hairs, which are also present on the flower-tube ....1. *C. montanus.*
    Backs of leaves subglabrous to densely pubescent but the hairs not concealing the veinlets; flower-tube thinly woolly at anthesis.
        Largest leaves less than 3 cm. long; flowers in a cluster 6 or fewer
                var. *glaber.*
        Some of leaves 3 cm. long or more; flowers sometimes up to 15 in a cluster but sometimes few or solitary ....var. *Blancheae.*
    Backs of leaves white-felty so that the veinlets are concealed; flower-tube densely white-woolly at anthesis ....var. *Traskiae.*
Backs of leaves yellowish green, minutely strigose along the veins, otherwise glabrous or nearly so; flower-tube microscopically strigose; curly hairs completely absent
                2. *C. minutiflorus.*

   1. **C. montanus** Raf. Mountain Mahogany. In open oak woods, or a component of chaparral. Local material is variable. On a somewhat arbitrary basis, our shrubs are referred to the following three varieties.

Var. **glaber** (Wats.) F. L. Martin. By far the commonest variety, if only because the name is allowed to cover such a wide range of variation, extending from low coastal hills inland to La Panza Range.

Var. **Blancheae** (C. K. Schn.) F. L. Martin. If defined on the basis of larger leaves with more numerous veins, this variety is best represented by an extensive stand on the hill west of Poly Canyon at San Luis Obispo. The shrubs at this locality have not been observed to flower. Fairly large leaves are occasionally found elsewhere, but never very far from the coast. Commonly var. *Blancheae* has more flowers in a cluster than var. *glaber,* although complete intergradation exists in both features. A collection from upper Arroyo Grande (*6296*) is more like var. *Blancheae* in both leaf size and number of flowers than var. *glaber* as found at most localities. I regard *C. betuloides* var. *multiflorus* Jepson as a synonym of var. *Blancheae.*

Var. **Traskiae** (Eastw.) F. L. Martin. Some of the shrubs on the ridge northwest of Cuesta Pass (*8736*) are referred to this variety because the calyx, flower-tube, and backs of the leaves are densely white-woolly. The leaves are smaller than in the type, however, and not so thickly tomentose on the back. Similar individuals have been found at Grover City and in the hills north of Arroyo Grande. Nothing exactly like the original material of var. *Traskiae* from Santa Catalina Island has been found elsewhere.

   2. **C. minutiflorus** Abrams. Occasional in Santa Lucia Mts.; above Cerro Alto Campground; ridge just west of Cuesta Pass. Specimens from these shrubs are identical in every detail with material from San Diego Co. despite the apparent absence of the species from the intervening area. The flowers in both San Diego Co. and San Luis Obispo Co. shrubs are in the size-range of *C. montanus* var. *glaber,* not distinctly smaller as described in the literature. At Salsipuedes Creek west of Pozo, some individuals approach *C. minutiflorus* closely, although the stand as a whole is referred to *C. montanus* var. *glaber.*

5. **Holodiscus** Maxim.

1. **H. discolor** (Pursh) Maxim. Cream Bush. Ocean Spray. Common in wooded canyons and on shaded slopes near coast, extending inland to hills west of Pozo. Whether any or all of our plants should be called var. *franciscanus* (Rydb.) Jepson is questionable. That variety seems to be recognized on the basis of its smaller, more densely hairy leaves, but in San Luis Obispo Co. there is no discoverable correlation between leaf-size and pubescence. Also, the plants do not fall into two readily distinguishable groups on the basis of any single character or any combination of characters.

6. **Adenostoma** H. & A.

Leaves mostly in axillary fascicles, stiff . . . . .1. *A. fasciculatum.*
Leaves not in fascicles, flexible . . . . .2. *A. sparsifolium.*

1. **A. fasciculatum** H. & A. Chamise. Abundant on rocky or sandy hills, near coast and eastward to La Panza Range and upper Cuyama Canyon. Absent from Cholame Hills and the region of Carrizo Plain. In some places this species forms a dense pure stand; or it may be mixed with other shrubs in varying proportions.

2. **A. sparsifolium** Torr. Ribbonwood. Formerly plentiful in upper Arroyo Grande watershed, on hills of limestone or calcareous sandstone, but in recent years great numbers of the shrubs have been destroyed. "Garcia, San Luis Obispo Co.," *Adolf Holm* in 1915 (possibly Garcia Mountain, between upper Salinas River and Stoney Creek); upper part of Cuyama Canyon. Spanish-speaking people call it "yerba pasmo" and believe it to have medicinal value.

7. **Prunus** L.

Flowers solitary or in small clusters, subsessile; fruit dry.
    Low spreading shrub with rigid branches; flowers not over 6 mm. in diameter
                                        1. *P. punctata.*
    Tree; flowers showy, 25 mm. or more in diameter . . . . .2..*P. Amygdalus.*
Flowers in racemes; fruit fleshy.
    Leaves thin, deciduous, finely serrate . . . . .3. *P. demissa.*
    Leaves leathery, evergreen, sharply dentate ("holly-like") . . . . . . .4. *P. ilicifolia.*

1. **P. punctata** (Jepson) Hoover. Sand Almond. Locally common in sand around south side of Morro Bay and on Nipomo Mesa. This low-growing coastal shrub seems sufficiently distinct from the taller (but widely branching) "desert almond" (*P. fasciculata*) and is well separated from it geographically. The petals are probably shorter and broader in the coastal plant, but fresh flowers of *P. fasciculata* from various desert localities should be compared before a confident statement to that effect is made.

2. **P. Amygdalus** Batsch. Almond. Sometimes growing spontaneously as scattered trees on open hillsides, as near San Luis Obispo and Atascadero.

3. **P. demissa** (Nutt.) Walp. Western Choke-Cherry. Moist places in sheltered canyons, occasional in Santa Lucia Range: Paso Robles Creek and tributaries; north base of Cuesta Pass; Lopez Canyon. An occurrence in "Rock Spring Canyon" of the Temblor Range is probably in Kern Co.

4. **P. ilicifolia** (Nutt.) Walp. Holly-Leaved Cherry. Sometimes on coastal sandhills (Hazard Canyon to Los Osos); more commonly in rocky places, extending inland to sandstone hills at western edge of Carrizo Plain. There is an isolated grove

on the east side of the Temblor Range in Kern Co. Unusually large individuals are seen along the Peachy Canyon Road near Paso Robles. The fruit, although having a thin flesh, is generally of fine flavor.

8. **Osmaronia** Greene

1. **O. cerasiformis** (T. & G.) Greene. Oso BERRY. Rather common on partly shaded, comparatively moist hillslopes near coast and eastward through Santa Lucia Mts.

9. **Photinia** Lindl.

1. **P. arbutifolia** (Ait.) Lindl. TOYON. CHRISTMAS BERRY. CALIFORNIA HOLLY. Common throughout the hill country from coast to La Panza Range. A notable display of berries can be seen in late fall and winter along the Peachy Canyon Road west of Paso Robles. This is one of the few native shrubs which have been used for ornamental plantings on a really extensive scale. A rare form with yellow berries was described as var. *cerina* Jepson, based on a collection from Templeton.

10. **Amelanchier** Medic. SERVICE BERRY

1. **A. pallida** Greene. Locally rare in woods or chaparral in Santa Lucia Range: ridge west of Cuesta Pass; 2 miles south of Atascadero; between Rocky Butte and Pine Mt. The flowers here are notably small as compared with the species elsewhere.

11. **Potentilla** L.

Stolons absent; stems branching, with flowers in cymes; leaflets green on both sides.
    Perennials with stout caudex; leaves pinnate.
        Basal leaves with 9 to 25 leaflets.
            Leaflets coarsely serrate or incised, not parted below the middle.
                Hypanthium deeply cup-shaped; bractlets equalling calyx-lobes, usually 3-toothed . . . . .1. *P. californica.*
                Hypanthium bowl-shaped; bractlets (at least slightly) smaller than calyx-lobes, entire.
                    Leaflets sparsely to densely glandular-pubescent, not silky.
                        Hypanthium densely hairy on inner surface; cyme often dense . . . . .2. *P. Kelloggii* var. *cuneata.*
                        Hypanthium glabrous to sparsely hairy on inner surface; cyme always loose . . . . .2..*P. Kelloggii* var. *puberula.*
                  Leaflets densely appressed-silky.
                    Leaflets pinnately incised or coarsely serrate; coastal
                              2. *P. Kelloggii* var. *Kelloggii.*
                    Leaflets fan-shaped, somewhat palmately cleft; interior mountains . . . . .3. *P. Bolanderi.*
            Leaflets parted nearly to base into narrow segments . .4. *P. Micheneri.*
        Basal leaves with 5 to 9 leaflets . . . . .5. *P. glandulosa.*
    Annual or biennial; leaflets 3 only . . . . .6. *P. rivalis.*
Stolons present; flowers solitary on scape-like peduncles; leaflets white on back
                                          7. *P. Egedii.*

1. **P. californica** (C. & S.) Greene. Along streamlets and on moist hillslopes, coast from Cambria northward.

2. **P. Kelloggii** Greene var. **Kelloggii.** Occasional in coastal sands: near Jack Lake (*9248,* the hair on the leaves so dense as to form a thick felt); Morro Bay.

Var. **cuneata** (Lindl.) Hoover. *Potentilla Lindleyi* Greene. Common near coast in dry to moist sandy soils or in moist clays. Intergrading with var. *Kelloggii* in the few places where that variety occurs.

Var. **puberula** (Greene) Hoover. In dry sandy places, upper Salinas Valley from vicinity of Atascadero southward, and in our southern coastal area from Indian Knob ridge (north of Pismo Beach) southward.

3. **P. Bolanderi** (Gray) Greene var. **Parryi** (Wats.) M. & J. Rare in La Panza Range: low wet meadow on northeast spur of Black Mountain (*Twisselmann 2929*); American Canyon (not found in flower but with leaves agreeing with the Twisselmann collection).

4. **P. Micheneri** Greene. Reportedly collected at "San Luis Obispo" by Curran in 1885 (Jepson, Fl. Cal. 2: 200). An error in data is suspected, because the species has never otherwise been found south of Marin Co.

5. **P. glandulosa** Lindl. On wooded slopes and along streams, common in western portion, occasionally extending inland to La Panza Range. Our plants are all of the form with creamy-white petals, described as *P. Wrangelliana*, but do not show the obtuse calyx-lobes which authors have associated with that name.

6. **P. rivalis** Nutt. var. **millegrana** (Engelm.) Wats. Estrella, *Jared* in 1895 (reported by Jepson, Fl. Cal. 2: 183, as *P. leucocarpa* Rydb.).

7. **P. Egedii** Wormsk. Common in moist soil along the coast on borders of salt-marshes and of fresh-water lakes and ponds, along streams, and in seepage spots on sea-cliffs. Presumably our plants belong to var. *groenlandica* (Tratt.) Polunin; the basis for distinguishing such a variety is not clear to me.

12. **Fragaria** L. Strawberry

Petioles and stolons stout; leaflets thick; flowers 25 to 30 mm. wide . . 1. *F. chiloensis.*
Petioles and stolons slender; leaflets thin; flowers 15 to 20 mm. wide 2. *F. californica.*

1. **F. chiloensis** (L.) Duch. Sand Strawberry. Sand of coastal dunes, known in the county only near Oso Flaco Lake.

2. **F. californica** C. & S. California Strawberry. In woods: Coon Creek in San Luis Range; common under pines around Cambria; occasional in Santa Lucia Mts. (e.g., San Carpoforo Creek, San Simeon Creek, Santa Rita Creek). Sometimes found on treeless north-facing slopes along the coast northward.

13. **Acaena** L.

1. **A. californica** Bitter. Open hills near the sea from near Piedras Blancas Point (*7663*) northward.

14. **Sanguisorba** L. Burnet

1. **S. minor** Scop. Occasionally escaping or persisting in formerly cultivated ground from plantings for forage: coast near Cambria; Atascadero; becoming established at Twisselmann Ranch according to Twisselmann.

15. **Alchemilla** L.

1. **A. occidentalis** Nutt. Sandy soils, often in cultivated fields: Arroyo Grande (reported as *A. arvensis* by Jepson, Fl. Cal. 2: 216). Probably common and generally distributed, but easily overlooked.

Mimosaceae. Mimosa Family

1. Acacia Willd.

1. **A. longifolia** Willd. Locally plentiful on sand-dunes south of Oceano. Presumably planted originally to stabilize the sand along the railway; now spreading westward toward the ocean.

Caesalpiniaceae. Senna Family

Flowers markedly irregular, appearing pea-like; leaves simple, cordate ....1. *Cercis.*
Flowers only slightly irregular; leaves pinnate or bipinnate.
   Leaves once pinnate ....2. *Cassia.*
   Leaves bipinnate.
      Shrub; stamens red, much longer than petals ....3. *Poinciana.*
      Herb; stamens shorter than petals ....4. *Hoffmannseggia.*

1. Cercis L. Redbud

1. **C. occidentalis** Torr. Western Redbud. Local on rocky slopes, upper Morro Creek. There is good reason to suspect that the species is introduced in this locality, as the nearest known occurrence is in the Sierra Nevada foothills, but the shrubs are well established and slowly increasing by seeding.

2. Cassia L. Senna

1. **C. tomentosa** L. f. Well established in the valley of lower Coon Creek, where there are several small trees, and numerous seedlings growing to maturity near them.

3. Poinciana Hook.

1. **P. Gilliesii** Hook. Bird-of-Paradise Tree. Cultivated ornamental shrub "commonly spontaneous about the ranchgrounds" at Twisselmann Ranch (Twisselmann (*8886*).

4. Hoffmannseggia Cav.

1. **H. Falcaria** Cav. Dorothy Twisselmann Ranch, "spreading slowly," *Twisselmann 4750* in 1958. Native to the California desert regions, eastward and southward.

Leguminosae. Pea Family

Leaves palmate, or with 3 leaflets only, or rarely simple.
   Thorny shrub; leaves simple or with 3 leaflets ....1. *Pickeringia.*
   Herbs or shrubs; if shrubby, without thorns; leaflets 3 or more.
      Plants not glandular-dotted.
         Leaflets more than 3 ....2. *Lupinus.*
         Leaflets normally 3 only.
            Shrubs ....3. *Cytisus.*
            Herbs, or stems slender and woody near base only.
               Stamens distinct; erect perennial with showy yellow flowers
                                  4. *Thermopsis.*
               Stamens diadelphous (9 of filaments united below, the tenth on upper side free).

Leaflets toothed; flowers in heads or short spikes or ra-
cemes (small umbels in some *Medicago* species).
Corolla deciduous after anthesis.
Style subulate; pod coiled or curved
5. *Medicago.*
Style filiform; pod straight . . . . .6. *Melilotus.*
Corolla withering-persistent . . . . .7. *Trifolium.*
Leaflets entire; flowers in umbels or solitary . . .8. *Lotus.*
Plants glandular-dotted . . . . .9. *Psoralea.*
Leaves pinnate, with more than 3 leaflets.
Tendrils absent.
Plants woody.
Petal 1, only the banner present .10..*Amorpha.*
Petals 5, differentiated as banner, wings, and keel . . . . . . .11. *Robinia.*
Plants herbaceous.
Plants not glandular-dotted; pods not prickly.
Flowers in umbels or solitary . . . . .8. *Lotus.*
Flowers in racemes or spikes (sometimes short head-like ones)
12. *Astragalus.*
Plants minutely glandular-dotted; pods with hooked prickles
13. *Glycyrrhiza.*
Terminal leaflets represented by tendrils.
Style with a ring of hairs near apex . . . . .14. *Vicia.*
Style flattened, hairy on inner side near apex . . . . .15. *Lathyrus.*

1. **Pickeringia** Nutt.

1. **P. montana** Nutt. Chaparral Pea. Common in chaparral, especially where
the soil is rocky and unfavorable for most plants, in Santa Lucia, San Luis, and La
Panza Ranges. No seeds or developing pods have been found in the county; yet the
wide distribution of the species is evidence of seed production in the past.

2. **Lupinus** L. Lupine.

Annuals with slender tap-root.
Flowers crowded at intervals on axis of raceme, thus verticillate or near-verticil-
late.
Pods ovate-acute, with 1 or 2 seeds.
Racemes remaining erect; flowers becoming suberect soon after anthesis, arranged symmetrically around axis.
Banner after drying 8 to 11 mm. wide; wings 5.5 to 8 mm. wide;
petals uniformly bright lilac-purple (fading in drying)
1. *L. horizontalis.*
Banner after drying 2 to 6 mm. wide; wings 3 to 4.5 mm. wide;
petals pale purple or pink to deep purple.
Flowers 13 to 15 mm. long (to tips of wings); banner after
drying 4 to 6 mm. wide; wings 3.5 to 4.5 mm. wide
2. *L. subvexus.*
Flowers 10 to 13 mm. long; banner after drying 2 to 4 mm.
wide; wings 3 to 4 mm. wide .3. *L. ruber.*
Racemes spreading or arching after anthesis, the flowers tending to
grow toward upper side . . . . .4. *L. densiflorus.*
Pods elongate, normally with 4 or more seeds.

Pedicels 3 to 8 mm. long; flowers 9 to 16 mm. long.
Stem stout, often hollow; keel ciliate on both margins near base
5. *L. succulentus.*
Stem slender; keel ciliate toward apex on upper margin only
6. *L. nanus.*
All leaflets less than 3 mm. wide; pods 5 mm. or less wide.
Flowers 9 to 11 mm. long; pods 4 to 5 mm. wide; seeds
usually 4 to 8 ....var. *nanus.*
Flowers 10 to 13 mm. long; pods 3.5 to 4.5 mm. wide;
seeds 8 to 12 ....var. *Menkerae.*
Some of leaflets over 3 mm. wide; pods more than 5 mm. wide.
Stems erect or ascending; pods 5 to 7 mm. wide
var. *latifolius.*
Stems, except for peduncle of earliest raceme, prostrate
or widely spreading; pods 7 to 8 mm. wide
var. *maritimus.*
Pedicels 1 to 3 mm. long; flowers 5 to 10 mm. long.
Keel non-ciliate; pods when immature fleshy, 6 to 9 mm. wide
7. *L. pachylobus.*
Keel ciliate or not; pods not markedly fleshy, 3 to 5 mm. wide
Keel acuminate; style long and slender ....8. *L. bicolor.*
Flowers 8 to 10 mm. long; coastal only ......var. *bicolor.*
Flowers 5 to 8 mm. long; widespread .. var. *microphyllus.*
Keel obtuse; style short and stout ....9. *L. polycarpus.*
Flowers in an even spiral around the axis, not in the least verticillate (rarely so
in *L. sparsiflorus,* but not throughout the raceme).
Stems erect or ascending; axis of raceme not inflated or curved at base.
Petals bright blue (varying to violet in an uncommon form of *L. con-
cinnus*).
Flowers 7 to 9 mm. long; plants usually with solitary flowers in
lower leaf-axils as well as in short racemes ......10. *L. concinnus.*
Plant grayish with copious spreading hair ....var. *concinnus.*
Plant sparsely hairy, green ....var. *Agardhianus.*
Flowers 10 to 15 mm. long; plants never with solitary flowers in
lower leaf-axils; racemes elongate.
Leaflets narrowly linear; bracts much exceeding the flower-
buds ....11. *L. Benthamii.*
Leaflets linear to oblanceolate; bracts slightly exceeding the
flower-buds ....12. *L. sparsiflorus.*
Petals red-purple (drying violet-purple).
Herbage appressed-pubescent; leaflets linear .....13. *L. truncatus.*
Herbage with stiff sharp bristles; leaflets oblanceolate to obovate
14. *L. hirsutissimus.*
Stems prostrate; axis of raceme swollen and inflated, abruptly curved up-
ward at base of raceme ....15. *L. nipomensis.*
Perennials, often more or less shrubby.
Herbs, often woody at the base but not with spreading woody stems.
Leaves bright green, sparingly appressed-hairy; keel ciliate on upper mar-
gin toward base; stems in tufts from a rather slender tough-fibrous tap-root
16. *L. latifolius.*
Leaves more or less grayish, with copious appressed to spreading hairs; keel
ciliate on upper margin toward apex, or non-ciliate; stems from a somewhat woody base or from rhizomes.

Keel non-ciliate; plants with rhizomes; leaves well spaced along the
stems . . . . .17. *L. formosus.*
Keel ciliate; plants without rhizomes, the stems tufted; leaves mostly
near base and long-petioled.
  Herbage villous and white-tomentose . . . . . . .18. *L. ludovicianus.*
  Herbage appressed-silky.
   Leaflets obovate or oblanceolate, mostly over 1 cm. wide
                    19. *L. cervinus.*
   Leaflets linear to narrowly oblanceolate, not over 5 mm. wide
               20. *L. austromontanus.*
Shrubs, or at least with spreading woody stems above ground.
 Lower leaves of each year's growth with petioles markedly longer than the
 leaflets . . . . .21. *L. albifrons.*
  Pubescence entirely appressed-silky.
   Bracts about equalling flower-buds . . . . .var. *albifrons.*
   Bracts conspicuously exceeding flower-buds . . . . .var. *Douglasii.*
  Pubescence somewhat spreading and of curved hairs . . .var. *Abramsii.*
 All leaves with petioles about equalling leaflets or slightly shorter.
  Leaflets densely silvery-silky; keel non-ciliate; petals bright to pale
  blue . . . . .22. *L. Chamissonis.*
  Leaflets appressed-hairy but green, or silvery only on young growth;
  keel ciliate; flower-color varied but seldom bright blue
                  23. *L. arboreus.*

1. **L. horizontalis** Heller. In disintegrated shale, sands, or barren clays, common
in Temblor Range and adjacent east side of Carrizo Plain, south to Cuyama Valley.
The size measurements of the flowers correspond to those given for var. *platypetalus*
C. P. Smith, which is based on a Mohave Desert plant but is not tenable on a geographic basis. Although *L. horizontalis* and *L. ruber* are closely related, the two in
the Temblor Range are quite distinct when they grow adjacent to each other.

2. **L. subvexus** C. P. Smith. Common in Salinas Valley; eastward to Cottonwood
Pass, La Panza district, and lower Cuyama Valley; often plentiful in clay soils and
firm sandy loams. On one hillside in the La Panza district, all the plants had flowers ascending at anthesis, but showing the size measurements of *L. subvexus,* while
other plants in the area were identical in every way except for having divergent
flowers. The conclusion is unavoidable that the key-character used by C. P. Smith
to separate *L. subvexus* (flowers spreading as contrasted with flowers ascending in
L. ruber) does not separate two really distinct species. As it occurs in this region,
L. subvexus is essentially merely a variant of *L. ruber* with somewhat larger flowers.
This may be a suitable instance for use of the subspecific category. There are no
definitely known occurrences of *L. subvexus* (or of *L. ruber*) on the west side of the
Santa Lucia Mts. (cf. Howitt and Howell, "Vascular Plants of Monterey County").
Those plants which are most conspicuously hairy represent var. *albilanatus* C. P.
Smith, of which the type locality is Paso Robles.

3. **L. ruber** Heller. Occasional in vicinity of Salinas River and Huasna River;
commoner eastward to Temblor Range, Caliente Mt., and Cuyama Valley. Largely
found in areas of lower rainfall than the territory of *L. subvexus,* but overlapping
it considerably. On an open clay hillside near the Salinas River between Santa Margarita and Atascadero were found some small pink-flowered plants, assumed to be
typical *L. ruber,* which looked distinct from the larger purple-flowered plants of *L.
subvexus* which were abundant in the vicinity. At no other locality has it been

noticed that the plants of this relationship are of two distinctly different kinds, and a large series of herbarium specimens is not readily separable into two groups. Both *L. ruber* and *L. subvexus* have at times been included in *L. microcarpus* Sims, a Chilean species. The limited material available from Chile suggests that true *L. microcarpus* has racemes which arch over after flowering with the pods all growing toward the upper side, as in the Californian *L. densiflorus* but unlike either *L. ruber* or *L. subvexus*.

4. **L. densiflorus** Benth. WHITE or PURPLE LUPINE. Common in western part, generally on clay banks or in sandy or gravelly flood-beds of streams. So far as at present known, the Salinas River marks the approximate eastern limit of the species in our area, although occurrences farther east are not improbable. Of the many named varieties of the species, our plants generally correspond best to the description of var. *palustris* (Kell.) C. P. Smith.

Var. **aureus** (Kell.) Munz. *L. Menziesii* Agardh. YELLOW ANNUAL LUPINE. Yellow-flowered plants are found sporadically within the range of the species, as near San Luis Obispo. The stems may be stouter than the average for the species, but some white-flowered plants also have notably stout stems.

5. **L. succulentus** Dougl. Common in soils containing considerable clay, or in crumbling shale, most abundant near coast, but also in scattered places through the interior. As Twisselmann has written, "it makes an excellent garden annual, and is useful for naturalizing in freshly graded areas or on fills, where it is a good soil builder."

6. **L. nanus** Dougl. var. **nanus.** Very common in "grassland" areas, usually in more or less sandy soils, in the Santa Lucia Range but especially in the Salinas Valley and hills eastward, where in spring large fields are solid blue with flowers. Plants found as far eastward as the La Panza district and Shandon hills can probably be considered typical *L. nanus,* although var. *Menkerae* is not sharply distinct in this area.

Var. **Menkerae** C. P. Smith. Along west base of Temblor Range from Crocker Grade to a few miles southward, where typical *L. nanus* is absent. Some of the plants found farther west, especially in the La Panza district, can probably be classified as var. *Menkerae* or as intergrades between it and var. *nanus.*

Var. **latifolius** Benth. Common in sandy soils near coast, at least from Cambria southward. Farther north, some of the plants, if not all of them, are more closely representative of the following variety. David B. Dunn's interpretation of the name var. *latifolius* is here followed, but it is possible that Bentham applied the name to a specimen of var. *maritimus.*

Var. **maritimus** Hoover, n. var. Caulibus (pedunculo racemae primae excepto) prostratis vel late patentibus; foliis villosis; foliolis 5 ad 9, eis foliorum inferiorum plerumque oblanceolatis, eis foliorum superiorum linearibus; floribus in 2 ad 5 verticellis, 10 ad 15 mm. longis; siliquis 7 ad 8 mm. latis.

Peduncle of earliest raceme erect; stems otherwise prostrate or widely spreading; leaves villous; leaflets 5 to 9, those of lower leaves mostly oblanceolate, of upper leaves mostly linear; flowers in 2 to 5 verticels, 10 to 15 mm. long; pods 7 to 8 mm. wide.

TYPE: near Arroyo del Oso, south of Arroyo de la Cruz, San Luis Obispo Co., *Hoover 9428.* An immature collection which evidently represents the same entity is from Bolinas Ridge, in open field, Marin Co., *Hoover 8680.* From these two records, it would appear that the variety is well distributed along the California coast. These plants have perhaps not been recognized as different from the much more widespread and plentiful form which David B. Dunn called subsp. *latifolius* (i.e., var. *latifolius* Benth.). The stems of var. *maritimus* have widely spreading, or even prostrate, branches from the base, the leaflets average narrower than in var. *latifolius* generally and are more conspicuously hairy, the racemes are short and relatively few-flowered, and the pods stouter. The stout pods, according to Dunn's key, are a feature of *L. affinis* Agardh, but the plants do not conform to that species in other respects.

7. **L. pachylobus** Greene. In open woods, known at only three localities in the county: between Rocky Butte and Pine Mt., Santa Lucia Range (*7893*); 17 miles east of Creston on La Panza Road (*7864*); summit between Pozo and Arroyo Grande. These occurrences represent the southern limit of the distribution of this rather rare Californian endemic.

8. **L. bicolor** Lindl. Common in all portions of the county in one form or another, mostly in sandy or gravelly soils. While it is true that there is considerable variation in certain obscure details of flower structure, the value of making each such variation the basis even of a variety, not to mention a subspecies, has not been convincingly demonstrated. A large proportion of the individual plants are intermediate between the published intraspecific entities rather than clearly referable to any one of them. Furthermore, if Dunn's key to the subspecies (El Aliso 3: 144) is taken at face value, a single plant can often be placed equally well in more than one of the "subspecies." The recognition of each minor variation is unquestionably of value to the geneticist but would be inconsistent with the treatment of other species in this work. Nevertheless, because plants of *L. bicolor* with relatively large flowers are absent from all our interior area, there is merit in making the following segregation on the grounds of flower size.

Var. **bicolor.** Plants with relatively large flowers, frequent near the coast only. Here included under this name are most, if not all, of the plants which Dunn has called subsp. *umbellatus* var. *umbellatus*. Dunn in his key stated no way in which subsp. *umbellatus* could be separated unequivocally from subsp. *bicolor*.

Var. **microphyllus** (Wats.) C. P. Smith. This name is here taken to include the small-flowered lupines which are found in such abundance throughout the interior region, but similar plants are also common near the coast. The named varieties *trifidus, Pipersmithii, tridentatus,* and *rostratus* are minor variations not correlated with the outward appearance of the plant and of only limited geographical significance. Two or more such variants can often be found among identical-looking plants in a single stand. If they are to be distinguished by name, I believe that the taxonomic category of *forma* would be most appropriate.

9. **L. polycarpus** Greene. *L. micranthus* Dougl. (1829), not Gussone (1828). 2 miles south of Atascadero, in fertile sandy loam (*9812*). Collections from Morro and Arroyo Grande are cited by Dunn (El Aliso 3: 147). It is likely that this species is often overlooked because of its very close superficial resemblance to small-flowered forms of *L. bicolor*.

10. **L. concinnus** Agardh var. **concinnus.** Common in gravelly and sandy places in the interior, from near Salinas River to western edge of Carrizo Plain. The presence of solitary flowers in the lower leaf-axils enables one at a glance to distinguish this from other lupines. Almost all San Luis Obispo Co. plants of *L. concinnus* show that feature, but this distinction evidently does not hold in other regions.

Var. **Agardhianus** (Heller) C. P. Smith. Cleared chaparral area on ridge northwest of Cuesta Pass; intergrades toward this variety occur within the main area of the species eastward.

11. **L. Benthamii** Heller. SPIDER LUPINE. Occasional on gravelly or stony banks: vicinity of Nacimiento River; upper Salinas River above Pozo; Agua Escondida Camp, and south to Cuyama Canyon and Cuyama Valley. Plants in the last named locality are larger in all parts and represent var. *opimus* C. P. Smith; which, however, is not geographically significant and is probably an expression of better environmental conditions.

12. **L. sparsiflorus** Benth. On steep gravelly and rocky slopes: upper Salinas River near Avenales Guard Station; Stoney Creek, and south to Cuyama Canyon and lower Cuyama Valley. These are the most northern and western occurrences of this species, which is a striking feature of the spring landscape in Arizona.

13. **L. truncatus** Benth. WOOD LUPINE. In gravelly and sandy soils in open woods or chaparral, or sometimes on sunny slopes, common in coastal area. Not yet definitely known to occur east of Santa Lucia Range.

14. **L. hirsutissimus** Benth. STINGING LUPINE. Of frequent occurrence, but rarely in any abundance, in gravelly soil among shrubs, or especially in disturbed soil, such as road-banks or burned areas, in western part, eastward as far as La Panza Range.

15. **L. nipomensis** Eastw. Dry sandy flats on Nipomo Mesa just back of the dunes. A remarkable rare and local species, entirely distinct in several respects from *L. concinnus*, with which it has been confused. Most notably, the stems are prostrate, the peduncle abruptly curved upward at the base of each raceme, and the axis of the raceme is thickened and inflated.

16. **L. latifolius** Agardh. Common in shady ravines and along streams near coast, extending inland to Rocky Canyon near Atascadero and to the mountains between Pozo and the headwaters of Arroyo Grande. No collections from La Panza Range have been seen.

17. **L. formosus** Greene var. **robustus** C. P. Smith. In sandy soils and disintegrated shale, common north and west of Paso Robles, eastward to vicinity of Estrella and Creston. The bright blue flowers are very showy. White-flowered plants occurring from Templeton to Atascadero are assumed to belong to the same species but call for more study.

18. **L. Ludovicianus** Greene. An endemic of San Luis Obispo Co., growing in sandy soil and centering in the upper Arroyo Grande watershed, extending to the hills north of Price Canyon and to the summit between Arroyo Grande and Huasna Valley. Also present in the upper Salinas River basin: 4 miles north of Pozo (*9847*). This has been designated as the official county flower.

19. **L. cervinus** Kell. In dry pine woods, an endemic of the Santa Lucia Range: ridge west of upper Lopez Canyon (*8928*); east side of Pine Mt. (*7903*). The species is somewhat more common in the higher parts of the Santa Lucia Mts. in Monterey Co.

20. **L. austromontanus** Heller. Atascadero, in stony soil around oaks (*6773*). This plant closely resembles the type collection of *L. austromontanus* from Tehachapi, Kern Co. Although it seems distinct from plants of *L. albifrons* occurring in the same area, it may be preferable to classify it, following Jepson (Fl. Cal. 2: 252), as a variety of *L. albifrons*.

21. **L. albifrons** Benth. Common in all except the most arid sections (except for

the sandy coastal area occupied by *L. Chamissonis*), on rocky or sandy hills and in flood-beds of streams. Reaches the coast near San Carpoforo Creek, where it is associated with a form of *L. arboreus*. Highly variable. Varieties can be distinguished only by emphasizing a single feature and ignoring the general appearance of the plant. Some of the shrubs near the western edge of Carrizo Plain are more green than silvery and so may be keyed out to *L. longifolius* (Wats.) Abrams, but I do not regard these shrubs as constituting a separate entity in this area. In most localities the plants are shrubs, but a form in which the woody stems spread out along the ground instead of growing upward is common in the Santa Lucia Mts., extending at least as far southward as the vicinity of San Luis Obispo. This variant is the basis for reports of *L. albifrons* var. *collinus* Greene in San Luis Obispo Co., but the typical form of that variety as found in the San Francisco Bay region differs from our plants in its short, spreading, densely leafy stems and short racemes on short peduncles. The following varietal names apply primarily to rather low-growing plants, woody only near the base, but there is no necessary correlation between growth-form and their distinguishing features.

Var. **Douglasii** (Agardh) C. P. Smith. Distinguishable only by having relatively long bracts, which are conspicuous when the flowers are in bud. Plants showing this feature occur along the coast northward and have been found in the Santa Lucia Mts. at least as far south as the ridge southeast of Cuesta Pass (*8648*). At this locality it was observed that the length of the bracts varies considerably from plant to plant in a single colony, and even on the same plant. In other words, the variety is rather arbitrary and not clearly defined.

Var. **Abramsii** (C. P. Smith) Hoover. Marked by having the hairs on the leaflets spreading rather than appressed; that is, the pubescence is woolly instead of silky. Restricted to the Santa Lucia Mts.: near Rocky Butte Fire Lookout (*9062*). This variant intergrades completely with the low woody-based form of *L. albifrons* which is common in the Santa Lucias.

22. **L. Chamissonis** Esch. Locally abundant in coastal sands around Morro Bay and from Pismo Beach southward. In the Los Osos the flowers are bright blue and form a beautiful display every spring. Most of the shrubs south of Pismo Beach have pale flowers and so are less attractive.

23. **L. arboreus** Sims. Common along the coast, usually in sand but sometimes on rocky slopes. There is some geographic segregation of the various forms of this species found in San Luis Obispo Co., but each of the forms is apparently represented here and there far to the north in California, if not beyond. South of Pismo Beach the flowers are all yellow and the leaves, being rather sparsely hairy, are bright green. A few shrubs of this yellow-flowered form grew for several years on roadbanks at the summit of Cuesta Pass, perhaps as the result of an intentional or an accidental introduction. From Arroyo Grande northward at least to Morro Bay, the shrubs have flowers in some shade of violet or blue. Intergrades between the color forms have been noticed near Arroyo Grande. On the steep sandy cliffs near Hazard Canyon, the plants, with violet flowers in that locality, are hardly shrubby but sprawl over the ground, and the flowers in each raceme are few. From Cambria northward to the Monterey Co. line, the species is represented entirely by shrubs in which the young growth is densely silvery-silky, much as in *L. albifrons*. The wing petals of this form are prevailingly light blue, and the banner white with yellow center, often shading into blue toward the margins. This "silvery" form has a dis-

tinctive aspect, but it is closely approached by some shrubs in the Los Osos district, as well as being duplicated apparently at several points farther north along the California coast.

3. Cytisus L.

1. **C. monspessulanus** L. French Broom. Occasionally established on openly wooded hills near coast: Cambria; north of Arroyo Grande.

4. Thermopsis R. Br.

1. **T. macrophylla** H. & A. Occasional on clay or rocky hillsides in western part, usually around springs or in beds of streamlets, but after once established thriving in quite dry places; often on serpentine: head of Perfumo Canyon in San Luis Range; just east of San Luis Obispo; upper Salinas valley between Santa Margarita and Pozo.

5. Medicago L.

Stems erect; petals violet-purple ....1. *M. sativa.*
Stems spreading to prostrate; petals yellow.
    Leaflets not spotted.
        Pods strongly curved but coiled, never spiny; usually perennial, with tough-fibrous root ....2. *M. lupulina.*
        Pods coiled, usually spine-margined; annuals with slender root.
            Plant glabrous or nearly so; stipules laciniate .....3. *M. polymorpha.*
            Plant soft-hairy; stipules entire ....4. *M. minima.*
    Leaflets each with central brown spot ....5. *M. arabica.*

1. **M. sativa** L. Alfalfa. Often coming up spontaneously by roadsides and on borders of fields, but rarely if ever becoming established in natural habitats.

2. **M. lupulina** L. Black Medick. Rather common as a weed of cultivation, especially in lawns; sometimes established in low valleys along streams, as in lower Lopez Canyon.

3. **M. polymorpha** L. Bur Clover. Very common as a garden weed and also spreading into open woods and open rangeland, especially in the areas of higher rainfall near the coast. Plants having reduced spines on the pods have been called *M. hispida* var. *confinis*; this form has rarely been found here.

4. **M. minima** (L.) Grufb. var. **pubescens** Webb. Plentiful locally in lower Lopez Canyon (*8324* in 1953) and probably elsewhere.

5. **M. arabica** (L.) Huds. Spotted Medick. Observed in Lopez Canyon in 1953 and in 1967, but no collection made.

6. Melilotus Juss.

Petals white ....1. *M. albus.*
Petals yellow.
    Flowers 5 to 7 mm. long ....2. *M. officinalis.*
    Flowers 2 to 3 mm. long ....3. *M. indicus.*

1. **M. albus** Desr. White Sweet-Clover. Particularly common in stream-beds, sometimes on road-banks and in seepage areas, widely distributed in both coastal and interior districts.

2. **M. officinalis** (L.) Lam. Sparingly introduced in eastern part of county: Gillis Canyon (*Twisselmann 2898*) and reportedly in lower Cammatta Canyon. A garden weed at Dorothy Twisselmann Ranch (*Twisselmann 5975* in 1960).

3. **M. indicus** (L.) All. Yellow Sweet-Clover. Common weed, especially in gardens and in cultivated ground generally, abundant near coast; in scattered localities toward the interior.

7. **Trifolium** L. Clover

Heads without involucre (in *T. Macraei* and *T. hirtum* closely subtended by enlarged bases of upper leaves).
 Calyx not inflated in age; heads without specialized sterile structures on top.
  Calyx-teeth glabrous or ciliate; flowers pedicelled, reflexed after anthesis.
   Plants annual from a slender tap-root; petals pink to rose-purple.
    Peduncles pilose ....1. *T. bifidum.*
     Leaflets deeply 2-lobed ....var. *bifidum.*
     Leaflets shallowly 2-lobed to merely notched at apex.
              var. *decipiens.*
    Peduncles glabrous.
     Calyx-teeth entire ....2. *T. gracilentum.*
     Calyx-teeth denticulate-ciliate ....3. *T. ciliolatum.*
   Plants perennial, the stems creeping; petals white ....4. *T. repens.*
  Calyx-teeth plumose; flowers sessile, not reflexed.
   Heads on long peduncles.
    Corolla distinctly longer than calyx.
     Heads cylindric; corolla red ....5. *T. incarnatum.*
     Heads ovoid; corolla purple, white-tipped 6. *T. dichotomum.*
    Corolla equalling calyx or shorter.
     Corolla equalling calyx ....7. *T. albopurpureum.*
     Corolla shorter than calyx, often much shorter
           8. *T. columbinum.*
   Heads closely subtended by enlarged bases of uppermost leaves.
    Heads solitary, 1.5 to 2.5 cm. in diameter ....9. *T. hirtum.*
    Heads mostly in pairs, not over 1.5 cm. in diameter 10. *T. Macraei.*
 Calyx-tube inflated in age, the lobes filiform; fruiting heads covered on top with reflexed branched structures (sterile pedicels or abortive flowers?)
           11. *T. subterraneum.*
Heads with an involucre which is usually lobed (in *T. depauperatum* a narrow unlobed disk).
 Involucre cup-shaped to bowl-shaped, usually lobed about to middle.
  Lobes of involucre toothed.
   Calyx-teeth longer than tube, plumose.
    Corolla shorter than calyx or barely equalling it, dark purplish red
           12. *T. barbigerum.*
    Corolla longer than calyx, bicolored, purple and white or lilac
    and cream-color ....13. *T. Grayi.*
   Calyx-teeth shorter than tube, denticulate-ciliate ....14. *T. microdon*
  Lobes of involucre entire ....15. *T. microcephalum.*
 Involucre flat or saucer-shaped, usually lobed nearly to base (except in *T. depauperatum*).
  Corolla not inflated after anthesis.
   Calyx-teeth gradually narrowed upward, triangular-subulate.
    Perennial; petals typically rose-purple ......16. *T. Wormskjoldii.*

Annual; petals usually dark purple (paler in *T. oliganthum*) and white.
  Corolla pale purple, only slightly exceeding the calyx
                                    17. *T. oliganthum.*
  Corolla dark purple, white-tipped, well exceeding the calyx
                                    18.. *T. variegatum.*
Calyx-teeth broad at base, abruptly narrowed upward and often with a tooth on each side.
  Herbage glabrous; petals rose-purple .... 19. *T. tridentatum.*
  Herbage glandular-pubescent; petals white .. 20. *T. obtusiflorum.*
Corolla inflated after anthesis.
  Involucre deeply lobed or of distinct bracts.
    Heads 2 to 4 cm. in diameter; petals creamy-white 21. *T. fucatum.*
      Lower calyx-teeth not exceeding calyx-tube .... var. *fucatum.*
      Lower calyx-teeth much longer than calyx-tube var. *flavulum.*
    Heads 1 cm. in diameter or less; petals white to purplish red
                                    22. *T. amplectens.*
  Involucre a narrow disk .... 23. *T. depauperatum.*

1. **T. bifidum** Gray var. **bifidum.** Occasional plants (near Cambria and southeast of Santa Margarita) seem referable to the typical form of the species; which is, however, rare as compared with the following variety.

Var. **decipiens** Greene. Frequent, usually in open woods, near coast, extending east at least to Salinas River.

2. **T. gracilentum** T. & G. Common throughout on open hills, or on plains in the interior. Plants under dry conditions are notably small. This size difference, together with a calyx almost as long as the corolla, is the basis for var. *inconspicuum* Fernald, which is not here regarded as warranting a separate name. The relative length of the calyx and corolla is not correlated with the size of the plant in San Luis Obispo Co. material, and neither feature is restricted to a particular geographic region.

3. **T. ciliolatum** Benth. Frequent on open or wooded hillsides and on fertile plains, widespread, but apparently absent from the more arid portions of the eastern part.

4. **T. repens** L. WHITE CLOVER. Often included in seed mixtures for lawns. Where grass alone is desired, the clover may be regarded as a weed, but here it shows hardly any tendency to spread beyond the lawns where it is sown.

5. **T. incarnatum** L. CRIMSON CLOVER. Grown as a hay crop near San Suis Obispo and apparently persisting after cultivation. It is often taken by students for a wild plant but probably never becomes permanently established.

6. **T. dichotomum** H. & A. Occasional on bare rocky or clay slopes in Santa Lucia Range: 2 miles northeast of Chimney Rock; Pine Top Mt.; road to Hi Mt. from Pozo-Arroyo Grande summit (*8772*, the southernmost known occurrence).

7. **T. albopurpureum** T. & G. Common in uncultivated areas throughout the interior, and occasional near the coast. The customary interpretation of *T. albopurpureum, T. dichotomum,* and *T. columbinum* (= *T. olivaceum*) is here followed. Actually, however, a continuous series can be traced from *T. dichotomum* with the largest corollas to *T. columbinum* with the smallest.

8. **T. columbinum** Greene. DOVE CLOVER. At scattered localities in upper Salinas Valley and La Panza Range. Jepson in 1901 first included *T. columbinum* and *T.*

olivaceum Greene in one species under the former name. The subsequent use of the name *T. olivaceum* for the inclusive species is therefore incorrect. Many of our plants have corollas intermediate in size between typical *T. columbinum* and *T. albopurpureum*. This form has been called *T. olivaceum* var. *griseum* Jepson.

9. **T. hirtum** All. Rose Clover. Sown for pasture, and for revegetation after fire. The plants may reseed themselves spontaneously to a limited extent but have not been observed to spread: Shandon, *Twisselmann 4545*.

10. **T. Macraei** H. & A. Occasional in sands or clays near coast, at least from Cambria northward.

11. **T. subterraneum** L. Cambria, in open pasture near ocean bluff (*8503*). Although the place where the plants were found was not under cultivation at the time, it is doubtful that the occurrence was a natural one.

12. **T. barbigerum** Torr. In moist places near coast from Cambria northward, locally plentiful in years of good rainfall; Price Canyon (*7849*), dwarf plants with very small heads, but otherwise showing the features of this species.

13. **T. Grayi** Loja. Moist soil, upper Salinas Valley from Atascadero to Pozo; also near the coast in Price Canyon.

14. **T. microdon** H. & A. Locally rare, mainly in stiff clay soils, near coast: Cambria to vicinity of San Luis Obispo.

15. **T. microcephalum** Pursh. Common, mostly in moist places, well distributed over most of the area but apparently absent from the most arid portions of the interior.

16. **T. Wormskjoldii** Lehm. In wet places, mainly near the coast; also at Atascadero and to be expected elsewhere in the upper Salinas Valley.

17. **T. oliganthum** Steud. San Luis Obispo and probably elsewhere in hill country of western part. Apparently rare, but more likely overlooked, as it is one of the less conspicuous clovers.

18. **T. variegatum** Nutt. Common in moist soils near coast, less so in Salinas Valley eastward to La Panza Range. More than in most of the clovers, the flower-heads of this species vary in size, so that, if abundant intermediates did not exist, one would surely conclude that the extremes were entirely different kinds of plants.

19. **T. tridentatum** Lindl. Frequently plentiful in "grassland" areas and in open woods, well distributed throughout the county, particularly common in Santa Lucia Mts.

20. **T. obtusiflorum** Hook. Creek Clover. Occasional along streams and around springs in Santa Lucia Mts. and upper Salinas basin: Calf Canyon; Wilson Canyon creek; west of Pozo; Lopez Canyon.

21. **T. fucatum** Lindl. var. **fucatum**. Coastal field west of Cambria (*10,570*).

Var. **flavulum** (Greene) Jepson. Locally common in clay soils, usually where moist in spring, in coastal area as far south as San Luis Obispo, throughout Santa Lucia Range, and in upper Salinas Valley; rare near Cholame. Most of our plants have the lower calyx-teeth elongated and simple, but sometimes one of the calyx-teeth is forked. This condition represents var. *Gambelii* (Nutt.) Jepson. Because var. *Gambelii* is merely a sporadic variant within the range of var. *flavulum,* it should, I believe, be included under the same name.

22. **T. amplectens** T. & G. In "grasslands" or open woods, common near coast, of more scattered occurrence in Salinas River watershed; apparently absent from eastern part.

23. **T. depauperatum** Desv. Known only in eastern part: mainly on Carrizo Plain, where locally abundant north of Soda Lake. This species and *T. amplectens,* which is closely similar, have been noticed growing together only in one place, 18 miles east of Creston on the La Panza Road. At this locality, strangely, *T. depauperatum* had deep red flowers and *T. amplectens* white flowers. Elsewhere the corollas of both species may be red, white, or intermediate shades.

8. Lotus L.

Perennials with woody or tough-fibrous roots, sometimes with lower part of stems woody.
    Stipules leaf-like to membranous in texture.
        Stipules similar to leaflets . . . . .1. *L. corniculatus.*
        Stipules membranous, much smaller than leaflets.
            Coarse plant with stiff erect or suberect stems; petals dull greenish yellow tinged with dark red or purple . . . . .2. *L. crassifolius.*
            Stems weak, trailing; banner bright yellow; wings rose-pink to white (deeper colored in drying) . . . . .3. *L. formosissimus*
    Stipules reduced to a minute prickle-like point or gland.
        Rachis of leaves 15 to 30 mm. long; leaflets mostly 7 to 9 . .4. *L. grandiflorus*
        Rachis of leaves less than 15 mm. long; leaflets 3 to 6.
            Calyx-teeth straight or sometimes curved outward, never recurved; petals yellow, sometimes turning reddish in age.
                Calyx-teeth subulate, or dilated at base only.
                    Herbage with spreading hairs, at least when young
                                    5. *L. Heermannii.*
                    Herbage at first villous, then glabrate; flowers 3 to 5 mm. long . . . . .var. *Heermannii.*
                    Herbage densely white-villous; flowers (in the common coastal form) 5 to 7 mm. long . . . . .var. *eriophorus.*
                Herbage glabrous to strigose.
                    Herbage densely strigose; flowers 6 to 8 mm. long
                                    6. *L. procumbens.*
                    Herbage glabrate (except in var. *Veatchii*); flowers 8 to 12 mm. long . . . . .7. *L. scoparius.*
                      Umbels sessile.
                          Herbage early glabrate . . . . .var. *scoparius.*
                          Herbage persistently appressed-pubescent
                                  var. *Veatchii.*
                    Umbels peduncled . . . . .var. *dendroideus.*
              Calyx-teeth triangular, always very short, obtuse or acute.
                Calyx-teeth acute, separated by rounded sinuses; pod partly exserted, its beak becoming coiled into a circle 8. *L. Biolettii.*
                Calyx-teeth obtuse or mucronulate, separated by narrow acute sinuses; beak of pod exserted, bent where emerging from calyx, the pods otherwise straight . . . . .9. *L. junceus.*
         Calyx-teeth recurved; petals white and rose-red . . . . .10. *L. Benthamii.*
Annuals with slender tap-root.
    Petals yellow, sometimes turning reddish in age.
        Flowers solitary or in small peduncled umbels of 2 to 5 flowers; pods straight, dehiscent.
            Flowers in peduncled umbels, or if solitary on a well-developed peduncle and usually with a bractlet.

Larger leaves with linear or oblanceolate leaflets mostly less than 1 cm. long (smaller leaves often with broader leaflets); wings much longer than keel; style hairy at base of stigma.

Leaflets slightly fleshy; flowers mostly in umbels of 2 or 3; seeds cubical . . . . 11. *L. strigosus.*

Leaflets markedly fleshy; flowers (locally) mostly solitary; seeds round . . . . 12. *L. tomentellus.*

Larger leaves with obovate leaflets 1 cm. long or more (smaller leaves often with narrower leaflets); wings slightly longer than keel; style glabrous . . . . 13. *L. salsuginosus.*

Flowers solitary, subsessile, without a bractlet.

Herbage sparsely strigose or nearly glabrous; calyx-teeth about equalling tube . . . . 14. *L. subpinnatus.*

Herbage densely hairy; calyx-teeth about twice as long as tube

15. *L. humistratus.*

Flowers several in sessile or subsessile umbels; pods curved, with beak hooked at tip, indehiscent . . . . 16. *L. hamatus.*

Petals white or pink.

Calyx-teeth shorter than tube; spring-flowering . . . . 17. *L. micranthus.*

Calyx-teeth longer than tube; summer-flowering . . . . 18. *L. Purshianus.*

1. **L. corniculatus** L. BIRD'S-FOOT TREFOIL. Escaped from cultivation along roadsides and in low moist places from Cambria to near San Luis Obispo.

2. **L. crassifolius** (Benth.) Greene. "Coast Ranges from San Luis Obispo Co. north," according to Munz (Cal. Fl. 844). No specimens confirming this report have been seen; but, as the species is known to be present in Monterey Co. and in Santa Barbara Co., its occurrence here is probable.

3. **L. formosissimus** Greene. In winter-wet places along the coast from Cambria northward.

4. **L. grandiflorus** (Benth.) Greene. Locally common in La Panza Range and hills to the west, in decomposing granite.

5. **L. Heermannii** (D. & H.) Greene var. **Heermannii.** The plants which differ most widely from var. *eriophorus* were found among the sand-dunes near Oso Flaco Lake (*9447*). Unfortunately, no specimens from the type locality (Tejon Pass, Los Angeles Co.) are available, so it is impossible to say whether typical *L. Heermannii* is of this extreme form or something more closely resembling var. *eriophorus*. All other specimens from the county, both coastal and interior, are here referred to var. *eriophorus* because the herbage is uniformly densely villous, although they vary in other respects.

Var. **eriophorus** (Greene) Ottley. Common in sandy soils along the coast. This coastal form represents the opposite extreme from *9447* cited above not only in being densely white-villous but also in having larger flowers, with especially long prominent calyx-teeth. At a few scattered localities in the interior are found plants which are as conspicuously villous (therefore included under var. *eriophorus*) but which have smaller flowers: 1.2 miles west of Bee Rock (*Twisselmann 3879*); upper Toro Creek north of Pozo (*7089*). Whether these small-flowered but densely hairy plants (rather than *9447*) represent the true *L. Heermannii* is yet to be determined.

6. **L. procumbens** (Greene) Greene. Occasional in dry places, mainly in hard gravelly soil, in Salinas River basin: hills west of Paso Robles, eastward and southward to La Panza Range.

7. **L. scoparius** (Nutt.) Ottley. DEER-WEED. Very common in sandy places near coast, and in hilly areas of both eastern and western parts; extending inland to Red Hills and Temblor Range, according to Twisselmann. Particularly abundant after brushy land has been burned or otherwise cleared. The species is in some respects highly variable (size and form of growth, number of flowers in a cluster, size of flowers, etc.).

Var. **Veatchii** (Greene) Ottley. Some plants along the coastline near Morro Bay represent intergrades toward this variety. The foliage is persistently appressed-hairy, not only on the young growth. No local specimens, however, approach the type collection (from Baja California) in degree of hairiness.

Var. **dendroideus** (Greene) Ottley. Plants in a few localities (near Morro Bay; ridge northwest of Cuesta Pass) have short-peduncled umbels and thus belong to (or approach) this variety, which occurs mainly on the southern California islands. A shrub with conspicuously peduncled umbels grew 5 miles east of Creston (*9826*), mixed with plants having the umbels sessile or occasionally short-peduncled. This occurrence is remarkable because the locality is remote from the coast.

8. **L. Biolettii** Greene var. **spiralis** Hoover. Chaparral areas in western part, known in only a few places but generally plentiful where it occurs: ridge southeast of Cuesta Pass; See Canyon; Oak Park district north of Arroyo Grande. One of the plants which are specially adapted to the conditions prevailing after fire.

9. **L. junceus** (Benth.) Greene. Common in rocky places from summit of Santa Lucia Mts. to the sea. This species and *L. scoparius* often grow side by side, yet in some places they show a notable difference in choice of habitat. So far as has been observed, *L. junceus* is absent from the coastal dunes and sand-plains where *L. scoparius* is most abundant. In the immediate vicinity of San Luis Obispo, though not elsewhere, *L. junceus* is restricted to serpentine.

10. **L. Benthamii** Greene. Frequent along coast from Cambria northward, mainly in rocky places, extending inland a few miles into Santa Lucia Mts. (Chris Flood Creek).

11. **L. strigosus** (Nutt.) Greene. In sand, gravel, or broken rock, scattered over virtually the entire county; commonest in Salinas River drainage.

12. **L. tomentellus** Greene. Bare sandy slopes, Temblor Range near north end of Elkhorn Plain (*10,305*). Hardly more than a geographic race of *L. strigosus*.

13. **L. salsuginosus** Greene. Mostly on freshly exposed clay or rocky slopes, less commonly in sandy soil, occasional from Monterey Co. line to San Luis Obispo, upper Arroyo Grande, southern Temblor Range, and Cuyama Valley. Apparently absent from central portion of county.

14. **L. subpinnatus** Lag. Common throughout the county in various soils, mostly in treeless and shrubless areas.

15. **L. humistratus** Greene. Frequent in hilly or mountainous areas in both eastern and western parts, in bare gravelly places and around rock outcrops.

16. **L. hamatus** Greene. Hard, dry gravelly or stony soil, locally rare: near Nacimiento River; near north end of La Panza Range; Stoney Creek.

17. **L. micranthus** Benth. Frequent in open woods in western part, inland to Salinas Valley.

18. **L. Purshianus** (Benth.) Clements & Clements. Common in dry places, mainly in sandy soils, coastal area eastward to La Panza Range. Not definitely known in driest portions of interior.

9. **Psoralea** L.

Leaflets 3 only.
    Stems erect or ascending, with several leaves, the peduncles arising from the upper axils.
        Plant usually less than 6 dm. tall; petals cream-white with purple shadings
                                                    1. *P. physodes.*
        Plant usually more than 1 m. tall; petals violet-purple . . 2. *P. macrostachya.*
    Leaves and peduncles arising from creeping rhizomes . . . . 3. *P. orbicularis.*
Leaflets more than 3 . . . . 4. *P. californica.*

    1. **P. physodes** Dougl. Cᴀʟɪꜰᴏʀɴɪᴀ Tᴇᴀ. Wooded hills from coast eastward to Salinas River, but of scattered occurrence and remarkably scarce in this area. The dried leaves are said to have been used to make an aromatic drink.

    2. **P. macrostachya** DC. Lᴇᴀᴛʜᴇʀ-Rᴏᴏᴛ. Frequent in stream-beds, usually among rocks, near coast: Santa Rosa Creek; San Bernardo Creek; Steiner Creek near San Luis Obispo. The long tough roots were a source of fiber for California Indians.

    3. **P. orbicularis** Lindl. Frequent in moist places near coast and in Santa Lucia Range.

    4. **P. californica** Wats. Bare rocky or clay slopes: not rediscovered in San Luis Obispo Co. since Dr. Edward Palmer long ago collected the type at "McGinnis' Ranch near head of Salinas River 25 miles from San Luis Obispo." Maps show a McGinnis Creek on the east slope of the La Panza Range, which would seem a likely locality for the occurrence of this species.

10. **Amorpha** L.

    1. **A. californica** Nutt. Locally common, mostly in open woods in decomposed granite, in region of Salinas River: east of Santa Margarita to vicinity of Pozo.

11. **Robinia** L.

    1. **R. Pseudo-Acacia** L. Lᴏᴄᴜsᴛ. Commonly planted in towns and around farmyards; apparently becoming established and spreading along streams: Salinas River at Paso Robles; California Polytechnic Campus, San Luis Obispo.

12. **Astragalus** L. Rᴀᴛᴛʟᴇᴡᴇᴇᴅ

Annuals; flowers in short spikes or subsessile in crowded racemes; pods small, neither inflated nor flattened, 2-celled, each cell 1-seeded.
    Pods deflexed, only the base enclosed by the calyx . . . . 1. *A. Gambelianus.*
    Pods ascending, mostly enclosed by the calyx . . . . 2. *A. didymocarpus.*
        Stems mostly spreading; flowers 3 to 5 mm. long . . . . . subsp. *didymocarpus.*
        Calyx black-hairy; herbage sparsely strigose or subglabrous.
            Calyx 2.3 to 4 mm. long . . . . var. *didymocarpus.*
            Calyx 4.5 to 5 mm. long . . . . var. *daleoides.*
        Calyx white-hairy; herbage densely strigose . . . . var. *dispermus*
        Stems mostly erect; flowers 7 to 9 mm. long . . . . subsp. *Milesianus.*
Perennials; flowers with evident pedicels; pods many-seeded, inflated or flattened.
    Pods linear-oblong or narrowly oblong, flattened, not inflated.
        Pods thick-coriaceous, 2-celled, wrinkled, curved upward, with short stout stipe . . . . 3. *A. pachypus.*
        Pods parchment-like, 1-celled, not wrinkled, pendent, with stipe much longer than calyx . . . . 4. *A. Antisellii.*
Pods subglobose-acuminate to semi-ovate, inflated (sometimes in *A. oxyphysus* very flat in age and not inflated).

Pods with a stipe.
    Petals yellowish; stipe 3 times as long as calyx ....5. *A. asymmetricus.*
    Petals whitish; stipe less than 3 times as long as calyx, often not exceeding it.
        Calyx-tube cylindric; pods compressed but also inflated at least when young; arid interior ....6. *A. oxyphysus.*
        Calyx-tube campanulate; pods not compressed; coastal.
            Pods on a "gynophore" from which the body can be rather readily detached when fully ripe ....7. *A. curtipes.*
            Pods with a true stipe from which the body is not readily detachable ....8. *A. leucopsis.*
Pods sessile within the calyx, not stipitate.
    Pods 1-celled.
        Leaflets mostly 25 or more; racemes dense; petals whitish; coastal.
            Herbage usually densely white-villous, sometimes only moderately so; calyx without black hairs ....9. *A. Nuttallii.*
            Herbage green, sparsely strigose or glabrate; calyx with black hairs ....10. *A. pomonensis.*
        Leaflets fewer than 25; racemes rather loose; petals yellowish; interior area.
            Herbage at first closely strigose, usually glabrate; calyx-teeth shorter than tube ....11. *A. Douglasii*
            Herbage ascending-villous, the leaflets in age sometimes sparsely so; calyx-teeth as long as tube ....12. *A. macrodon.*
    Pods 2-celled ....13. *A. lentiginosus.*
        Herbage white-hairy; petals light yellow ....var. *nigricalycis.*
        Herbage green, sparsely strigose; petals violet-purple var. *idriensis.*

1. **A. Gambelianus** Sheldon. Common in grasslands and rocky places, often on serpentine, in western and central portions.

2. **A. didymocarpus** H. & A. var. **didymocarpus.** Common in herbaceous plant communities of the interior, in sandy soils, friable clays, and crumbling shale.

Var. **daleoides** Barneby. Occasional within the area of var. *didymocarpus* in the county: San Miguel; Cholame; La Panza district; west base of Temblor Range to western Kern Co.

Var. **dispermus** (Gray) Jepson. In calcareous sand, east of San Juan River on La Panza-Simmler road (*7615*). This collection is somewhat intermediate toward var. *didymocarpus.* The extreme form of var. *dispermus* (plants densely white-hairy throughout) is not known within our limits.

Subsp. **Milesianus** (Rydb.) Hoover. Clay soils, usually if not always in areas of serpentine, from Morro Bay to San Luis Obispo.

A. didymocarpus var. *obispensis* (Rydb.) Jepson was based on a plant reportedly collected at San Luis Obispo. This record is almost certainly an error resulting from confusion of labels, as stated by Barneby (El Aliso 2: 213).

3. **A. pachypus** Greene. "Dry hillsides, in stiff alkaline clay soils": not yet detected in San Luis Obispo Co. but known in Cuyama Valley, Santa Barbara Co. Included here because of the high probability of finding it on or near Caliente Mt.

4. **A. Antisellii** Gray. Rocky slopes or stream-beds, uncommon in our area: lower end of Rocky Canyon east of Atascadero; Huasna River; north side of lower Cuyama Valley.

5. **A. asymmetricus** Sheldon. In herbaceous plant communities, commonly in

sandy loams: northern part of Temblor Range; Red Hills; Shandon-Creston road; Salinas Valley near Paso Robles, and northward.

6. **A. oxyphysus** Gray. Common on plains and open hills in eastern part: Shandon Hills, southward through region of Carrizo Plain to Cuyama Valley. Twisselmann says that "it has been rigorously controlled, and in some areas exterminated," but, at least until the spring of 1967, plants could be found in abundance.

7. **A. curtipes** Gray. Rocky slopes, most commonly on serpentine, from southern Monterey Co. to vicinity of San Luis Obispo; near lower Cuyama River into Santa Barbara Co. Plants from San Benito Co. have been included in this species but have remarkably large pods, are probably spreading rather than erect, grow in a different climatic area, and so constitute "probably a distinct entity," as Barneby says.

8. **A. leucopsis** (Torr.) T. & G.? A collection from between Suey Creek and Cuyama River (*7593*) keys to *A. trichopodus* var. *lonchus* (a synonym of *A. leucopsis*) in Barneby's monograph of the genus, but differs in no obvious respect from *A. curtipes*. The difference used as a basis for separating the two is very obscure and apparently not geographically consistent.

9. **A. Nuttallii** (T. & G.) J. T. Howell. Ocean bluffs and hollows among dunes: Monterey Co. line southward to near Cambria; vicinity of Morro Bay to Pt. Buchon; from Oceano southward. The species includes an assortment of races, of which some are more or less restricted geographically. Plants found from Cambria northward are low and spreading, generally densely white-hairy, and with exceptionally large pods. Similar plants, though perhaps less conspicuously hairy, occur on the bluffs south of Islay Creek. The pods are smaller in plants found farther south and also formerly among the dunes north of Morro Bay. Plants among the Nipomo Dunes (near Jack Lake) are tall and nearly erect, thus resembling var. *virgatus* of the San Francisco Bay region. (Their hairy leaves exclude them from var. *virgatus*.) Plants in which the stipules are persistently and rather densely hairy may be identified from the current literature as var. *miguelensis* (Greene) Jepson, of the southern California islands. That variety, however, differs from all mainland plants in the degree to which the stems are persistently, finely, and densely white-velvety. The statement that the stipules of "*A. miguelensis*" are "herbaceous" and those of *A. Nuttallii* "scarious" does not correspond to any difference in the texture of the stipules in actual specimens. This species, as well as all others having inflated pods, is often called "rattleweed."

10. **A. pomonensis** Jones. Common in sand around south side of Morro Bay; also on Nipomo Mesa. There is some indication in the latter district of intergradation with *A. Nuttallii*.

11. **A. Douglasii** (T. & G.) Gray. Common in the interior, generally in sandy soils but also sometimes in friable clays: Salinas Valley and eastward; upper Arroyo Grande.

12. **A. macrodon** Gray. Usually in fertile but dry soils derived from decomposed shale, less commonly in sandstone areas, at scattered localities but sometimes locally abundant: vicinity of Paso Robles and Atascadero; foothills northeast of Simmler on west side of Temblor Range and on higher summits of that range; notably plentiful on Caliente Mt.

13. **A. lentiginosus** Dougl. var. **nigricalycis** Jones. Sandy soils or disintegrated shale, southern part of Temblor Range and Carrizo Plain to Cuyama Valley. Twisselmann states that "it is excellent feed for livestock in early spring."

Var. **idriensis** Jones. Grassy areas in higher parts of Temblor Range, apparently uncommon.

13. Glycyrrhiza L. Licorice

1. **G. lepidota** Pursh. Occasional in sandy soil, usually in low valleys and along streams, sometimes on open hillsides: Salinas Valley eastward to La Panza Range; Arroyo Grande to Cuyama River and Santa Maria Valley.

14. Vicia L. Vetch

Flowers few to many in a raceme.
    Plants perennial, sparsely pubescent or apparently glabrous.
        Leaflets mostly fewer than 15; flowers 3 to 8 in a raceme ...1. *V. americana.*
        Leaflets mostly 15 or more; flowers many ....2. *V. gigantea.*
    Plants annual, densely pubescent ....3. *V. benghalensis.*
Flowers solitary or in pairs.
    Flowers subsessile in the axils ....4. *V. sativa.*
    Flowers on slender peduncles ....5. *V. exigua.*

1. **V. americana** Muhl. var. **truncata** (Nutt.) Brewer f. **oblonga** Hoover. Frequent in woods from coast eastward through Santa Lucia Mts. Leaflets oblong, elliptic, or lanceolate, acute to obtuse. The great majority of plants are referable to this *forma.*

Forma **truncata.** Leaflets truncate at apex, mostly relatively broad. Found with *f. oblonga:* Santa Rita Creek west of Templeton, *Hoover 8831* in part.

Forma **angusta** Hoover. Associated with *f. oblonga* at Cambria and doubtless elsewhere. Leaflets linear or narrowly lanceolate, acute or acuminate.

2. **V. gigantea** Hook. Frequent in moist places, usually in woods and thickets, near coast: Black Lake; Oceano; San Luis Range; shores of Morro Bay and northward.

3. **V. benghalensis** L. Purple Vetch. Often escaping from cultivation and becoming a weed along roadsides and in low ground generally, particularly abundant from Cambria to San Luis Valley.

4. **V. sativa** L. Common Vetch. Common weed in cultivated ground in western part, often spreading into undisturbed areas in woods and grasslands.

5. **V. exigua** Nutt. California Vetch. Frequent on wooded hills in western part, and in fertile soils (mostly clay) in interior.

15. Lathyrus L.

Native perennials in wooded and brushy areas; leaflets 6 or more; flowers 2 cm. long or less.
    Stems not wing-margined ....1. *L. vestitus.*
    Stems distinctly wing-margined on the angles ....2. *L. Jepsonii.*
Introduced annuals by roadsides and in old fields; leaflets (in addition to tendrils) 2 only; flowers more than 2 cm. long.
    Herbage pubescent; leaflets broadly elliptic or ovate ....3. *L. odoratus.*
    Herbage glabrous; leaflets narrow ....4. *L. tingitanus.*

1. **L. vestitus** Nutt. Among trees, and in relatively moist brushy areas, very common near coast, of more scattered occurrence inland to La Panza Range. Several specimens have been named by C. L. Hitchcock subsp. *puberulus,* which seems

hardly different from typical *L. vestitus* and is certainly not distinctive enough to rank as a subspecies. An occasional plant which is nearly glabrous corresponds to the inconsequential form called *L. Bolanderi* Wats. (Chorro Creek, *Eastwood 16,860*; "Poly Canyon," *Hoover 6616*).

2. **L. Jepsonii** Greene var. **californicus** (Wats.) Hoover. In woods, usually along streams, Salinas Valley near Templeton and Paso Robles.

3. **L. odoratus** L. Sweet Pea. Sometimes escaping from cultivation near fields where grown for seed, especially near Morro Bay and San Luis Obispo, and continuing to appear spontaneously at least for several years.

4. **L. tingitanus** L. Tangier Pea. Probably a garden escape, rare and found only near coast: weedy roadside, Cambria, *Howell 40,813* in 1964.

Geraniaceae. Geranium Family

Calyx without a spur; pedicels not appearing jointed.
    Fertile stamens 10 . . . . .1. *Geranium.*
    Fertile stamens 5 . . . . .2. *Erodium.*
Calyx with a slender hollow spur fused with the pedicel, the pedicel therefore appearing jointed at base of spur . . . . .3. *Pelargonium.*

1. Geranium L. Cranesbill

Annuals with slender tap-root; petals about equalling sepals.
    Lobes of leaves acute; petals rose-purple . . . . .1. *G. dissectum.*
    Lobes of leaves mostly obtuse; petals pink . . . . .2. *G. carolinianum.*
Perennial with stout fleshy tap-root; petals longer than sepals . . . .3. *G. glabratum.*

1. **G. dissectum** L. Common near coast, at least from San Luis Obispo northward, and occasional in Santa Lucia Mts., usually in moist places.

2. **G. carolinianum** L. "San Luis Obispo Co.," according to Jepson in Fl. Cal. It is, in any event, much less common in this region than *G. dissectum*.

3. **G. glabratum** (Hook.) Small. Near Piedras Blancas Point, in swampy area (*6960*), and probably elsewhere in wet places near coast. California collectors have called similar plants *G. retrorsum* L'Hér., but the plants do not conform to published descriptions of that species.

2. Erodium L'Hér.

Leaves simple, palmately veined, the blades cordate at base, little if at all longer than wide.
    Stems leafy above base; leaves ovate in outline, 3-lobed . . . . .1. *E. texanum.*
    Peduncles and leaves all arising from base; leaves round in outline, crenate or shallowly lobed . . . . .2. *E. macrophyllum.*
Leaves pinnate or pinnately lobed, the blades much longer than wide.
    Leaves simple, except sometimes the upper ones; beak of fruit more than 5 cm. long.
        Concavities at top of carpel subtended by 2 folds, the upper part of carpel glabrous; beak 9 to 12 cm. long . . . . .3. *E. Botrys.*
        Concavities at top of carpel subtended by 1 fold, the upper part of carpel pubescent; beak 5 to 9 cm. long .4. *E. obtusiplicatum.*
    Leaves pinnate; beak of fruit less than 5 cm. long.
        Stems and petioles pale green; leaflets serrate-incised . . . . .5. *E. moschatum.*
        Stems and petioles usually reddish; leaflets deeply lobed or divided
                                        6. *E. cicutarium.*

1. **E. texanum** Gray. Sandy soil: San Miguel (*Chester Dudley* in 1935 and 1937). To be expected also in eastern part of county, since it is known in the hills along the western margin of the San Joaquin Valley.

2. **E. macrophyllum** H. & A. Frequent on open hillsides in the interior, generally in friable (probably calcareous or gypseous) clay soils.

3. **E. Botrys** (Cav.) Bertol. In soils ranging from sand to gravelly clay, very common over most of interior (but not reported by Twisselmann from Temblor Range or Carrizo Plain), less so near coast.

4. **E. obtusiplicatum** (M., W. & W.) J. T. Howell. Interior portion of county, sometimes locally abundant: Palo Prieto Creek and Red Hills to hills south of Creston. Probably more widely established and continually spreading.

5. **E. moschatum** (L.) L'Hér. Widespread as a weed: common near coast (though less so than *E. cicutarium*); at scattered places in interior.

6. **E. cicutarium** (L.) L'Hér. Filaree. Very plentiful everywhere except in heavily wooded or densely brush-covered areas. No other introduced species except *Bromus rubens* has proved so widely adaptable to varying conditions. Although an objectionable weed in gardens, filaree is considered an excellent forage plant.

3. Pelargonium L'Hér.

Petals showy, the longer over 1 cm. long ....1. *P. capitatum.*
Petals inconspicuous, the longer three 5 to 6 mm. long ....2. *P. grossularioides.*

1. **P. capitatum** (L.) Ait. Escaped from cultivation at Cambria (*Howell 36,505*).
2. **P. grossularioides** (L.) Ait. Occasional weed in gardens and by roadsides: Cambria; San Luis Obispo.

Oxalidaceae. Oxalis Family

Stems arising from a corm, not forming stolons but bearing numerous small corms underground; petals 15 to 20 mm. long ....1. *O. Pes-caprae.*
Stems at first from a tap-root, later spreading and forming stolons which root at the nodes; petals 4 to 12 mm. long.
Hairs on stems and pedicels upwardly appressed; petals 4 to 8 mm. long
2. *O. corniculata.*
Hairs on stems reflexed, on pedicels spreading; petals 8 to 12 mm. long
3. *O. pilosa.*

1. **O. Pes-Caprae** L. Bermuda Buttercup. Common weed near coast, as at Cambria, Morro Bay, San Luis Obispo, and Arroyo Grande.

2. **O. corniculata** L. Yellow Wood-Sorrel. Common as a pernicious weed in gardens and lawns. A considerable proportion of the plants have purple leaves and so are var. *atropurpurea* Planch.

3. **O. pilosa** Nutt. Woods or open hillsides, from coast into Santa Lucia Mts. Not all specimens of this native species are readily distinguishable from the European weed *O. corniculata*. Both show a tendency for the stems to take root at the lower nodes and spread horizontally.

Tropaeolaceae. Nasturtium Family

1. Tropaeolum L.

1. **T. majus** L. Nasturtium. Occasionally naturalized along creek-banks and at bases of rocky cliffs near the coast: Cambria; San Luis Obispo; near Avila; Pismo Beach.

Linaceae. FLAX FAMILY

1. Linum L. FLAX

Petals 10 to 15 mm. long, blue, rarely varying to white; styles 5. . . . . . .1. *L. Lewisii.*
Petals 2 to 3 mm. long, white or pale pink; styles 3 . . . . .2. *L. micranthum.*

1. **L. Lewisii** Pursh. BLUE FLAX. Rare in hills bordering Salinas Valley: 4 miles east of Templeton (*Ferris 9754*) and Atascadero. To be expected on Caliente Mt., as it occurs in Kern Co. on the road to Mt. Abel.

2. **L. micranthum** Gray. Occasional on rocky wooded or chaparral-covered hills, in our region generally in areas of serpentine: Santa Lucia Range and foothills; Perfumo Canyon in San Luis Range.

Zygophyllaceae. CALTROPS FAMILY

1. Tribulus L.

1. **T. terrestris** L. PUNCTURE VINE. Occasional weed along roads or railways and in sandy fields: San Luis Obispo; between Edna and Pismo Beach; Salinas Valley and eastward; Cuyama Valley.

Simaroubaceae. QUASSIA FAMILY

1. Ailanthus Desf.

1. **A. altissima** (Miller) Swingle. TREE OF HEAVEN. Persisting from previous planting at scattered places through the interior; noticeably spreading only along Salinas River at Paso Robles.

Polygalaceae. MILKWORT FAMILY

1. Polygala L.

1. **P. californica** Nutt. North-facing slopes near coast, southward at least to Arroyo de la Cruz and probably farther south.

Euphorbiaceae. SPURGE FAMILY

Staminate flowers with calyx; flowers not in a cyathium.
    Small plants with simple entire leaves.
        Herbage with spreading bristly hairs as well as stellate-pubescent; leaves broadly ovate . . . . .1. *Eremocarpus.*
        Herbage with closely appressed stellate hairs only; leaves oblong-elliptic or narrower . . . . .2. *Croton.*
    Large shrub with palmately lobed leaves . . . . .3. *Ricinus.*
Flowers without calyx, borne in a cup-like structure (cyathium) . . . . .4. *Euphorbia.*

1. Eremocarpus Benth.

1. **E. setigerus** (Hook.) Benth. TURKEY MULLEIN. In uncultivated ground and fallow fields, and especially abundant by roadsides, commonest toward interior but also present in warmer parts of coastal area.

2. Croton L.

1. **C. californicus** Muell. Arg. Very common in sandy soils near coast. An occasional plant has leaves which are green instead of the usual silvery-white, because

the hairs are more widely spaced. This form is notably plentiful on the dunes south of Pismo Beach.

3. **Ricinus L.**

1. **R. communis** L. Castor Bean. Established and evidently long-persistent at several places near coast, inland to Cuesta Pass.

4. **Euphorbia L.**

Stems erect.
    Leaves all opposite, rounded at base ....1. *E. Lathyrus.*
    Leaves below inflorescence alternate, narrowed toward base.
        Leaves serrulate; capsule tuberculate ....2. *E. spathulata.*
        Leaves entire; capsule not tuberculate.
            Stem without sterile branches; floral leaves distinct; capsule with longitudinal ridges ....3. *E. Peplus.*
            Stem usually with sterile branches below, which bear crowded leaves; floral leaves connate at base; capsule without ridges ...4. *E. crenulata.*
Stems prostrate.
    Herbage glabrous; leaves without red spot.
        Leaves entire ....5. *E. ocellata.*
        Leaves serrulate toward apex ....6. *E. serpyllifolia.*
    Herbage pubescent; each leaf with central red spot ....7. *E. supina.*

1. **E. Lathyrus** L. Caper Spurge. Occasionally appearing as a scarce weed in towns but not noticeably spreading: Morro Bay; San Luis Obispo.

2. **E. spathulata** Lam. Widespread in clay soils of open fields and pastures or treeless hills, both coastal and interior. Found at many localities but usually in small numbers.

3. **E. Peplus** L. Petty Spurge. Common weed in gardens, particularly plentiful at San Luis Obispo.

4. **E. crenulata** L. Under trees or in recently cleared places near coast and in Santa Lucia Mts. In this part of California, this species is, oddly, represented by only one or a few plants in each of the places where it has been found. The facts that some of the plants do not flower in any given year, and that the flowering ones often have sterile shoots below, indicate that *E. crenulata* is a biennial or a short-lived perennial.

5. **E. ocellata** D. & H. Common in summer-dry, usually sandy soil, throughout the interior.

6. **E. serpyllifolia** Pers. Occasional in low ground, usually in or near dried stream-beds, near coast and in Salinas Valley.

7. **E. supina** Raf. A rapidly spreading weed in gardens and irrigated places generally, though apparently not yet widespread here: San Luis Obispo; Twisselmann Ranch. Probably this is the species reported as *E. maculata* L. by Twisselmann in his Temblor Flora.

Callitrichaceae. Water-Starwort Family

1. **Callitriche L.**

Fruits peduncled ....1. *C. marginata.*
Fruits sessile or nearly so ....2. *C. heterophylla.*

1. **C. marginata** Torr. Places which are wet in winter: Cambria; Los Osos Valley; Price Canyon; Carrizo Plain north of Soda Lake. Probably overlooked rather than rare.

2. **C. heterophylla** Pursh. var. **Bolanderi** (Hegelm.) Fassett. Cambria, among pines north of town, on pond in marshy meadow (*9351*); Stoney Creek (*8036,* immature but apparently belonging to this variety).

Limnanthaceae. MEADOW-FOAM FAMILY

1. Limnanthes R. Br.

1. **L. Douglasii** R. Br. var. **nivea** C. T. Mason. MEADOW-FOAM. Locally common in low winter-wet places, upper Salinas Valley between Atascadero and Santa Margarita, to Rinconada Creek, the southern limit for the species.

Anacardiaceae. SUMAC FAMILY

Shrubs or vines; leaves with 3 leaflets . . . . 1. *Rhus.*
Tree; leaves pinnate, with many leaflets . . . . 2. *Schinus.*

1. Rhus L. SUMAC

Upper leaf-surfaces shining; flowers pale green or whitish, pedicelled; fruit white, glabrous . . . . 1. *R. diversiloba.*
Leaves not shining; flowers yellowish, sessile; fruit red, viscid-pubescent
2. *R. trilobata.*

1. **R. diversiloba** T. & G. POISON-OAK. Very abundant in wooded areas and on north-facing slopes near coast; more widely scattered inland to La Panza Range. An isolated occurrence on the east slope of the Temblor Range is probably in Kern Co. In some places, especially near Cambria, a large proportion of the plants are *f. radicans* McNair, an ivy-like climber. Repeated experiences on field trips have shown that a larger proportion of people than is commonly supposed is immune to the injurious effects of contact with this plant.

2. **R. trilobata** Nutt. SQUAW BUSH. Common in interior hills from east slope of Santa Lucia Mts. to La Panza Range.

2. Schinus L.

1. **S. Molle** L. PEPPER TREE. Established and reproducing spontaneously on a hillslope above Highway 101 on the west side of San Luis Obispo.

Aceraceae. MAPLE FAMILY

1. Acer L. MAPLE

Leaves simple, palmately lobed . . . . 1. *A. macrophyllum.*
Leaves compound, with 3 leaflets . . . . 2. *A. Negundo.*

1. **A. macrophyllum** Pursh. BIG-LEAF MAPLE. Common in canyons or on north-facing slopes in Santa Lucia Mts., and locally in San Luis Range (See Canyon and Diablo Canyon).

2. **A. Negundo** L. var. **californicum** (T. & G.) Sarg. BOX-ELDER. Stream banks and bottoms, mainly along Salinas River; also occasionally west of Santa Lucia Range, as along San Luis Creek and Arroyo Grande.

Hippocastanaceae. BUCKEYE FAMILY

1. Aesculus L.

1. **A. californica** (Spach) Nutt. CALIFORNIA BUCKEYE. Just north of summit on Old Creek road; California Polytechnic campus at San Luis Obispo. Both these occurrences are undoubtedly relics from early plantings, but the trees are long established and reproduce spontaneously. The nearest known native buckeyes are on Parkfield Grade, southwestern Fresno Co., and along the upper Nacimiento River, Monterey Co.

Rhamnaceae. BUCKTHORN FAMILY

Flowers green, with inconspicuous petals or none; fruit a berry ....1. *Rhamnus.*
Flowers white to blue or violet, with spoon-shaped petals; fruit a capsule
2. *Ceanothus.*

1. Rhamnus L.

Leaf-blades mostly entire, over 3 cm. long; berries dark purple or black.
    Backs of leaves bright green and glabrous ....1. *R. californica.*
    Backs of leaves finely and closely white-woolly ....2. *R. tomentella.*
Leaf-blades dentate or denticulate, less than 3 cm. long; berries red.
    Shrub less than 2 m. tall, with rigid branches, often thorny; leaf-blades (except on juvenile shoots) denticulate, 10 to 17 mm. long; petioles 1 to 3 mm. long
3. *R. crocea.*

    Large shrub or small tree, often with a distinct trunk, over 2 m. tall when full grown, with flexible branches; leaf-blades sharply dentate, 18 to 30 mm. long; petioles 3 to 8 mm. long ....4. *R. ilicifolia.*

1. **R. californica** Esch. CALIFORNIA COFFEE-BERRY. Common in western part, extending inland to east slope of Santa Lucia Mts., upper Arroyo Grande, and lower Cuyama Canyon.

2. **R. tomentella** Benth. Toward the interior, from east slope of Santa Lucia Mts. to La Panza Range and Cuyama Canyon. In the region of overlap between this and *R. californica,* no intergrades or other evidence of interfertility have been seen. *Rhamnus tomentella* seems identical with *R. californica* except for the finely white-woolly lower surfaces of the leaves, but in the related genus *Ceanothus* even smaller differences are universally accepted as marks of recognition for species, sometimes even in the presence of intergradation. It would be wholly illogical to follow the lead of authors dealing with *Rhamnaceae* who base species in *Ceanothus* on the most trivial and inconstant characters, while at the same time denying emphasis to comparable differences in *Rhamnus.*

3. **R. crocea** Nutt. REDBERRY. Common on rocky hills near coast, extending inland to crest of Santa Lucia Range (northwest of Cuesta Pass), Huasna district, and lower Cuyama Canyon; often a component of chaparral. It may extend farther inland. Twisselmann, while listing "var. *ilicifolia*" separately, says of typical *R. crocea*: "occasional in the chaparral in the Temblor Mountains. A few shrubs grow in lower Palo Prieto Canyon." No specimens in verification of this statement have been seen. Vigorous juvenile shoots have leaves somewhat imitating those of *R. ilicifolia.* Such shoots perhaps are the basis for statements that the two intergrade. In San Luis Obispo Co., at least, *R. crocea* and *R. ilicifolia* are clearly distinct.

4. R. ilicifolia Kell. Holly-Leaf Redberry. Common toward interior except in most arid parts: Santa Lucia Mts. (mostly east slope but also in Arroyo Grande watershed including Lopez Canyon) to La Panza Range; Diablo Canyon in San Luis Range. Often in open woods, and rarely if ever in dense chaparral. Small trees, exceptionally large for the species, occur on the east slope of the Temblor Range in Kern Co.

2. Ceanothus L.

Leaves alternate.
    Leaves on flowering branches with 3 principal veins from base.
        Branchlets not longitudinally ridged or angled.
            Branchlets white; leaves usually glabrous . . . . .1. *C. leucodermis.*
            Branchlets green to brownish; leaves usually appressed-pubescent on lower surface (sparsely so in *C. integerrimus*).
                Evergreen; leaves denticulate . . . . .2. *C. oliganthus.*
                    Branchlets and upper leaf-surfaces somewhat spreading-pubescent . . . . .var. *oliganthus.*
                    Branchlets puberulent or glabrate; upper leaf-surfaces glabrous . . . . .var. *sorediatus.*
                Deciduous; leaves usually entire, sometimes denticulate near apex
                          6. *C. integerrimus.*
        Branchlets longitudinally ridged or angled.
            Leaves glabrous, or sparsely appressed-hairy beneath on the veins only, the margins not revolute .3. *C. thyrsiflorus.*
            Leaves densely appressed-hairy on lower surface, the margins narrowly revolute between the teeth . . . . .4. *C. griseus.*
    Leaves on flowering branches pinnately veined (largest leaves on vigorous shoots sometimes with 3 prominent veins, especially in *C. spinosus*).
        Leaves bright green on lower surface (although lighter than upper surface) and glabrous or nearly so, with flat margins, not glandular-dentate or very sparingly so.
            Evergreen, usually with some thorny branches . . . . .5. *C. spinosus.*
            Deciduous, without thorny branches . . . . .6. *C. integerrimus.*
        Leaves either glaucous or densely pubescent on lower surface, with more or less revolute margins or, if flat, strongly glandular-dentate.
            Leaves glandular-papillate on upper surface.
                Main branches erect to spreading; leaves rounded or bluntly pointed at apex, not notched . . . . .7. *C. papillosus.*
                    Leaf-blades broadly linear to lanceolate or oblong, mostly over 3 mm. wide . . . . .var. *papillosus.*
                    Leaf-blades narrowly linear, not over 3 mm. wide
                          var. *Roweanus.*
                Main branches prostrate, mat-forming, some of the flowering branches ascending; leaves truncate or broadly rounded at apex, usually involute so as to appear notched . . . . . .8. *C. Hearstiorum.*
            Leaves not glandular-papillate.
                Leaves when flat orbicular to oval, but the shape often obscured by the strongly revolute margins; usually tall shrubs
                          9. *C. impressus.*
                Leaves oblong, the margins not strongly revolute but markedly glandular-dentate and usually undulate; low spreading shrubs
                        10. *C. foliosus.*

Leaves opposite.

    Main branches erect or ascending, or rarely depressed and spreading; leaves very slightly if at all revolute, densely puberulent between the veinlets beneath but not white-woolly . . . . .11. *C. cuneatus.*

        Internodes of flowering branches mostly 1 cm. long or more   var. *cuneatus.*

        Internodes of flowering branches mostly less than 1 cm. long.

            Leaves, except on juvenile shoots, entire . . . . .var. *fascicularis.*

            Leaves on mature shrubs sharply few-toothed . . . . .var. *rigidus.*

    Main branches prostrate, often with some flowering branchlets ascending; leaves with revolute margins, white-woolly beneath, the veinlets concealed

                                               12. *C. maritimus.*

1. **C. leucodermis** Greene. White-Bark Ceanothus. Common on rocky or sandy hills back from the coast: east slope of Santa Lucia Mts. and upper Arroyo Grande to La Panza Range.

2. **C. oliganthus** Nutt. var. **oliganthus.** Strictly "typical" *C. oliganthus,* with relatively large leaves (on the average) somewhat hairy on the upper surface, and branchlets conspicuously villous, does not seem to be represented in San Luis Obispo Co., although McMinn has reported it ("Ceanothus," p. 204). Our shrubs vary considerably in leaf size and to some degree in pubescence, but nevertheless are sufficiently uniform to be all included in the following variety.

Var. **sorediatus** (H. & A.) Hoover. Mainly on north-facing or east-facing slopes, in open woods or a component or chaparral, common from coast inland to La Panza Range.

3. **C. thyrsiflorus** Esch. Coast from near Arroyo de la Cruz northward. Dense thickets existed in this area until a few years ago. Most of the shrubs have now been destroyed.

4. **C. griseus** (Trel.) McMinn. Localized in San Luis Range, particularly in the canyon of Coon Creek. Here many of the individuals are small trees comparable in size to the island species *C. arboreus,* and probably are the largest examples of *C. griseus* in existence. The display of blue flowers in the spring is a memorable sight. This species, including both erect and prostrate forms, is one of the few native shrubs which have become widespread in cultivation.

5. **C. spinosus** Nutt. Green-Bark Ceanothus. Occasional on hills in southwestern part of county: upper Pennington Creek; near Cerro Romauldo; San Luis Range. In past years this species presented a beautiful spectacle in Perfumo Canyon, where hillsides were covered with the shrubs, many of them filled with bright blue flowers in April and May. Great numbers have been destroyed. Within its area, this is the only species of *Ceanothus* which to a marked degree crown-sprouts after fire or cutting.

6. **C. integerrimus** H. & A. "San Luis Obispo Co.," according to McMinn in "Illustrated Manual of California Shrubs," without mention of a definite locality. The species is known in the Monterey Co. portion of the Santa Lucia Range, but records from south of the county line are lacking.

7. **C. papillosus** T. & G. var. **papillosus.** At scattered localities in Santa Lucia Range and San Luis Range. The "most typical" *C. papillosus*; that is, the shrubs with the largest leaves, differing most widely from var. *Roweanus,* are found on Pine Top Mt. and in Perfumo Canyon. Otherwise, the varieties merge so gradually into each other that the decision to call the shrubs at certain localities var. *Roweanus* can only be arbitrary.

Var. **Roweanus** McMinn. Nothing exactly like the shrubs at the type locality, Mt. Tranquillon in Santa Barbara Co., has been found elsewhere. If the variety is extended to include all shrubs of *C. papillosus* with comparatively small leaves, then it is commoner than typical *C. papillosus* in the Santa Lucia Range and also occurs sparingly in the San Luis Range (Islay Creek and Coon Creek). Seedlings seldom appear except after fire. Probably for that reason, the species, although known at many localities, is usually represented by only a few individuals.

8. **C. Hearstiorum** Hoover & Roof. Low hills near coast north and south of Arroyo de la Cruz. The stems are prostrate or sprawl over low vegetation. For horticultural purposes, this plant makes an excellent ground-cover, being highly ornamental at all times but especially when in bloom.

9. **C. impressus** Trel. Sandstone hills between San Luis Valley and Arroyo Grande, and in sand on Nipomo Mesa. Our shrubs are on the average taller than those found in Santa Barbara Co. The largest are slender trees about 12 feet tall. The variety *nipomensis* McMinn seems, however, too vaguely defined to warrant the use of a separate name. The leaves of many shrubs on Nipomo Mesa are exactly like those of supposedly typical *C. impressus* from Santa Barbara Co., while the leaves of some Santa Barbara Co. specimens are as described by McMinn for var. *nipomensis* (larger, lighter green, and less revolute).

10. **C. foliosus** Parry var. **medius** McMinn. In Santa Lucia Mts. from Cuesta Pass northwestward to about Tassajera Peak; Hi Mountain; La Panza Range. A shrub found northwest of Cuesta Pass is evidently a hybrid between *C. foliosus* var. *medius* and *C. oliganthus* var. *sorediatus*. The shrubs near Cuesta Pass, the type locality of the variety, have the lower leaf-surfaces densely white-hairy. Elsewhere the leaves are more sparsely hairy, thus varying toward typical *C. foliosus*.

11. **C. cuneatus** (Hook.) Nutt. var. **cuneatus**. BUCKBRUSH. Our commonest *Ceanothus*, widespread on rocky hills throughout Santa Lucia Range, San Luis Range, and eastward to La Panza Range. Also present in Temblor Range, but perhaps only in Kern Co. The species is not yet definitely known in the Diablo Range (vicinity of Cottonwood Pass), Cholame Hills, or Caliente Range. *Ceanothus Greggii* Gray var. *vestitus* (Greene) McMinn has been reported from San Luis Obispo Co., but the specimen on which this record is based (Rinconada Creek, *Armstrong 1157*) differs in no significant way from the prevailing form of *C. cuneatus* in the area where it was collected. The name *C. ramulosus* Greene is customarily used for shrubs of *C. cuneatus* with blue flowers. In many localities the flower-color within a stand of shrubs varies from pure white to medium blue. East of the summits of the Santa Lucia Mts., the flowers only rarely have any blue tinge.

Var. **fascicularis** (McMinn) Hoover. Sandy hills and plains: south side of Morro Bay, southward along coast to near Islay Creek; hills on south side of San Luis Valley to Nipomo Mesa. Recognizable by its short internodes and often narrower leaves. (Plants from the interior with rather narrow leaves are excluded from this variety by their long internodes). As in var. *cuneatus*, the flowers vary from white to blue or lilac. Some individuals have flowers in particularly attractive shades and are well worth propagating for horticultural use. In 1948 it was noted that the sepals have a median longitudinal ridge on the upper side in var. *cuneatus* but not in var. *fascicularis*. Extended observations should be made to determine whether this difference holds true generally.

Var. **rigidus** (Nutt.) Hoover. On coastal sand-hills near Hazard Canyon, where var. *fascicularis* is the prevailing form, some shrubs resemble very closely, if they do not exactly duplicate, the Monterey Co. plants which have been called *C. rigidus.* The distinguishing mark of this variety is that the leaves on full-grown shrubs are dentate, like miniature holly leaves. There is complete intergradation to var. *fascicularis* in this area. Some of the shrubs near Hazard Canyon are low and mat-forming, a form of growth which has rarely been noted in *C. cuneatus* elsewhere.

12. **C. maritimus** Hoover. Hill summit south of Arroyo de la Cruz and extending about 2 miles northward, in one place reaching the edge of the ocean bluff. The flowers, usually blue, are sometimes white. The species, highly promising as an ornamental, has been cultivated successfully in climates quite different from that of its native habitat.

Vitaceae. Grape Family

1. Vitis L.

1. **V. californica** Benth. California Grape. San Luis Obispo, *Miles,* cited by Jepson, Fl. Cal. 2: 481. Since it has been impossible to confirm this occurrence, I suspect that the record is due to a confusion of labels. *Vitis vinifera* L., the familiar cultivated grape of Old-World origin, is sometimes seen growing wild at the sites of abandoned vineyards or homes in the hills.

Malvaceae. Mallow Family

Shrubs.
    Bractlets distinct; corolla rose-pink to white . . . . .1. *Malacothamnus.*
    Bractlets united below; corolla rose with purple veins . . . . .2. *Lavatera.*
Herbs (*Sidalcea Hickmanii* woody only at base).
    Herbage green, sparsely to rather densely pubescent; corolla rose-color to white; rhizomes short or absent.
        Stigmas capitate . . . . .3. *Eremalche.*
        Stigmas not capitate, the style-branches longitudinally stigmatic.
            Flowers in racemes (in *S. diploscypha* short umbel-like ones), much surpassing the reduced bracts . . . . .4. *Sidalcea.*
            Flowers in small clusters or solitary in the axils, surpassed by the subtending leaves . . . . .5. *Malva.*
    Herbage densely white-hairy; corolla cream-color; low spreading perennial with extensive rhizomes . . . . .6. *Sida.*

1. Malacothamnus Greene. Bush Mallow

Flowers in dense terminal head-like clusters surrounded by leafy bracts
                                      1. *M. Palmeri.*
    Leaves stellate-pubescent on both surfaces .var. *Palmeri.*
    Leaves glabrate on upper surface . . . . .var. *involucratus.*
Flowers in panicles or scattered along branches.
    Pubescence in part spreading, villous; leaves 3 to 7 cm. wide   2. *M. orbiculatus.*
    Pubescence of leaves and branches closely appressed, tomentose; leaves 1 to 4 cm. wide.
        Calyx heavily white-tomentose, the surface of its lobes concealed by the hair.
            Inflorescence subspicate, the flowers 1 or few in the upper axils, short-pedicelled or subsessile . . . . .3. *M. Jonesii.*

Inflorescence paniculate, some of the flowers on lateral branches and
often long-pedicelled ....4. *M. niveus*.
Calyx stellate-pubescent but hardly tomentose, the surface of the lobes not
concealed ....5. *M. gracilis*.

1. **M. Palmeri** (Wats.) Green var. **Palmeri**. Occasional on rocky slopes in Santa
Lucia Mts. from the Atascadero-Morro Bay road northward, mostly near summits
but occasionally extending down canyons to near the sea, as at Cambria, the type
locality.

Var. **involucratus** (Rob.) Kearney. Cuesta Pass (*Chester Dudley* in 1937). An error
in the label is suspected. This well-marked variety has not otherwise been found
outside Monterey Co.

2. **M. orbiculatus** (Greene) Greene? Specimens collected at Bee Rock (north of
Nacimiento Reservoir near Monterey Co. line), though not typical for the species,
are tentatively called *M. orbiculatus,* as they resemble quite closely some specimens
so named. The bractlets in these collections are remarkably short, as in *M. David-
sonii* of the San Fernando Valley. Also referred to *M. orbiculatus* are plants from
the Temblor Range in Kern Co. and the Mt. Abel region. Similar plants may be
expected on Caliente Mt.

3. **M. Jonesii** (Munz) Kearney. Dry hills, mainly in disintegrated shale, localized
west of Paso Robles (the type locality) to near Nacimiento Dam.

4. **M. niveus** (Eastw.) Kearney. In rocky or sandy soil, from Atascadero eastward
to Gillis Canyon near Shandon, La Panza Range, and lower Cuyama Valley. Closely
related to *M. Jonesii* and probably not really distinct. Some specimens of *M. niveus*
have reduced inflorescences much like those of *M. Jonesii*. The type locality of *M.
niveus* is the Eldorado School north of Pozo.

5. **M. gracilis** (Eastw.) Kearney. Dry rocky slopes from upper Arroyo Grande to
lower Cuyama Canyon; entirely restricted to this small area.

2. Lavatera L.

1. **L. assurgentiflora** Kell. Tree Mallow. Occasional along the coast, as around
Morro Bay and south of Pismo Beach. It is generally assumed that the species is
native only on the southern California islands, all mainland occurrences representing escapes from cultivation. No reason is evident, however, why it could not have
been present in some spots along the mainland shore before it was planted.

3. Eremalche Greene

Calyx at anthesis 5 mm. long; petals 5 to 6 mm. long ....1. *E. exilis*.
Calyx 5 to 12 mm. long; petals 10 to 25 mm. long ....2. *E. Parryi*.

1. **E. exilis** (Gray) Greene. Sandy soils, occasional in southeastern part of county:
just east of San Juan River, La Panza district (*9350*); southern Temblor Range to
Cuyama Valley.

2. **E. Parryi** (Greene) Greene. Desert Mallow. Common through the interior in
calcareous sands and clays, or often in crumbling shale, from Paso Robles east to
Cottonwood Pass and Temblor Range, south to Cuyama Valley. Absent from granite areas (La Panza Range and westward). In the southeastern part of the county,
plants with small flowers but otherwise identical are often found mingled among
the typical large-flowered form of the species. In addition, at several places in the

Temblor Range the plants vary in flower-color from pure white to the usual rose-purple. No correlation between the size and the color of the flowers is evident. Plants with small and pale flowers resemble specimens from the type locality of *E. kernensis* Wolf in the adjacent San Joaquin Valley. The nature of their occurrence, mixed in highly variable colonies and with possible distinguishing features combined in every conceivable way, leads to the conclusion that *E. kernensis* is merely a localized form of *E. Parryi.*

4. Sidalcea Gray

Annual; racemes short, few-flowered ....1. *S. diploscypha.*
Perennial; racemes elongate.
    Plants with short rhizomes bearing tough fibrous roots; stems not or very rarely branched above base; bractlets none ....2. *S. malvaeflora.*
    Plants with woody roots; vigorous stems branched above base; bractlets 1 to 3
3. S. Hickmanii.

1. **S. diploscypha** (T. & G.) Gray. Occasional in clay soils near coast from San Luis Valley northward.

2. **S. malvaeflora** (DC.) Gray. Common on open hills in western part, and on ocean bluffs northward. The few specimens now available do not segregate well into the numerous subspecies and varieties which have been named by C. L. Hitchcock. Extensive field studies and copious collections are needed before these names can be properly evaluated with respect to San Luis Obispo Co. An exception to this statement is provided by plants found in a few widely scattered moist places in the interior: Nacimiento River, Creston, and San Juan River. These plants owe their distinctive appearance to long many-flowered racemes of small flowers. Hitchcock has called these plants subsp. *laciniata* var. *sancta,* but exactly similar plants from the upper Cuyama River (outside our limits) are called subsp. *sparsifolia* by the same authority.

3. **S. Hickmanii** Greene ssp. **anomala** C. L. Hitchcock. Restricted to the area of serpentine rock northwest of Cuesta Pass, where it is locally plentiful along the ridge, especially in cleared spots.

5. Malva L. Mallow

Corolla about twice as long as calyx.
    Bractlets ovate to oblong; carpels 7 to 9 ....1. *M. nicaeensis.*
    Bractlets linear; carpels 13 to 15 ....2. *M. neglecta.*
Corolla scarcely longer than calyx ....3. *M. parviflora.*

1. **M. nicaeensis** All. Common weed in cultivated or formerly cultivated ground and by roadsides, especially around San Luis Obispo; scarce toward the east except in irrigated spots.

2. **M. neglecta** Wallr. Twisselmann Ranch; Atascadero; perhaps overlooked elsewhere for *M. nicaeensis,* which it resembles in aspect.

3. **M. parviflora** L. Cheeseweed. In fertile soils, probably even more abundant than *M. nicaeensis.*

6. Sida L.

1. **S. hederacea** (Dougl.) Torr. Alkali Mallow. In low places, generally in saline soils of the interior; infrequent near the coast, as in Los Osos Valley.

Sterculiaceae. Sterculia Family

1. Fremontodendron Cov.

1. **F. californicum** (Torr.) Cov. var. **obispoense** (Eastw.) Hoover, n. comb. *Fremontia obispoensis* Eastw., Leafl. West. Bot. 1: 140. 1934. Fremontia. Rocky soil in a few scattered spots: upper Sycamore Canyon in San Luis Range (once called Pettitt's Canyon, the type locality); in Santa Lucia Mts. near Cerro Alto and northwest of Cuesta Pass; Calf Canyon; east slope of Pine Mt. in La Panza Range. There has never been a verifiable report of the shrubs here forming seeds, but their wide dispersal is clear proof of seed production in the past. The original collection of *F. californicum* came from the Sierra Nevada. Many specimens from that region have large palmately lobed leaves quite unlike those of San Luis Obispo Co. shrubs. On the other hand, reduced leaves similar to those of var. *obispoense* are by no means restricted to this part of California. A collection from a solitary large shrub found just north of the summit on the Cambria-Adelaida road shows a closer resemblance to Sierra Nevada material than does anything else in this region. Although this specimen has mostly 3-lobed leaves, its closest relationship is believed to be with var. *obispoense*.

Guttiferae. St. John's-Wort Family

1. Hypericum. St. John's-Wort

Flowering stems erect; petals showy, much longer than sepals . . . . . . 1. *H. formosum.*
Stems creeping; petals about as long as sepals . . . . 2. *H. anagalloides.*

1. **H. formosum** H. B. K. var. **Scouleri** (Hook.) Coulter. Near a spring on east slope of Pine Mt. above San Simeon; to be looked for elsewhere in Santa Lucia Range.

2. **H. anagalloides** L. Uncommon in permanently moist spots near coast.

Elatinaceae. Waterwort Family

Stems creeping on mud or floating; leaves entire . . . . 1. *Elatine.*
Stems ascending; leaves serrate toward apex . . . . 2. *Bergia.*

1. Elatine L.

Flowers sessile; seeds straight or slightly curved . . . . 1. *E. brachysperma.*
Flowers short-pedicelled; seeds strongly curved . . . . 2. *E. californica.*

1. **E. brachysperma** Gray. Beds of vernal pools, near Estrella (*Eastwood & Howell 4202*) and probably elsewhere.

2. **E. californica** Gray. Laguna near San Luis Obispo (*7922*). Both this species and the preceding are very inconspicuous, so probably are overlooked rather than actually rare.

2. Bergia L.

1. **B. texana** (Hook.) Seub. Scarce in dried pool beds in eastern part: Twisselmann Ranch and summit of Palo Prieto Canyon.

Frankeniaceae. Frankenia Family

1. Frankenia L.

Upper stems, leaf-bases, and calyces hirsutulous (vegetative branches produced in rainy season apparently glabrous); petals rose-purple, the blade oblong, obtuse, exserted 2 to 4 mm. from calyx . . . . 1. *F. grandifolia.*

Upper stems, leaf-bases, and calyces apparently quite glabrous; petals lilac-pink, the blade linear, acute, exserted 1.5 to 2 mm. from calyx . . . . .2. *F. campestris.*

1. **F. grandifolia** C. & S. Common in moist saline soil along coast; notably abundant in the salt-marshes around Morro Bay and at Avila.

2. **F. campestris** (Gray) Hoover, n. comb. *F. grandifolia* var. *campestris* Gray, Syn. Fl. 11: 208. 1895. Plentiful on alkaline plains and in alkaline hollows in eastern part, especially Cholame Valley and Carrizo Plain.

Tamaricaceae. TAMARISK FAMILY

1. Tamarix L. TAMARISK

Flowers in lateral racemes from one-year-old branches; sepals and petals 4
1. *T. tetrandra.*
Flowers in terminal panicles on new growth; sepals and petals 5 . .2. *T. pentandra.*

1. **T. tetranda** Pall. *T. parviflora* DC. Common in cultivation; occasionally escaped in moist places near our eastern border, as at Grant Lake.

2. **T. pentandra** Pall. Three plants were noted by Twisselmann as growing in 1952 at O'Brien Lake at the summit of Palo Prieto Pass. In 1964 he reported the shrubs as "now all dead."

Cistaceae. ROCKROSE FAMILY

Leaves opposite; flowers more than 4 cm. in diameter . . . . .1. *Cistus.*
Leaves alternate; flowers less than 2 cm. in diameter . . . . .2. *Helianthemum.*

1. Cistus L. ROCKROSE

1. **C. ladaniferus** L. In dry sand 2 miles south of Los Berros on Nipomo Mesa, the shrubs locally plentiful and freely reproducing. Presumably escaped from cultivation, although the species has not been observed to be now cultivated in the area. It is uncertain which of the many named varieties of this variable species is represented here. The petals are white with a red-purple spot on the lower part.

2. Helianthemum Mill.

Inner sepals 4 to 6 mm. long; petals 6 to 8 mm. long . .*H. scoparium* var. *scoparium.*
Inner sepals 3 to 3.5 mm. long; petals 4 to 6 mm. long . . . . .var. *vulgare.*

1. **H. scoparium** Nutt. var. **scoparium.** CALIFORNIA RUSHROSE. Common in sandy soils near coast, at least from Morro Bay southward; occasional inland to Salinas Valley (Atascadero to near Pozo). Not noted as occurring in our northern coastal area. The species is variable. On Nipomo Mesa some of the plants are white-hairy as described for *H. suffrutescens* Schreiber, while others growing with them, regarded as the typical form of the species, are green.

Var. **vulgare** Jepson. Rocky hills in chaparral areas, especially abundant after fires, throughout Santa Lucia Range and eastward to La Panza Range.

Violaceae. VIOLET FAMILY

1. Viola L. VIOLET

Leaves twice pinnately divided into narrow segments, all basal . . . .1. *V. Douglasii.*
Leaves coarsely toothed to entire, some of them on stem above the base.
    Plants with a deeply buried short rhizome bearing fibrous roots and slender ascending underground stems which continue into the flowering branches; leaf-blades mostly about as wide as long . . . . .2. *V. pedunculata.*
    Plants with a relatively stout tap-root; bases of stems hardly if at all underground; leaf-blades mostly longer than wide.

Cauline leaves more than 2, with 5 or 6 teeth on each margin; petals 10
to 12 mm. long . . . . .3. *V. quercetorum.*
Cauline leaves usually 2, with not more than 4 teeth on each margin; petals 8 to 10 mm. long . . . . .4. *V. purpurea.*

1. **V. Douglasii** Steudel. Rare in uncultivated land toward the interior: 18 miles
east of Creston on La Panza road, where plentiful locally in red gravelly-clay soil
near oaks; formerly near Paso Robles; roadside ½ mile west of Templeton (*Wiggins
5835* in 1932).

2. **V. pedunculata** T. & G. Wild Pansy. Johnny-Jump-Up. Common in firm soils
in treeless areas or open woods, from coast eastward to Choice Valley hills; apparently absent from vicinity of Carrizo Plain. Baker and Clausen published a subspecies *tenuifolia,* which supposedly includes the plants of eastern San Luis Obispo
County, but some coastal specimens are identical with them in every detail.

3. **V. quercetorum** Baker & Clausen. Frequent in hilly regions of the interior in
sandy or gravelly soils, usually in open oak woods or under digger pines. A collection from the summit of the Temblor Range (*Twisselmann 1975*), in Kern Co. but
very near the San Luis Obispo Co. line, has woolly leaves with prominent teeth and
conforms to descriptions of *V. aurea* Kell. Because this form merges gradually into
V. quercetorum, the inclusion of both in the same species would doubtless be a
more usable classification.

4. **V. purpurea** Kell. Plants found in the *Pinus ponderosa* forest on Pine Top
Mt. south of San Carpoforo Creek clearly belong to the typical form of this species,
according to Munz's key. The distinctness of *V. quercetorum* from *V. purpurea* may
properly be questioned, in view of many plants which resemble one about as closely
as the other. Such plants have been found between Rocky Butte and Pine Mt.

Loasaceae. Blazing Star Family

1. Mentzelia L.

Petals narrowed toward both ends, at least 4 cm. long . . . . .1. *M. laevicaulis.*
Petals broadly rounded at apex or abruptly pointed, not over 22 mm. long.
    Petals showy, 10 to 22 mm. long.
        Petals yellow, often with orange base, 10 to 16 mm. long; capsule 15 to 25
        mm. long . . . . .2. *M. gracilenta.*
        Petals bright orange with copper-red base, 12 to 22 mm. long; capsule 20 to
        35 mm. long . . . . .3. *M. pectinata.*
    Petals 3 to 10 mm. long.
        Bracts ovate-lanceolate or narrower, shorter than the flowers.
            Leaves sparsely scabrous-puberulent, bright green; bracts entirely
            green; seeds grooved on one or more of the angles.
                Calyx-lobes 3 to 5 mm. long; petals 5 to 10 mm. long; leaves pin-
                nately lobed, less commonly merely toothed or the uppermost en-
                tire . . . . .4. *M. affinis.*
                Calyx-lobes scarcely 1 mm. long; petals 3 to 4 mm. long; leaves
                usually coarsely toothed to entire but sometimes lobed

                                  5. *M. dispersa.*
        Leaves densely scabrous-puberulent, dark under the grayish pubes-
        cence; bracts usually with whitish or pale center; seeds not grooved
        on any angle . . . . .6. *M. albicaulis.*
    Bracts deltoid-ovate or oval, equalling the flowers or longer

                               7. *M. micrantha.*

1. **M. laevicaulis** (Dougl.) T. & G. In stream-beds or on barren slopes which are damp during part of the year, known definitely within our limits only in Cuyama Canyon.

2. **M. gracilenta** T. & G. In crumbling shale gravel, calcareous clays, or calcareous sandy loams, usually on steep slopes, frequent from Cottonwood Pass to near Creston and Caliente Mt., but absent from Temblor Range. The yellow petals often, but not always, have an orange base. A collection from Caliente Mt. (*8224*) has the flowers rather crowded and the subtending leaves unusually broad. These features suggest *M. congesta* (Nutt.) T. & G., but the plants otherwise, particularly in their large flowers, agree with *M. gracilenta*. Intensive field studies on Caliente Mt. are needed to determine whether both species may occur there and may hybridize.

3. **M. pectinata** Kell. Gypseous clay soils, Temblor Range, Caliente Range, and hills bordering Cuyama Valley. The plants usually grow on slopes which are otherwise nearly barren, and in a year of good rainfall are abundant and showy. The petals are vivid orange, always with a copper-red base. The nature of the relationship between this species and *M. gracilenta* is an unsolved enigma. Flower color seems to be the only difference which is completely reliable; yet the two kinds of plants are completely distinct ecologically, geographically, and apparently genetically. Even in the Caliente Range, which is the only known area where they occur relatively near each other, there are no mixed colonies, intergrades as to flower color, evident hybrids, or any sort of evidence of interfertility. Unfortunately, it is usually impossible to determine the flower color from herbarium specimens.

4. **M. affinis** Greene. Rather common in sandy soils in eastern part.

5. **M. dispersa** Wats. Locally plentiful after fires in La Panza Range; also found as a scarce wheat-field weed 2 miles east of Soda Lake (*Twisselmann 1331*) and said by Twisselmann to be occasional throughout the Temblor Range.

6. **M. albicaulis** Dougl. ex T. & G. Occasional on sandy slopes in interior; in places abundant, as on Caliente Mt. Not reported from Temblor Range. Some of our plants may, on the basis of their somewhat larger petals, represent var. *Veatchiana* (Kell.) Urban & Gilg, but that name does not correspond to a clearly defined entity in our area.

7. **M. micrantha** (H. & A.) T. & G. Wooded and brushy areas, in sandy or rocky soil, in hill country from near the coast eastward to La Panza Range. Mostly occurring after fires.

Datiscaceae. DATISCA FAMILY

1. Datisca L.

1. **D. glomerata** (Presl) Baillon. DURANGO ROOT. Common in rocky stream-beds in Santa Lucia Mts., extending inland to La Panza Range.

Cactaceae. CACTUS FAMILY

1. Opuntia Mill.

Joints of stems flattened.
    Stems tending to grow upright, becoming over 1 m. tall; joints often over 3 dm. long ....1. *O. Ficus-indica*.
    Stems spreading and rooting along the ground, less than 1 m. tall; joints less than 3 dm. long ....2. *O. phaeacantha*.
Joints of stem cylindrical ....3. *O. echinocarpa*.

1. **O. Ficus-indica** (L.) Mill. Prickly Pear. Mission Cactus. Well established in various places in the hills around San Luis Obispo. Scattered clumps are widespread through our coastal area as relics from prior cultivation.

2. **O. phaeacantha** Engelm. Mostly on south or west slopes of rocky hills, from San Luis Obispo and the summit between Pozo and Arroyo Grande southward. A large and vigorous stand occurs on the sandy slope bordering the Santa Maria Valley southwest of Nipomo. This is the species which previous references have called *O. occidentalis*. According to David Walkington, this application of that name is erroneous, and the species in question should be called *O. phaeacantha*.

3. **O. echinocarpa** Engelm. & Bigelow. Staghorn Cholla. Cuyama Canyon, ¼ mile east of Big Rocks, noted as found at "four different locations back in the hills," *Hendrix 1064* in 1939.

Elaeagnaceae. Oleaster Family

1. Shepherdia Nutt.

1. **S. argentea** Nutt. Buffalo Berry. Along streams. The only local record is Cuyama Ranch, *Peterson 325* in 1936.

Lythraceae. Loosestrife Family

1. Lythrum L.

Erect annual, or becoming perennial and then with sprawling stems; petals 1 to 3 mm. long . . . . .1. *L. Hyssopifolia.*
Erect perennial; petals 4 to 6 mm. long . . . . .2. *L. californicum.*

1. **L. Hyssopifolia** L. Common in places which are wet or even flooded in winter, but not known with certainty from eastern portion of county. Plants growing in permanently damp soil seem to become perennial and correspond to *L. adsurgens* Greene, but it is doubted that such plants differ genetically from the prevailing form of *L. Hyssopifolia.*

2. **L. californicum** T. & G. California Loosestrife. In moist low valleys, rare in this region: Creston; San Luis Obispo. Not recently observed and perhaps exterminated here.

Myrtaceae. Myrtle Family

1. Eucalyptus L'Hér.

1. **E. Globulus** Labill. Bluegum. Great numbers of trees were once planted in the coastal region, particularly south of Morro Bay and on Nipomo Mesa. These planted stands are extending themselves by spontaneous seeding, and this species easily crowds out native trees when in competition with them. Seedlings of some other species of *Eucalyptus* may sometimes appear spontaneously, but none except bluegum has really become established.

Onagraceae. Evening Primrose Family

Petals 5; calyx-lobes persistent after petals fall . . . . .1. *Jussiaea.*
Petals 4; calyx deciduous with the rest of the flower from the developing capsule.
    Calyx-lobes erect or ascending.
        Seeds bearing a tuft of hairs.
            Flowers funnel-shaped, bright red (rarely white), the calyx colored like
            the petals . . . . .2. *Zauschneria.*

Flowers not funnel-shaped; petals rose-purple to white, colored differ-
ently from the calyx ....3. *Epilobium.*
Seeds without a tuft of hairs ....4. *Boisduvalia.*
Calyx-lobes reflexed, or the tips united and turned to one side.
Anthers attached near base; petals rose-red, purplish, pink, or white
5. *Clarkia.*
Anthers attached near middle; petals yellow or white, sometimes turning
pink or red in age ....6. *Oenothera.*

1. Jussiaea L.

1. **J. repens** L. var. **peploides** (H. B. K.) Griseb. Oso Flaco Lake. Rarely flowering
here.

2. Zauschneria Presl

Leaves linear ....*Z. californica* var. *californica.*
Leaves narrowly elliptic to lanceolate ....var. *mexicana.*

1. **Z. californica** Presl var. **californica.** CALIFORNIA FUCHSIA. Occasional on rocky
slopes from coast eastward to Santa Lucia Mts. As Raven has demonstrated (Aliso
5: 215. 1962), the typical phase of the species is the plant with small leaves which
has been called var. *microphylla* Gray or, when the leaves are conspicuously white-
hairy, *Z. cana* Greene. None of the San Luis Obispo Co. plants shows the extremely
reduced leaves seen in specimens from coastal Monterey Co. and from Santa Barbara Co. southward, but some are intermediate. A collection with conspicuously
hairy herbage (Yaro Creek, *9902*) may represent var. *villosa* (Greene) Jepson.

Var. **mexicana** (Presl) Hoover, n. comb. *Z. mexicana* Presl, Rel. Haenk. 12: 29.
1835. Far more common locally than the preceding variety and extending inland
to La Panza Range; also found on east slope of Temblor Range in Kern Co. No
matter how species and subspecies in *Zauschneria* have been defined, the majority
of plants have remained of uncertain identity. The most helpful solution to the
problem seems to me to be the inclusion of all in one species with the more notable
variants designated as varieties. Even the designation of subspecies, I believe, would
place too much emphasis on what seem to be only slight genetic differences which
are often masked by the effect of the environment.

3. Epilobium L. WILLOW-HERB

Annuals, growing in dry places.
Stems 3 to 10 dm. tall, glabrous throughout or puberulent only in inflores-
cence; calyx-tube at least 2 mm. long ....1. *E. paniculatum.*
Stems 5 to 30 cm. tall, puberulent throughout; calyx-tube very nearly obsolete
2. *E. minutum.*
Perennials, growing in wet places ....3. *E. Watsonii.*
Petals 6 to 10 mm. long.
Upper leaves and stems velvety-pubescent ....var. *Watsonii*
Herbage moderately puberulent to glabrous ....var. *franciscanum.*
Petals 3 to 6 mm. long.
Inflorescence puberulent to glabrous, not glandular ....var. *Parishii.*
Inflorescence glandular-puberulent ....var. *occidentale.*

1. **E. paniculatum** Nutt. Frequent throughout the county, except in the driest
parts; commonest in Santa Lucia Mts. and upper Salinas Valley. Forms which have

been identified as present include f. *paniculatum* (pedicels slightly glandular-puberulent), f. *adenocladon* Hausskn. (pedicels densely glandular-puberulent), and f. *sub-
ulatum* Hausskn. (pedicels glabrous).

2. **E. minutum** Lindl. Occasional in woods or chaparral, generally after fires, in
Santa Lucia and La Panza Ranges. Probably more common than the very few records would indicate.

3. **E. Watsonii** Barbey var. **Watsonii.** The typical form of the species (upper
leaves and stems velvety; flowers relatively large) is a rare plant of our northern
coast (Piedras Blancas Point, in swampy area, *6963*). The species as a whole, however, as represented by the following varieties, is common in wet places. Perhaps
this entire group should be included in *E. mexicanum* Schlecht. Apparently no
author has discussed the question of how *E. Watsonii* var. *franciscanum* and var.
Parishii can be differentiated from that species. Available specimens from Mexico
are too few and too fragmentary for an adequate study of the problem at this time.

Var. *franciscanum* (Barbey) Jepson. Common in wet places along coast. This
name is used for plants with the large flowers of typical *E. Watsonii* but with the
pubescence sparse, not velvety. A large proportion of the plants agrees equally well
with descriptions of var. *franciscanum* and of var. *Parishii*. Plants identifiable as
var. *franciscanum* on the basis of having rather large flowers are sometimes found
as far inland as Tassajera Creek on the east slope of the Santa Lucia Range (*8826*).

Var. **Parishii** (Trel.) C. L. Hitchc. *E. californicum* Hausskn. Very similar to var.
franciscanum but with smaller flowers. Occasional near coast, mainly as intergrades
with var. *franciscanum,* but also extending inland at least to upper Salinas River.

Var. **occidentale** (Trel.) C. L. Hitchc. Distinguished by the presence of minute
glandular hairs in the inflorescence. Known in the county only in the vicinity of
Oceano and the dune lakes.

4. **Boisduvalia** Spach

Capsule terete.
 Plants spreading-pubescent.
  Bracts ovate or lanceolate ....1. *B. densiflora.*
  Bracts linear ....2. *B. stricta.*
 Plants obscurely puberulent or glabrous ....3. *B. glabella.*
Capsule sharply 4-angled ....4. *B. cleistogama.*

1. **B. densiflora** (Lindl.) Wats. Frequent in moist or drying beds of streams and
around springs, Santa Lucia Mts. eastward to La Panza Range.

2. **B. stricta** (Gray) Greene. Occasional in the upper Salinas basin. The only definite record now available is north base of Cuesta Pass (*9660*), but the species has
been seen elsewhere.

3. **B. glabella** (Nutt.) Walp. Beds of vernal pools: rare in coastal area (Los Osos
Valley, *9031*), common in upper Salinas Valley, occasional eastward to Palo Prieto
Pass and Carrizo Plain.

4. **B. cleistogama** Curran. Beds of vernal pools, 4 miles from Paso Robles on
Estrella road, *4078* in 1939. Probably now exterminated at this locality. Otherwise
found only in the Great Valley and bordering low foothills.

5. **Clarkia Pursh**

Flower-buds erect.
 Petals 15 to 25 mm. long.

Capsule inconspicuously ribbed, 2 to 2.5 mm. thick; red spot on petal, if present, near middle or below.

Petals 20 to 25 mm. long, bright rose-purple with red base

1. *C. rubicunda.*

Petals 15 to 20 mm. long, variously colored and marked but, if rose-purple, not red at base ....2. *C. speciosa.*

Petals wholly dark purple-red, or lighter and with red spot near middle or below ....subsp. *speciosa.*

Petals rose-purple with white base ....subsp. *immaculata.*

Capsule prominently ribbed, 3 to 4.5 mm. thick; red spot on petal, if present, near apex ....3. *C. purpurea* var. *Arnottii.*

Petals 5 to 12 (rarely to 15) mm. long.

Stems erect or with ascending branches; leaves acute or narrowed toward apex.

Calyx-lobes distinct and reflexed under the opened petals; capsule prominently ribbed, 3 to 4.5 mm. thick ....3. *C. purpurea.*

Upper leaves and flowers crowded ....var. *purpurea.*

Upper leaves and flowers well spaced ....var. *hirsuta.*

Calyx-lobes remaining united and turned to one side under the opened petals; capsule inconspicuously ribbed, 1.5 to 3 mm. thick   4. *C. affinis.*

Stems with widely spreading branches, or if simple horizontally curved near base; leaves obtuse, oblong or slightly widened toward apex

5. *C. Davyi.*

Flower-buds deflexed or pendent until just before opening.

Petals not lobed; stamens 8.

Petals gradually narrowed toward base, the claw, if present, short and poorly defined.

Petals 15 to 30 mm. long.

Axis of inflorescence at first drooping, becoming erect as the buds open; flower-tube with ring of hairs inside near middle

6. *C. cylindrica.*

Axis of inflorescence erect, the buds deflexed; flower-tube with ring of hairs inside near its mouth ....7. *C. deflexa.*

Petals 5 to 12 mm. long.

Petals rose-purple or pink, rhombic or with acute apex.

Leaves ovate or elliptic; claw of petal with a tooth on each side near base ....8. *C. rhomboidea.*

Leaves linear to narrowly lanceolate; claw of petal not toothed

9. *C. modesta.*

Petals white, broadly rounded at apex ....10. *C. epilobioides.*

Petals abruptly narrowed into a narrow claw about as long as the blade

11. *C. unguiculata.*

Petals deeply 3-lobed; stamens 4 ....12. *C. concinna.*

1. **C. rubicunda** (Lindl.) Lewis & Lewis. Coastal bluffs between San Carpoforo Creek and Arroyo de la Cruz, in heavy soil. A collection from this locality (*7387*) was cited as subsp. *Blasdalei* by Lewis (U. C. Publ. Bot. 20: 271), but the plants show no clearly defined difference from specimens identified by Lewis as typical *C. rubicunda*. A study of available material does not support subsp. *Blasdalei* as a distinguishable entity.

2. **C. speciosa** Lewis and Lewis subsp. **speciosa.** Common and often plentiful from Santa Lucia Mts. eastward to La Panza district; seldom extending beyond Salinas River drainage. Flower color is remarkably variable. At some localities the petals

are uniformly deep purple-red, while elsewhere wide differences in intensity of color and degree of spotting can be observed within a small area. The following sub-species, geographically separated from the remainder of the species, is quite constant in its color pattern.

Subsp. **immaculata** Lewis & Lewis. Local in the sandy hills between San Luis Valley and Arroyo Grande, inland to Huasna district.

3. **C. purpurea** (Curtis) Nels. & Macbr. var. **purpurea**. The typical phase of the species, if defined somewhat arbitrarily by a combination of crowded flowers and comparatively broad leaves, is rare in the county. It has been reported from dunes north of Guadalupe.

Var. **Arnottii** (T. & G.) Hoover, n. comb. *Oenothera Arnottii* T. & G. Fl. N. Am. 1: 503. 1840. *Oenothera lepida* var. *Arnottii* Wats. (1873). *Godetia viminea* (Dougl.) Spach. (See footnote under var. *hirsuta).* Sandy hills at edge of Nipomo Mesa, 2 miles south of Arroyo Grande *(7065)*; head of See Canyon (*B. Smith* in 1966). This rather large-flowered plant merges gradually into the following variety.

Var. **hirsuta** (Kell.) Hoover, n. comb. *Oenothera quadrivulnera* Dougl. var. *hirsuta* Kell., Proc. Cal. Acad. 5: 45. 1873.[3] *Godetia quadrivulnera* (Dougl.) Spach. Because the name is allowed to cover a wide range of variation, this variety is much the commonest. It is found plentifully from the Santa Lucia Mts. eastward probably to the Temblor Range. It is absent from the coastline and only occasional on the west slope of the Santa Lucia Range.

4. **C. affinis** Lewis & Lewis. Frequent from Salinas Valley eastward to Choice Valley and La Panza district. Very similar in general aspect to *C. purpurea* var. *hirsuta* and growing in much the same kinds of habitats; doubtfully distinct from that variety, although most specimens are readily identifiable as one or the other.

5. **C. Davyi** (Jepson) Lewis & Lewis. Ocean bluffs and borders of beaches at Morro Bay (*Condit* in 1911, but not since collected there) and from Cambria northward. These plants have been called *C. prostrata* Lewis & Lewis and said to be hexaploid, whereas true *C. Davyi,* found in Santa Barbara Co. and from Monterey Co. northward, is tetraploid. Apparently there is no reliable difference between the two in outward appearance, as implied (though not explicitly stated) by Lewis and Raven (Leafl. West. Bot. 9: 94). The flowers of *C. prostrata* are not consistently larger, nor do all plants in San Luis Obispo Co. show the spot on each petal which is alleged to be an identifying mark of *C. prostrata. Clarkia prostrata* is said to be derived from hybridization between *C. Davyi* and *C. speciosa,* but no feature of *C. speciosa* is evident in it. Not only do some individual plants of *C. prostrata* exactly resemble some which are called genuine *C. Davyi,* but they also grow under the same ecologic conditions, quite different from the habitat of *C. speciosa.* The purposes of classification therefore are best served by regarding *C. prostrata* as a chromosomal race of *C. Davyi.* Furthermore, in view of the obvious impossibility of determining the chromosome number of every individual plant, it is entirely possible that some of the San Luis Obispo Co. plants may be tetraploid *C. Davyi* rather than the hexaploid race. The presence of supposedly tetraploid *C. Davyi* in both Monterey Co. and Santa Barbara Co. (Leafl. West. Bot. 9: 94 and 9: 132) gives strong support to such a suggestion.

---

3. An earlier name which may apply to this variety is *Oenothera viminea* var. *intermedia* Kell. Jepson cited it under *Godetia quadrivulnera,* but Lewis and Lewis believed it synonymous with *Clarkia purpurea* subsp. *viminea* (=var. *Arnottii).*

6. **C. cylindrica** (Jepson) Lewis & Lewis. Common from east slope of Santa Lucia Range eastward in the hill country. The species is notably abundant and showy in the region of Polonio Pass in a good year. Near our eastern border, the flowers are larger than farther west, although no plants in the county match in this respect plants found in the hills near Bakersfield. The common name "Farewell to Spring" is used for this, as for most species of *Clarkia*. One plant found west of Paso Robles, mixed with pink-flowered plants, had deep red petals. Some of the plants in the Santa Lucia Mts. of northern San Luis Obispo Co., assumed to be *C. cylindrica,* may actually belong to the closely related *C. Bottae* (Spach) Lewis & Lewis, a coastal species which, as far as known, is localized in Monterey Co. Plants of this relationship call for much more extensive field study in that region.

7. **C. deflexa** (Jepson) Lewis & Lewis. Common on shaded slopes in western part: Santa Lucia Range, San Luis Range, and south to Huasna district and lower Cuyama River.

8. **C. rhomboidea** Dougl. See Canyon, *Condit,* according to Jepson (Fl. Cal. 2: 574). Undoubtedly more widespread in the county than this single record would indicate.

One peculiar form, which in vegetative features resembles *C. rhomboidea,* calls for special mention. Eben McMillan collected plants 2 miles northeast of Cholame which have very small narrow dull-colored petals. This same form was seen near the west end of Caliente Mt. Vasek identified the McMillan collection as *C. tembloriensis,* but the leaves are too broad for that species as differentiated in Vasek's key (Madroño 17: 219. 1964), and in other respects also it differs from other specimens placed in *C. tembloriensis.* The plants in question perhaps represent a species not yet named, but more adequate material is needed for study before their classification can be considered satisfactory.

9. **C. modesta** Jepson. In chaparral or oak woods, occasional on east side of Santa Lucia Range, as near Atascadero, and eastward through the hill country to La Panza Range, where locally common. Especially vigorous and plentiful in 1952 following a fire the previous year.

10. **C. epilobioides** (Nutt.) Nels. & Macbr. Wooded areas in western part, rare or overlooked: San Luis Mt. (acc. to Jepson, Fl. Cal. 2: 585); Tar Spring Creek east of Arroyo Grande; Suey Creek (*Eastwood 409*).

11. **C. unguiculata** Lindl. CLARKIA. Common on hillslopes in loose or rocky soil. Plants found west of the summit of the Santa Lucia Range, regarded as typical *C. unguiculata,* have relatively broad leaves, relatively broad petal-blades, and copious long hair on the ovary and calyx. All plants of the interior show one or more of the supposedly distinctive features of *C. tembloriensis* Vasek, which in its extreme form has narrow leaves, small petal-blades, and no long hairs on the ovary or calyx. Relatively few of the plants, however, combine all the features of *C. tembloriensis.* Most are intermediate. If two species actually exist in the area, it is impossible to establish a geographic boundary between them except on a highly arbitrary basis. As an example, it might be true that all plants east of the San Juan River have no long hairs whatever, but it has been noted that these hairs may be very sparse even as far west as the Santa Lucia Range.

12. **C. concinna** (F. & M.) Greene. RED RIBBONS. On crumbling rocky banks, rare in Santa Lucia Range: near Hearst Castle (*Susan Allison* in 1966); Lopez Canyon (*Sharon Patterson* in 1966).

6. **Oenothera** L. Evening Primrose

Petals 2 to 4 cm. long; stigma with 4 linear lobes; flowers opening in evening, closing in hot sunshine.
    Petals yellow . . . . .1. *O. Hookeri.*
        Calyx-lobes 3 to 3.5 cm. long, the projecting tips in bud 3 to 6 mm. long
                                                                            var. *Hookeri.*
        Calyx-lobes 2 to 3 cm. long, the projecting tips in bud 1 to 2.5 mm. long
                                                                        var. *montereyensis.*
    Petals white, often aging pink.
        Plants with tap-root; herbage sparsely pilose . . . . .2. *O. deltoides.*
        Plants with deep spreading rhizomes; herbage densely white-pubescent
                                                                        3. *O. californica.*
Petals 2 to 20 mm. long; stigma round; flowers opening in sunshine.
    Flower-tube long and very slender, appearing like a peduncle; ovaries largely concealed among basal leaves.
        Perennial with fleshy tap-root; leaf-blades oblong to narrowly ovate
                                                                            4. *O. ovata.*
        Annual with slender root; leaf-blades linear to narrowly oblanceolate.
            Leaves and flowers closely tufted, the stem rarely with evident short branches; petals 8 to 10 mm. long . . . . .5. *O. graciliflora.*
            Stem usually with stout spreading branches up to 4 cm. long; petals 3 to 5 mm. long . . . . .6. *O. Palmeri.*
Flower-tube obconic or funnelform, much shorter than the ovary; ovaries not concealed by basal leaves.
    Upper leaves well developed, the flowers axillary.
        Capsule 4-angled, 2 to 3 mm. thick.
            Petals 2 to 6 mm. long . . . . .7. *O. micrantha.*
                Stems prostrate or widely spreading . . . . .var. *micrantha.*
                Stems erect.
                    Plants hairy . . . . .var. *Jonesii.*
                    Plants subglabrous . . . . .var. *ignota.*
            Petals 6 to 14 mm. long.
                Upper leaves oblong or obovate, sometimes cordate; perennial
                                                                    8. *O. cheiranthifolia.*
                    Leaves hairy.
                        Leaves whitish with dense appressed hairs, mostly narrowed toward base . . . . .var. *cheiranthifolia.*
                        Leaves green, pilose, mostly auriculate at base
                                                                        var. *viridescens.*
                    Leaves glabrous . . . . .var. *nitida.*
                Upper leaves lanceolate to oblanceolate; annual 9. *O. bistorta.*
        Capsule terete, 1 mm. thick or less.
            Petals 5 to 15 mm. long . . . . .10. *O. dentata.*
                Petals 5 to 8 mm. long.
                    Stems short-villous .var. *campestris.*
                    Stems appressed-puberulent to glabrous . . . .var. *Parishii.*
                Petals 8 to 15 mm. long .var. *Johnstonii.*
            Petals 2 to 4 mm. long . . . . .11. *O. contorta.*
    Upper leaves reduced to small bracts shorter than the flowers.
        Flowers crowded in the spike; petals white, often turning pink or red in age; capsules curved outward . . . . .12. *O. decorticans.*

        Flowers widely spaced; petals yellow; capsules straight or nearly so, deflexed .... 13. *O. leptocarpa.*

1. **O. Hookeri** T. & G. var. **Hookeri.** EVENING PRIMROSE. Common in wet places near coast, especially in hollows among dunes.

Var. **montereyensis** (Munz) Hoover, n. comb. *O. Hookeri* subsp. *montereyensis* Munz, El Aliso 2: 14. 1949. Probably as common as var. *Hookeri* along this part of the coast: south side of Morro Bay; with var. *Hookeri* at Oso Flaco Lake.

2. **O. deltoides** Torr. & Frem. var. **cognata** (Jepson) Munz. Sandy flats and hills along Cammatti Creek and its tributary Shell Creek. The plants bloom the first year from seed but are not necessarily annual as described in the literature, as many of them develop a stout fleshy tap-root and become perennial. The same fact holds true for the variety near the Merced River in the San Joaquin Valley.

3. **O. californica** Wats. In deep sand, about 6 or 7 miles east of Creston (*7100*); Santa Maria River (*Eastwood 305*).

4. **O. ovata** Nutt. SUN CUPS. GOLDEN EGGS. Firm sandy-loam or clay soils: Los Osos Valley, where surviving only along roadsides in 1948 and by now perhaps completely exterminated; abundant along coast from Cambria northward.

5. **O. graciliflora** H. & A. "Grassland" areas throughout the interior, widespread but the plants never plentiful in any one place.

6. **O. Palmeri** Wats. Coarse sandy or gravelly soils, Panorama Hills (*7820*) to Cuyama Valley (*8126*).

7. **O. micrantha** Hornem. var. **micrantha.** Very common in sand near coast; occasional inland as far as 10 miles east of Creston (*7099*).

Var. **Jonesii** (Lev.) Munz. Chaparral areas back from the coast, especially common following fires.

Var. **ignota** Jepson. Dry sandy or rocky slopes from Santa Lucia Range to La Panza Range and hills to the north.

8. **O. cheiranthifolia** Hornem. var. **cheiranthifolia.** BEACH EVENING PRIMROSE. Common in sand along the coast. The species is variable in several respects.

Var. **viridescens** (Lehm.) Hoover, n. comb. *O. viridescens* Lehm. in Hook., Fl. Bor.-Am. 1: 214. 1834. Near south end of Morro Bay and Hazard Canyon, on sandy cliffs. Not to be confused with var. *suffruticosa,* a more woody plant with silvery-white leaves and much larger flowers, although recent authors have included both under the same name.

Var. **nitida** (Greene) Munz. A rare form found sparingly with var. *cheiranthifolia* at Point Sierra Nevada.

9. **O. bistorta** Nutt. In sandy field, Los Osos (*8970*). The plants have broader leaves than other specimens of *O. bistorta* and may actually represent an annual form of *O. cheiranthifolia.*

10. **O. dentata** Cav. var. **campestris** (Greene) Jepson. Common in sandy soils, mainly in the interior, but approaching the coast in the Arroyo Grande and Santa Maria River watersheds.

Var. **Parishii** (Abrams) Munz. Rather poorly differentiated from var. *campestris* in this region and occupying about the same area, but decidedly less common.

Var. **Johnstonii** Munz. In sand, east side of Carrizo Plain south to Cuyama Valley. In years of favorable rainfall, this large-flowered variety forms a significant part of the spectacular flower displays.

11. **O. contorta** Dougl. Common in all parts of the county, usually in sandy soils. The varieties *strigulosa* (F. & M.) Munz and *epilobioides* (Greene) Munz are not here geographically segregated and have even been mixed in the same collection. Because they are exactly alike in general appearance, I believe they would be better classified as *formae,* as has been done in *Epilobium paniculatum,* a species which varies in a parallel manner.

12. **O. decorticans** (H. & A.) Greene. Frequent from San Juan River eastward, on steep banks of calcareous sand or clay and crumbling shale, and in sandy-clay washes. Some of the plants, in which the capsules are strongly curved downward, could be interpreted as intergrades toward var. *desertorum* Munz.

13. **O. leptocarpa** Greene. Frequent in eastern part on sandy or gravelly slopes which are otherwise barren.

Araliaceae. Ginseng Family

1. Aralia L.

1. **A. californica** Wats. Elk Clover. California Spikenard. Along streams and around hillside springs: frequent in Santa Lucia Range; See Canyon in San Luis Range.

Umbelliferae. Parsley Family

Flowers in umbels (or in *Sanicula* in heads arranged in umbels); bracts, bractlets, and sepals not spine-tipped; petals larger than sepals, or sepals absent.
  Fruit not bearing prickles, stiff bristles or tubercles.
    Leaves simple, round in outline, palmately lobed or crenate; umbels simple, few-flowered or reduced to a solitary flower,
      Aquatic or marsh plants; leaves and peduncles arising from stolons or rhizomes . . . . .1. *Hydrocotyle.*
      Terrestrial plants, with palmately lobed opposite leaves and short axillary peduncles . . . . .2. *Bowlesia.*
    Leaves compound, often finely divided into small segments; umbels compound.
      Fruit not flattened.
        Compound umbels, or some of them, sessile; involucels none.
          Leaves pinnate, with incised or deeply lobed leaflets
                              3. *Apium.*
          Leaves ternately subdivided into small narrow segments
                            4. *Apiastrum.*
        Compound umbels long-peduncled; involucels usually present (absent in *Foeniculum*).
          Petals yellow.
            Plant over 1 m. tall; leaf-divisions filiform
                          5. *Foeniculum.*
            Plant less than 1 m. tall; leaf-divisions ovate to oblong
                        6. *Tauschia.*
          Petals white.
            Leaves bipinnate or more finely divided.
              Stems not purple-dotted.
                Stylopodium prominently conical; leaf-divisions, or at least some of them, narrowly linear
                        7. *Perideridia.*

Stylopodium flattened or low-conical; leaf-divisions broad, serrate.
Stem stiffly erect; style less than half as long as fruit ....8. *Cicuta.*
Stem weak, tending to bend at nodes and grow horizontally; styles about half as long as fruit ....9. *Oenanthe.*
Stems purple-dotted ....10. *Conium.*
Leaves simply pinnate, with serrate-incised leaflets
11. *Berula.*
Fruit strongly flattened.
Plant over 1 m. tall; stem stout; petals white, those on outer margin of umbel enlarged ....12. *Heracleum.*
Plant less than 1 m. tall; stem less than 1 cm. thick; petals variously colored, not markedly enlarged on outer margin of umbel
13. *Lomatium.*
Fruit bearing stiff bristles or hooked prickles (or in *Sanicula tuberosa* bearing tubercules without prickles).
Fruit or its beak at least 3 times as long as thick.
Annual; petals white; fruit long-beaked ....14. *Scandix.*
Perennial; petals green; fruit slender, with short beak or none
15. *Osmorhiza.*
Fruit only slightly, if at all, longer than thick.
Umbels compound.
Involucre absent or represented by 1 bract ....16. *Anthriscus.*
Involucre present, of reduced but leaf-like bracts.
Fruit ribbed, with barbed or hooked prickles on the ribs.
Sepals absent; prickles of fruit barbed ......17. *Daucus.*
Sepals present; prickles of fruit hooked ....18. *Caucalis.*
Fruit not ribbed, tuberculate, the tubercles usually ending in a hooked prickle ....19..*Sanicula.*
Umbels simple, borne opposite the leaves ....20. *Torilis.*
Flowers in cymosely arranged heads; bracts, bractlets, and sepals spine-tipped; petals smaller than sepals ....21. *Eryngium.*

1. Hydrocotyle L.

Leaves peltate, the petiole attached near center of blade; blade crenate
1. *H. verticillata.*
Leaves not peltate, the petiole attached in a notch on one side of blade; blade palmately lobed, the lobes crenate ....2. *H. ranunculoides.*

1. **H. verticillata** Thunb. In wet soil or in water at margins of lakes, ponds, and streams near coast; often abundant.

2. **H. ranunculoides** L. f. In the same habitats as the preceding species; often growing with it.

2. Bowlesia R. & P.

1. **B. incana** R. & P. Mainly in shaded places, in loose soil, widespread but seldom collected in the county: Paso Robles; Cayucos; between San Juan River and Carrizo Plain, under junipers.

3. Apium L.

1. **A. graveolens** L. Celery. Well established in wet places near coast, from Price Canyon near Pismo Beach to Oso Flaco Lake.

4. Apiastrum Nutt.

1. **A. angustifolium** Nutt. In sandy or gravelly soils, at scattered localities in both eastern and western parts of county, but not generally common: Temblor Range; Paso Robles; Los Osos; Black Lake.

5. Foeniculum Adans.

1. **F. vulgare** Mill. Fennel. In clay or firm gravelly soils or on rocky banks, abundantly naturalized near coast, especially around San Luis Obispo; less common in Salinas Valley.

6. Tauschia Schlecht.

Leaves simply pinnate . . . . .1. *T. arguta.*
Leaves ternate, the primary divisions again divided . . . . .2. *T. Hartwegii.*

1. **T. arguta** (T. & G.) Macbr. Shady places in sandy or rocky soil: La Panza Range; apparently also in upper Arroyo Grande basin.

2. **T. Hartwegii** (Gray) Macbr. Shady places, usually in rocky soil, Santa Lucia Range eastward to Salinas River.

7. Perideridia Reichb.

Leaf-divisions all narrow.
    Tuberous roots mostly solitary or in pairs; petioles dilated at base but not inflated; fruit 2 to 3 mm. long . . . . .1. *P. Gairdneri.*
    Tuberous roots usually several, rarely solitary; petioles and midribs somewhat inflated; fruit 4 to 6 mm. long . . . . .2. *P. Pringlei.*
Leaf-divisions dimorphic, some linear and entire, others broader and toothed
                                                      3. *P. californica.*

1. **P. Gairdneri** (H. & A.) Mathias. Coastal grasslands and openings in pine woods, in soils which are wet in winter but may become very dry, from Cambria northward.

2. **P. Pringlei** (C. & R.) Nels. & Macbr. In clay soils: serpentine areas around San Luis Obispo; Highland district (about 10 miles east of Creston); abundant on summits of Caliente Range.

3. **P. californica** (Torr.) Nels. & Macbr. Rocky stream-beds, found in two separate areas of the county: Santa Lucia Range north of a line from Paso Robles to Cambria (San Carpoforo Creek, San Simeon Creek, Las Tablas Creek, etc.); Stoney Creek in Huasna River watershed.

8. Cicuta L. Water Hemlock

Ribs of fruit broader than intervals; seeds not very oily . . . . .1. *C. Douglasii.*
Ribs of fruit narrower than intervals; seeds very oily . . . . .2. *C. Bolanderi.*

1. **C. Douglasii** (DC.) C. & R. Rare in wet places: Atascadero; Black Lake Canyon.

2. **C. Bolanderi** Wats. Lopez Canyon (*Tracey & Viola Call 2546*). The collection, identified by Lincoln Constance, is not markedly different from specimens of *C. Douglasii,* and *C. Bolanderi* elsewhere is said to be confined to salt marshes. Consequently there exists some doubt that two different species are present here.

9. **Oenanthe** L.

1. **O. sarmentosa** Presl. Common in wet places throughout our coastal region, often very plentiful.

10. **Conium** L.

1. **C. maculatum** L. Poison Hemlock. Abundantly established in coastal area and along Salinas River and its tributaries.

11. **Berula** Hoffm.

1. **B. erecta** (Huds.) Cov. In shallow water or wet soil, common near coast; rare in interior (Trout Creek near Santa Margarita; Ortega Spring near Palo Prieto Canyon).

12. **Heracleum** L.

1. **H. lanatum** Michx. Cow-Parsnip. Rare along coastal streams northward: Santa Rosa Creek at Cambria; Little Pico Creek.

13. **Lomatium** Raf.

Leaves repeatedly divided into segments less than 2 cm. long.
  Bractlets obovate to broadly oblong.
    Leaves and peduncles all arising directly from the root-crown
                                             1. *L. caruifolium.*
    Cauline leaves as well as basal ones present . . . . .2. *L. utriculatum.*
  Bractlets linear to narrowly lanceolate.
    Plants more or less pubescent.
      Petals glabrous; fruit oblong, glabrate . . . . .3. *L. macrocarpum.*
      Petals pubescent; fruit oval or nearly circular, hairy.
        Herbage densely and persistently gray-pubescent; fruit finely pubescent . . . . .4. *L. mohavense.*
        Herbage at first woolly, often somewhat glabrate; fruit thinly tomentose to glabrate . . . . .5. *L. dasycarpum.*
    Plants glabrous or seemingly so.
      Leaf-segments small and very narrow . . . . .5. *L. dasycarpum.*
      Leaf-segments with broad pointed teeth . . . . .6. *L. parvifolium.*
Leaves once or twice divided into broad leaflets 2 to 5 cm. long, the leaflets coarsely toothed or incised . . . . .7. *L. californicum.*

1. **L. caruifolium** (H. & A.) C. & R. In clay soils, common in coastal region from San Luis Valley northward; Salinas Valley east of Paso Robles to Carrizo Plain and low hills near north end of Temblor Range.

2. **L. utriculatum** (Nutt.) C. & R. Common in grasslands or open woods throughout. Some specimens from the interior show minute sepals on the young fruits and in that respect correspond to *L. Vaseyi* C. & R. This feature is, however, not usually shown by all the fruits even on a single plant and can not be correlated with any other feature. It is concluded, therefore, that the plants previously called *L. Vaseyi* should be included in *L. utriculatum.*

3. **L. macrocarpum** (H. & A.) C. & R. In sandy soils or disintegrating shale, frequent in the area around northern end of La Panza Range.

4. **L. mohavense** C. & R. Caliente Mountain, locally plentiful on sandstone (*8215*).

5. **L. dasycarpum** (T. & G.) C. & R. Rocky hills near coast and eastward about to

Salinas River. Scarce in Temblor Range, according to Twisselmann; to be expected in La Panza Range. *Peucedanum Jaredi* Eastw., from near Estrella, is included here. A form found locally northwest of Cuesta Pass is unusual in having glabrous leaves but is duplicated in specimens from Marin Co.

6. **L. parvifolium** (H. & A.) Jepson. Rocky hills around San Luis Obispo, where usually growing in serpentine; north slope near mouth of Arroyo de la Cruz. In the plants near San Luis Obispo, the leaves are notably fleshy and more or less glaucous, corresponding to var. *pallidum* (C. & R.) Jepson. The variety does not, however, differ consistently otherwise.

7. **L. californicum** (Nutt.) M. & C. Wooded hills, localized in two separate areas: east side of Santa Lucia Range west of Paso Robles; Stoney Creek in Huasna River basin.

14. **Scandix** L.

1. **S. Pecten-Veneris** L. SHEPHERD'S NEEDLE. Introduced sparingly in western part, especially around San Luis Obispo.

15. **Osmorhiza** Raf.

Bractlets absent; pedicels mostly equalling fruit or longer . . . . .1. *O. chilensis.*
Bractlets several; pedicels much shorter than fruit . . . . .2. *O. brachypoda.*

1. **O. chilensis** H. & A. Common in shady woods in the hills of western part.

2. **O. brachypoda** Torr. Shaded places in Santa Lucia Range, apparently not quite so common as *O. chilensis.*

16. **Anthriscus** Hoffm.

1. **A. scandicina** (Weber) Mansfeld. BUR CHERVIL. Wooded slopes in Santa Lucia Range west of Paso Robles, where well established and abundant.

17. **Daucus** L.

Annual with slender root; bracts with short linear or lanceolate divisions
                                                                      1. *D. pusillus.*
Biennial with fleshy root; bracts with long filiform divisions . . . . .2. *D. Carota.*

1. **D. pusillus** Michx. Frequent in sandy or gravelly soils in all except driest parts of the county. A maritime race of the species found north of San Simeon show a tendency for the branches to spread horizontally.

2. **D. Carota** L. CARROT. Rarely escaping from cultivation and probably not becoming permanently established here: between Atascadero and Morro, *Winblad* in 1937.

18. **Caucalis** L.

1. **C. microcarpa** H. & A. Common in chaparral and wooded areas in all except the most arid localities.

19. **Sanicula** L.

Lower leaves toothed to deeply lobed or, if compound, the divisions sessile or decurrent.
    Flowers yellow, greenish, or orange; basal leaves not pinnatifid.
        Bractlets yellow, longer than flowers . . . . .1. *S. arctopoides.*
        Bractlets green, shorter than flowers.

Basal leaves palmately lobed or compound, the lobes or divisions acute
to acuminate.
Middle leaf-division little if at all longer than the others.
Stems mostly ascendingly branched above; lower leaves with
lobes or divisions toothed, rarely laciniate.
Basal leaves simple, palmately lobed; flowers yellow
2. *S. crassicaulis.*
Basal leaves with blades divided quite to the base into 3
leaflets; flowers orange or perhaps sometimes "greenish-
yellow" ....3. *S. Hoffmannii.*
Stems mostly divergently branched near base; lobes of leaves
deeply laciniate, or rarely merely serrate ......4. *S. laciniata.*
Middle leaf-division distinctly longer than the others, narrowed
below to a toothed rachis.
Bractlets linear-acuminate; flowers deep yellow . . 5. *S. arguta.*
Bractlets oblong, obtuse; flowers light yellow . . 6. *S. simulans.*
Basal leaves obtuse, simple and toothed, or divided into 3 toothed
leaflets ....7. *S. maritima.*
Flowers deep purple; basal leaves mostly pinnatifid ....8. *S. bipinnatifida.*
Lower leaves compound, at least the primary divisions with petiolules, not decurrent.
Root elongate; ultimate divisions of lower leaves lanceolate to ovate; fruit with
hooked prickles ....9. *S. bipinnata.*
Root tuberous, subglobose; leaves repeatedly divided into fine divisions; fruit
tuberculate but the tubercles not extended into hooked prickles 10. *S. tuberosa.*

1. **S. arctopoides** H. & A. Coastal grassland areas from San Simeon (*6657*) northward.

2. **S. crassicaulis** Poepp. Common, usually in shaded places, in western part, extending inland through La Panza Range.

3. **S. Hoffmannii** (Munz) Bell. Shady places in winter-moist clay soil from near
Monterey Co. line on San Carpoforo Creek (*8992*) to Reservoir Canyon and Rinconada Mine. Often, but not always, in serpentine areas.

4. **S. laciniata** H. & A. Locally abundant in portions of coastal area in more or
less sandy soils: hills between San Luis Valley and Arroyo Grande; pine woods
around Cambria, and northward. Some of the plants have basal leaves which are
not deeply parted or laciniate, resembling a leaf type which is common in *S. crassicaulis.* Conversely, there are plants in the Santa Lucia Mts. which seem to belong
to *S. crassicaulis* but in which the leaves suggest *S. laciniata.* Such plants could be
of hybrid origin, but more probably each species varies independently.

5. **S. arguta** Greene. Common in coastal region from San Simeon southward, extending inland to Cuesta Pass, upper Arroyo Grande, and lower Cuyama Canyon.

6. **S. simulans** Hoover. Open hills near San Luis Obispo; ocean bluff between
Morro Bay and Cayucos. Perhaps overlooked elsewhere because of its close resemblance to *S. arguta.*

7. **S. maritima** Kell. In more or less moist clay soils on low hills and in valleys
west of San Luis Obispo and near Arroyo de la Cruz. Otherwise known only in the
San Francisco Bay region, where long ago exterminated.

8. **S. bipinnatifida** Dougl. Widespread and often plentiful in grassland or sparsely
wooded areas, mainly in the interior but reaching the coast at San Simeon. Absent
from region of Carrizo Plain.

9. **S. bipinnata** H. & A. Common in woods or on open north slopes throughout

the interior. We have no record of this species in the coastal area, though it may occur there.

10. **S. tuberosa** Torr. From summit of Santa Lucia Mts. eastward to low hills bordering La Panza Range, widely scattered and sometimes locally plentiful but not generally common.

20. **Torilis** Adams.

1. **T. nodosa** (L.) Gaertn. KNOTTED HEDGE-PARSLEY. Apparently sparingly introduced: Arroyo Grande (Jepson, Fl. Cal. 2: 673).

21. **Eryngium** L.

Bracts and bractlets spine-tipped but entire, without marginal spines 1. *E. armatum.*
Bracts (surrounding heads) bearing marginal spines, at least near base.
    Bractlets (within heads) entire, or with a pair of small teeth at base
                                                  2. *E. aristulatum.*
    Bractlets with marginal spines . . . . .3. *E. Vaseyi.*
        Heads above bracts less than 20 mm. in diameter; sepals entire
                                              var. *Vaseyi.*

        Heads above bracts mostly 20 mm. in diameter or more at maturity; sepals, or some of them, with spiny teeth on margin . . . . .var. *globosum.*

1. **E. armatum** (Wats.) C. & R. Coastal plains, open woods, or winter-flooded places: Laguna near San Luis Obispo; common on coast from Cambria northward. San Simeon is the type locality of *E. longistylum* C. & R., which is here included in *E. armatum.*

2. **E. aristulatum** Jepson. Locally found only around the Laguna near San Luis Obispo (*6265, 6328*).

3. **E. Vaseyi** C. & R. var. **Vaseyi.** In low ground, often in depressions which are flooded after rains, Palo Prieto Pass to near Creston, Paso Robles Airport, and northward. A collection from Choice Valley (*E. McMillan 154*) is, in the size of its heads, intermediate toward var. *globosum.*

Var. **globosum** (Jepson) Hoover. Local on Carrizo Plain south of Simmler.

Cornaceae. DOGWOOD FAMILY

1. **Cornus** L. DOGWOOD

Leaves with 3 or 4 veins on each side of midrib; twigs brownish or grayish
                                                1. *C. glabrata.*
Leaves with 4 to 7 veins on each side of midrib; twigs red . . . . .2. *C. stolonifera.*
    Lower leaf-surfaces sparsely appressed-pubescent with nearly straight hairs
                                          var. *stolonifera.*
    Lower leaf-surfaces densely covered with curved or curly hairs var. *occidentalis.*

1. **C. glabrata** Benth. Along streams or dry stream-beds, frequent in Santa Lucia Mts. and eastward to near Pozo.

2. **C. stolonifera** Michx. var. **stolonifera.** RED-OSIER DOGWOOD. Shrubs found along Tassajera Creek (*9677*) fit the description of typical *C. stolonifera*. Because the occurrence here of only var. *occidentalis* has been assumed, it may well be that other occurrences have been overlooked.

Var. **occidentalis** (T. & G.) Hoover, n. comb. *C. sericea* var. *occidentalis* T. & G.,

Fl. N. Amer. 1: 652. 1840. Western part of county, in moist places, probably much commoner than var. *stolonifera,* but extensive collections have not been made.

Lennoaceae. Lennoa Family

1. Pholisma Nutt.

1. **P. arenarium** Nutt. In sand near south end of Morro Bay and on Nipomo Dunes. Here it is parasitic, so far as is definitely known, on the roots of *Ericameria ericoides,* but various other hosts have been recorded elsewhere. Coastal plants have been named *P. depressum* Greene or, more recently, *P. paniculatum* Templeton, and may actually be distinct from the desert plants which have been included in the same species. An ample series of living plants from both coast and desert would be needed for a proper decision on this question. Available material is far from adequate.

Ericaceae. Heath Family

Leaves serrulate; calyx in fruit becoming fleshy, forming a dark blue berry-like fruit
1. *Gaultheria.*
Leaves (except on juvenile shoots) entire or rarely sparingly dentate; calyx small, inconspicuous at the base of the red, brown, or white berry.
    Surface of berry granular; tree becoming over 10 m. tall . . . . .2. *Arbutus.*
    Surface of berry not granular; shrubs or small trees rarely over 5 m. tall
3. *Arctostaphylos.*

1. Gaultheria L.

1. **G. Shallon** Pursh. Salal. Rare on north slopes near coast: Islay Creek in San Luis Range; Cerro Alto, *Condit & Waters.* The presence of this species on the mountain now called Cerro Alto is improbable; that name may perhaps have been applied to Hollister Peak.

2. Arbutus L.

1. **A. Menziesii** Pursh. Madrone. Santa Lucia Mts., mainly near the summits, southward to Lopez Canyon; sparingly in Coon Creek basin of San Luis Range. A stray sapling was seen in the pine forest at Cambria.

3. Arctostaphylos Adans. Manzanita

Leaf-blades cuneate to rounded or truncate at base, sometimes subcordate, but then the rounded basal lobes much shorter than the petioles.
    Low spreading to erect shrubs or small trees, never creeping; leaves acute to obtuse; flowers in a panicle.
        Bark rough and shaggy, brown to gray.
            Leaf-blades with both surfaces alike, bright green, glabrous or glabrate; bracts 2 to 4 mm. long . . . . .1. *A. rudis.*
            Leaf-blades with surfaces unlike, the lower surface paler and usually somewhat tomentose; lower bracts of panicle 5 to 15 mm. long.
                Shrub not forming a burl . . . . .2. *A. morroensis.*
                Shrub usually forming a basal burl . . . . .3. *A. tomentosa.*
        Bark smooth, dark red.
            Not forming a burl.
                Branchlets with scattered long spreading hairs ("bristles" of authors) as well as pubescent.

Berries glabrous, not viscid ....4. *A. pilosula.*
Berries strongly glandular-viscid ....13. *A. Hooveri.*
Branchlets pubescent or glabrous, without long hairs.
Leaf-blades at first white-pubescent, often more or less gla-
brate in age; berry containing separate nutlets, not glandular
5. *A. obispoensis.*
Leaf-blades glabrous from the first; berry with a solid stone,
glandular ....6. *A. glauca.*
Branchlets glabrous ....var. *glauca.*
Branchlets finely pubescent ....var. *puberula.*
Stems forming a large woody swelling (burl) at or just below the
ground, from which vigorous shoots arise after burning or cutting.
Both surfaces of leaves essentially alike, bright green to glaucous
or white-puberulent.
Branchlets with long spreading hairs ....7. *A. glandulosa.*
Branchlets and inflorescences glandular ..var. *glandulosa.*
Branchlets and inflorescences pubescent but not glandular
var. *Campbelliae.*
Branchlets pubescent, without long spreading hairs
8. *A. Cushingiana.*
Lower leaf-surface paler or duller than the upper, bearing most or
all of the stomata.
Branchlets and inflorescences glandular .....9. *A. subcordata.*
Branchlets and inflorescences not glandular ..10. *A. crustacea.*
Prostrate mat-forming creeper; leaves acuminate; flowers 7 or fewer in a simple
raceme ....11. *A. Hearstiorum.*
Leaf-blades auriculate at base, sessile or with petioles mostly shorter than the basal
lobes.
Branchlets with long spreading hairs as well as finely puberulent.
Leaf-blades bright green, glabrous or slightly pubescent at base only; ber-
ries glabrous ....12. *A. pechoensis.*
Leaf-blades at first hairy, glabrate and glaucous or persistently white-puberulent; berries usually strongly glandular, sometimes varying to merely pubescent or nearly glabrous ....13. *A. Hooveri.*
Branchlets densely pubescent, with no long spreading hairs.
Shrubs erect or somewhat spreading; leaves whitish, glaucous or persistently
white-puberulent; berries 9 to 12 mm. in diameter ....14. *A. luciana.*
Shrubs mound-shaped or with sprawling stems; leaves bright green at least
in age, glabrous to partly glabrate; berries 6 to 9 mm. in diameter
15. *A. cruzensis.*

1. **A. rudis** Jeps. & Wies. Shagbark Manzanita. In sand on Nipomo Mesa; formerly plentiful but rapidly being destroyed by clearing operations. Repeated un-
successful search has been made for the basal burl which is said in the original description to be present. Apparently the presence or absence of that structure is not
a distinctive feature of this species.

2. **A. morroensis** Wies. & Schr. In sand, south side of Morro Bay, extending to
just south of Hazard Canyon. A highly localized species. Great numbers of shrubs
have been destroyed, but it still is plentiful on the slope south of Los Osos. The
species is easily adaptable to cultivation. An occasional shrub with rose-pink flowers
represents a form of superior ornamental value. A shrub from Los Osos Valley with
auriculate leaves (*6600*) may be a hybrid with *A. cruzensis.*

3. **A. tomentosa** (Pursh) Lindl. Infrequent near coast: Arroyo de la Cruz; pine woods at Cambria; sandstone hills on northeast side of Los Osos Valley, the southernmost locality for true *A. tomentosa*.

4. **A. pilosula** Jeps. & Wies. Restricted to this county, where localized in three separate areas: higher elevations in La Panza Range from near summit on Pozo Road southward an undetermined distance (head of American Canyon, the type locality); east base of Santa Lucia Range between Santa Margarita and Atascadero; from San Luis Range (upper Coon Creek, *6574*) to upper Arroyo Grande and Nipomo. Most plentiful on sandstone hills between San Luis Valley and the coast.

Forma **microphylla** Hoover, n. f. Frutex erectus ad 4 dm. altus; foliis ovatis ad orbicularibus, 7 ad 15 mm. longis. Dwarfed, up to 4 dm. tall; leaves only 7 to 15 mm. long. Summit of La Panza Range east of Pozo, *Hoover 8553*. This unique shrub represents the extreme in a series of progressively smaller-leaved variants.

5. **A. obispoensis** Eastw. Common, often locally abundant, in Santa Lucia Range from Cuesta Pass northwestward into Monterey Co.; often, but by no means always, in areas of serpentine rock. The type locality is upper Chorro Creek. A solitary shrub occurs in the pine woods at Cambria, and another grew in lower Lopez Canyon before the construction of the dam. The leaves are highly variable in size, shape, pubescence, and color. Doubtfully distinct from *A. canescens* Eastw., which shows a comparable range of variation. Individual shrubs with comparatively narrow leaves represent *A. silvicola* Jeps. & Wies., which has been supposed to be restricted to Santa Cruz Co. No adequate basis for separating it from *A. obispoensis* can now be found.

6. **A. glauca** Lindl. var. **glauca**. Common on east side of Santa Lucia Range and eastward to La Panza Range. Some individuals become remarkably large. An unusual occurrence is on a serpentine outcrop near the Pozo-Arroyo Grande summit. Here the leaves on some shrubs vary to a yellow-green color, giving them an appearance very different from the normal form of the species.

Var. **puberula** J. T. Howell. Associated with var. *glauca* in La Panza Range and in Cuyama Canyon; apparently replacing it entirely in the upper Arroyo Grande basin, where extending westward to lower Lopez Canyon, and, according to Twisselmann, on east slope of Temblor Range. The exceptional trees which once existed in the upper Arroyo Grande region, among the largest manzanitas found anywhere, all seem to have been destroyed in a state-sponsored "range improvement" program.

7. **A. glandulosa** Eastw. var. **glandulosa**. Rare in the county, at least in typical form: Santa Lucia Mts. (Pine Top Mt.; near Rocky Butte; near summit on Paso Robles-Adelaida road); summits of La Panza Range. An apparent intergrade toward *A. Cushingiana* has been named var. *Howellii* (Eastw.) Adams (*A. Howellii Eastw.*).

Var. **Campbelliae** (Eastw.) Adams. This variant has been previously reported only from the Mt. Hamilton Range of Santa Clara Co., but shrubs which "key out" to it were found on the ridge opposite County Park on upper Arroyo Grande (*6552*), and near Cerro Alto (*9334*).

8. **A. Cushingiana** Eastw. *A. glandulosa* var. *crassifolia* Jepson. Very common in Santa Lucia Range, often growing with other manzanitas, sometimes forming dense pure stands; La Panza Range; ridge south of Los Osos. Hybridization with *A. glandulosa* unquestionably occurs, but not to any greater extent than between many other pairs of species in the genus. Inclusion of *A. Cushingiana* in *A. glandulosa* would

therefore logically call for the combining of other species in a way that contemporary students of the genus would find unacceptable. Within certain well-defined limits, *A. Cushingiana* is highly variable. Some shrubs in the La Panza Range with small almost glabrous leaves probably furnish the basis for reports of *A. pungens* H. B. K. in this region (Jepson, Fl. Cal. 3: 44). Vegetative branches and even flowering branches show a close resemblance to Mexican specimens of *A. pungens,* but our shrubs differ from *A. pungens* in having a burl at the base.

9. **A. subcordata** Eastw. North-facing slope among live oaks, on sandstone, Price Canyon between Edna and Pismo Beach (*6614, 6767, 7305*). The types of both *A. subcordata* and *A. confertiflora* Eastw., which I believe to be specimens of the same species, happen to have subcordate leaves. Other specimens from the type localities, Santa Cruz and Santa Rosa Islands, show leaves with the blades merely rounded at base and do not differ in any noticeable way from the mainland shrubs cited here.

10. **A. crustacea** Eastw. In this county apparently restricted to San Luis Range, where found in the watersheds of Coon Creek and Islay Creek.

11. **A. Hearstiorum** Hoover & Roof. Local on coastal hills north and south of Arroyo de la Cruz. An evident hybrid between this species and *A. cruzensis* was seen where the two grew together. Records of *A. Hookeri* G. Don in the county refer actually to this species.

12. **A. pechoensis** Dudley. Localized in western part of San Luis Range, from head of See Canyon down to the coast near mouth of Coon Creek. The type locality is "head of Wild Cherry Canyon." Reports of the occurrence of the species outside this small area, so far as I can learn, are all due to incorrect application of the name. *Arctostaphylos viridissima* (Eastw.) McMinn, an endemic of Santa Barbara Co., has been said to occur in San Luis Obispo Co., but such reports can be traced back to specimens of genuine *A. pechoensis,* or perhaps sometimes of *A. cruzensis.*

13. **A. Hooveri** Wells. Santa Lucia Mts. toward the north, commonly an associate of *Pinus ponderosa,* often locally abundant: rhyolite hill between Rocky Butte and Pine Mt. to Pine Top Mt. and northward into Monterey Co. On Pine Top Mt. was collected a series of specimens in which the berries range from nearly glabrous to hairy but not glandular and to densely glandular-hairy (the typical condition). In view of the range of variation, it is difficult to see how this species can be kept separate from either *A. glutinosa* Schreiber or *A. Andersonii* Gray, both of which have been regarded as endemics of the Santa Cruz Mts. Perhaps these plants will all eventually be included under *A. Andersonii* (the earliest published name), but the problem is one which calls for much more extensive field study, with special emphasis on the extent of variation in the shrubs at a particular locality.

14. **A. luciana** Wells. Rocky slopes in Santa Lucia Mts.; restricted to this county. Locally abundant and in places forming pure stands in the white shale area southeast of Cuesta Pass; rare to the northward: "road from Morro to Atascadero," *Eastwood 15,009;* Klau Mine, *Twisselmann 2508.* A collection from the summit of Cuesta Pass (*8629*) is intermediate toward *A. obispoensis.*

15. **A. cruzensis** Roof. Infrequent on coastal hills: sandstone ridge on northeast side of Los Osos Valley; sparingly in pine woods near Cambria; near Arroyo de la Cruz, the type locality. Rumors of the presence of a manzanita on Black Mountain near Morro Bay and on Hollister Peak may be based on this species, but no specimens from those localities have been seen.

Vacciniaceae. Huckleberry Family

1. Vaccinium L.

1. **V. ovatum** Pursh. Huckleberry. In scattered localities near coast, mainly on north-facing slopes: Arroyo de la Cruz; San Luis Range (forming a dense pure stand along lower Coon Creek) and eastward to sandstone hills bordering San Luis Valley on the south.

Primulaceae. Primrose Family

Flowers in umbels or solitary.
    Leaves not in a basal rosette.
        Leaves closely spaced in an apparent whorl at top of stem, the flowers on filiform pedicels in a sessile umbel . . . . .1. *Trientalis.*
        Leaves well distributed along the stems, the flowers axillary.
            Leaves opposite.
                Calyx white, appearing like a corolla; corolla absent . . .2. *Glaux.*
                Calyx green; corolla orange-red . . . . .3. *Anagallis.*
            Leaves alternate; corolla very small and inconspicuous.
                                4. *Centunculus.*
    Leaves in a basal rosette.
        Corolla inconspicuous, its lobes ascending or spreading . . . . .5. *Androsace.*
        Corolla showy, its lobes reflexed . . . . .6. *Dodecatheon.*
Flowers in racemes . . . . .7. *Samolus.*

1. Trientalis L.

1. **T. latifolia** Hook. Starflower. Frequent in humus-rich soil under trees near coast from San Luis Range northward, and in Santa Lucia Range as far south as Lopez Canyon.

2. Glaux L.

1. **G. maritima** L. Sea Milkwort. Oceano Beach, at edge of marsh behind sand-dunes (*7726*); Pismo.

3. Anagallis L.

1. **A. arvensis** L. Pimpernel. Common weed near coast, especially in gardens; occasionally found in interior.

4. Centunculus L.

1. **C. minimus** L. In moist sandy openings among the pines at Cambria; probably overlooked elsewhere.

5. Androsace L.

1. **A. acuta** Greene. Grassland areas of eastern part, uncommon or overlooked: Cottonwood Pass to Temblor Range and Carrizo Plain.

6. Dodecatheon L.

1. **D. Clevelandii** Greene. Shooting Stars. In grasslands or open woods, occasional near coast from east side of San Luis Range northward, and common from

summits of Santa Lucia Range to western edge of Carrizo Plain. According to H. J. Thompson, the coastal plants are subsp. *sanctarum,* and those of the interior are subsp. *insulare.* Dried specimens do not separate readily into two groups, and there are greater differences within each subspecies than between them.

7. Samolus L.

1. **S. floribundus** H. B. K. In wet places: Oceano; upper Arroyo Grande.

Plumbaginaceae. PLUMBAGO FAMILY

Leaves of rosette broad; flowers in an openly branched inflorescence .. 1. *Limonium.*
Leaves of rosette narrowly linear; flowers in a head .... 2. *Armeria.*

1. Limonium Mill.

Leaves entire; stems terete .... 1. *L. commune.*
Leaves sinuate-pinnatifid; stems winged on the angels .... 2. *L. sinuatum.*

1. **L. commune** S. F. Gray. var. **mexicanum** (Blake) Jepson. Locally common in salt-marshes bordering Morro Bay and at Avila.

2. **L. sinuatum** (L.) Mill. Garden flower native to Mediterranean region, established at Wellsona Crossing between Paso Robles and San Miguel, where persistent for several years; also escaping from gardens in coastal region (Cayucos).

2. Armeria Willd.

1. **A. maritima** (Mill.) Willd. var. **california** (Boiss.) Lawr. SEA-PINK. THRIFT. Common on coastal bluffs from vicinity of Cambria northward.

Styracaceae. STORAX FAMILY

1. Styrax L.

1. **S. officinalis** L. var. **fulvescens** (Eastw.) M. & J. SNOWDROP BUSH. 1¼ miles northwest of Colwell Ranch (Stoney Creek area), H. C. LEE.

Oleaceae. OLIVE FAMILY

Leaves pinnate; fruit a samara .... 1. *Fraxinus.*
Leaves simple; fruit a drupe .... 2. *Forestiera.*

1. Fraxinus L. ASH

1. **F. dipetala** H. & A. FLOWERING ASH. Foothills west of Paso Robles; Cuyama Canyon on Santa Barbara Co. side, and probably entering the county on the north side of the canyon.

2. Forestiera Poir.

1. **F. neo-mexicana** Gray. DESERT OLIVE. Rare along streams or around springs in interior: Paso Robles (Jepson, Fl. Cal. 3: 82); Temblor Range in Kern Co., and perhaps entering San Luis Obispo Co. in that area.

Gentianaceae. GENTIAN FAMILY

Corolla-lobes 4, yellow .... 1. *Microcala.*
Corolla-lobes 4 or 5, rose-red to white .... 2. *Centaurium.*

1. Microcala Hoffmgg. & Link

1. **M. quadrangularis** (L.) Griseb. Seen in a moist sandy opening among pines at Cambria. No specimens from that locality can now be found.

2. Centaurium Hill.

Pedicels mostly equalling flowers or shorter; corolla-lobes 4 to 6 mm. long
1. *C. Davyi.*
Pedicels mostly longer than flowers; corolla-lobes 3 to 4 mm. long   2. *C. exaltatum.*

1. **C. Davyi** (Jepson) Abrams. Occasional near coast, probably in sandy soils: Cambria, under pines (*Eastwood 15,153*); road from Arroyo Grande to Huasna (Eastwood *14,989*). Jepson's report of *C. Muhlenbergii* in San Luis Obispo Co. (Fl. Cal. 3: 85) is based on specimens which probably are *C. Davyi.*

2. **C. exaltatum** (Griseb.) Wight. Occasional in moist places in interior: Cottonwood Pass (*7272*); Mariana Creek in La Panza Range; Wilson Canyon Creek.

Apocynaceae. Dogbane Family

Stems long-trailing; flowers solitary from the axils; corolla blue . . . . .1. *Vinca.*
Stems erect; flowers in cymes; corolla white to pale green or pink . . . .2. *Apocynum.*

1. Vinca L.

1. **V. major** L. Periwinkle. Planted as a ground cover; often naturalized on shaded stream-banks near coast.

2. Apocynum L. Dogbane

Plant 6 to 20 dm. tall; stem with ascending branches above.
  Blades of upper leaves narrowed toward base into petioles 2 mm. long or more
1. *A. cannabinum.*
  Blades of upper leaves rounded at base, sessile or with petioles less than 2 mm. long . . . . .2. *A. sibiricum.*
Plant 3 dm. tall or less; stem with spreading branches above . . . . . . .3. *A. pumilum.*

1. **A. cannabinum** L. var. **glaberrimum** A. DC. Indian Hemp. Low places, usually along streams, frequent in western portion, extending inland at least to Salinas River.

2. **A. sibiricum** Jacq. Specimens identified as this species have been found near San Luis Obispo and at Sandiego Joe's Spring in the Temblor Range. They do not seem to me, however, to be clearly distinguishable from *A. cannabinum.* Californian plants of this species are generally called var. *salignum* (Greene) Fernald, but plants found at the Stalnaker Orchard near San Luis Obispo (*Sam Wright* in 1966) look exactly like var. *cordigerum* (Greene) Fernald, mainly of the eastern states and not previously reported from California.

3. **A. pumilum** Greene. Rare in Santa Lucia Mts., in open woods: Pine Mt. (*8019*). Plants were not found in flower or fruit but are readily recognizable by their vegetative features.

Asclepiadaceae. Milkweed Family

1. Asclepias L. Milkweed

Leaves woolly, often glabrate in age, 2 or 3 at a node, ovate or oblong.
  Hoods dark red, without horns . . . . .1. *A. californica.*
  Hoods white to pale yellow or purple-tinged, each with a horn inside.

Stems decumbent; lateral umbels usually sessile, sometimes short-peduncled
2. *A. vestita.*
Stems erect; lateral umbels peduncled.
Some of leaves usually whorled, persistently woolly; horns scarcely exserted from hoods . . . . .3. *A. eriocarpa.*
Leaves all opposite, glabrate; horns well exserted from hoods
4. *A. erosa.*
Leaves glabrous, 2 to 6 at a node, mostly linear to oblong-lanceolate
5. *A. fascicularis.*

1. **A. californica** Greene. Rare on dry ridges in Santa Lucia Mts.: between Rocky Butte and Pine Mt. (*7894*).

2. **A. vestita** H. & A. In sandy or rocky soils, common in Salinas Valley and eastward to La Panza district; occasional in Temblor Range.

3. **A. eriocarpa** Benth. Common in dry soil in Santa Lucia Range; occasional near coast; eastward to Shandon Hills and La Panza Range; also Temblor Range, according to Twisselmann.

4. **A. erosa** Torr. DESERT MILKWEED. Pass at southeast end of Carrizo Plain (*10,148*), the plants scattered, most of them on a north-facing slope.

5. **A. fascicularis** Dec. Common in low ground and along streams in the hills, sometimes becoming a weed in cultivated land; commonest westward, but rarely if ever on the coastline; in eastern part only around springs and along streams.

Convolvulaceae. MORNING-GLORY FAMILY

Plants with leaves; stems green or white-woolly.
Stems creeping, mat-forming; fruit deeply 2-lobed . . . . .1. *Dichondra.*
Stems trailing, twining, or ascending, not mat-forming; fruit not 2-lobed.
Corolla 15 mm. long or more; leaf-blades normally with 2 basal lobes, or at least truncate at base . . . . .2. *Convolvulus.*
Corolla 5 to 6 mm. long; leaf-blades without basal lobes.
Annual; stems trailing; leaf-blades gradually narrowed into a petiole; style 1 . . . . .2. *Convolvulus simulans.*
Perennial with extensive rhizomes; stems ascending; leaf-blades sessile or abruptly narrowed into a very short petiole; styles 2 . . . . .3. *Cressa.*
Leafless parasites; stems yellow or orange . . . . .4. *Cuscuta.*

1. **Dichondra** Forst.

Leaf-blades broader than long, almost glabrous on lower surface   1. *D. Donnelliana.*
Leaf-blades orbicular in outline, about as broad as long, thinly strigose on lower surface . . . . .2. *D. carolinensis.*

1. **D. Donnelliana** Tharp & Johnston. Dry, open hillside facing ocean just south of Arroyo de la Cruz. Probably overlooked elsewhere.

2. **D. carolinensis** Michx.? The Dichondra which is commonly used for lawns, of which the exact identity is uncertain, may sometimes be regarded as a weed in lawns or gardens where it is not wanted.

2. **Convolvulus** L.

Perennials; leaf-blades usually sagittate, hastate, or reniform, or sometimes merely truncate at base; corolla at least 15 mm. long.
Leaves succulent, reniform, broader than long . . . . .1. *C. Soldanella.*
Leaves not succulent, not broader than long.

Flower subtended by a pair of bracts which closely envelop the calyx;
bracts about as long as calyx-lobes or only a little shorter.
Plants glabrous to pubescent, not woolly.
Plants glabrous or very nearly so . . . . .2. *C. cyclostegius*.
Bracts orbicular to oblong-oval, broader than calyx-lobes
var. *cyclostegius*.
Bracts lanceolate to ovate-lanceolate, similar to the calyx-lobes
var. *intermedius*.
Plants pubescent, though often minutely so.
Stems usually showing some tendency to twine; leaf-margins
nearly straight or with a concave curve on either side
3. *C. aridus*.
Stems with little or no tendency to twine; leaf-margins each
forming a strongly convex curve . . . . .4. *C. subacaulis*.
Plants white-woolly . . . . .5. *C. malacophyllus*.
Flower with an evident pedicel above the pair of bracts; bracts much nar-
rower than calyx-lobes and usually shorter.
Corolla 20 to 40 mm. long; leaves acute or acuminate, the margins
straight or concavely curved on either side.
Basal lobes of leaves broad, often toothed; corolla 3 to 4 cm. long
6. *C. occidentalis*.
Basal lobes of leaves narrow, entire; corolla 2 to 3 cm. long
7. *C. longipes*.
Corolla 15 to 20 mm. long; leaves often obtuse, the margins convexly
curved on either side . . . . .8. *C. arvensis*.
Annual; leaf-blades narrowed toward base; corolla 6 mm. long . . . . .9. *C. simulans*.

1. **C. Soldanella** L. BEACH MORNING-GLORY. In sand at edge of beach near south
end of Morro Bay (*7297*); dunes south of Oso Flaco Lake.

2. **C. cyclostegius** House var. **cyclostegius.** Very common near coast, especially in
rocky soil among shrubs; inland at least to summits of Santa Lucia Mts. Coastline
plants from Cambria northward have short stems without much tendency to twine,
and the flowers are generally more rosy. Careful comparisons of living plants might
reveal a distinguishable entity in that region. Collectors, regrettably, have often
ignored morning-glories, so that it is difficult to determine the range of variation
in any species, or the precise area occupied by it.

Var. **intermedius** (Abrams) Hoover, n. comb. *C. aridus* Greene subsp. *intermedius*
Abrams, Contr. Dudley Herb. 3: 357. 1946. Plants which key out to *C. aridus* subsp.
intermedius Abrams, found on the ridge northwest of Cuesta Pass and in Cuyama
Canyon, do not seem adequately separable from *C. cyclostegius*. There is a continuous gradation from nearly orbicular to lanceolate bracts, and width of the bracts is
not clearly correlated with any other feature.

3. **C. aridus** Greene. A plant which may belong to this species, but having exceptionally long and narrow bracts, was found near Creston (*Jasper Ferrero 25* in 1953.)
Unidentified plants seen, but not collected, in the hills north of Pozo are also to
be compared with this species.

4. **C. subacaulis** Greene. In grasslands, generally in clay soils: San Luis Valley and
bordering hills; upper Salinas Valley near Santa Margarita, and northward.

5. **C. malacophyllus** Greene. WOOLLY MORNING-GLORY. Frequent in hilly country
in interior, mostly in rocky soils: Upper Arroyo Grande; upper Salinas Valley eastward to La Panza Range. On geographic grounds our plants should be subsp. *pedi-*

cellatus (Jepson) Abrams, but no consistent difference from the typical phase of the species can be found.

6. **C. occidentalis** Gray. Occasional in Santa Lucia Range: serpentine outcrop near Pozo-Arroyo Grande summit; Paso Robles Creek near York Mountain; hills west of Paso Robles. The few available specimens do not fall readily into the named varieties.

7. **C. longipes** Wats. Dry hills among shrubs, usually in decomposed granite or sandstone: hills bordering Salinas River eastward to La Panza Range.

8. **C. arvensis** L. BINDWEED. FIELD MORNING-GLORY. Low ground, commonly in fertile clay soils, mostly in cultivated areas and by roadsides. That the native morning-glories almost never invade cultivated ground, while this European introduction is an aggressive weed, is a difficult fact to explain.

9. **C. simulans** Perry. Open treeless areas in northeastern part of county, growing in friable clay: Cottonwood Pass to Estrella. The species is found in rather few widely separated places.

3. Cressa L.

1. **C. truxillensis** H. B. K. Common in low saline places throughout eastern part.

4. Cuscuta L. DODDER

Corolla-lobes reflexed; capsule depressed-globose . . . . .1. *C. californica.*
Corolla-lobes spreading or ascending; capsule ovoid.
    Corolla-lobes about half as long as corolla-tube . . . . .2. *C. Ceanothi.*
    Corolla-lobes about as long as the tube . . . . .3. *C. salina.*

1. **C. californica** H. & A. Widespread over most of county: Avila (*Eastwood 13,772*); common in interior, where it often parasitizes *Eriogonum fasciculatum.* Other genera noted as hosts include *Hemizonia, Pickeringia, Lotus, Lupinus, Navarretia,* and *Trichostema.* Dodder is abundant, but collections are few.

2. **C. Ceanothi** Behr. *C. subinclusa* D. & H. Wooded and brushy areas, mainly along streams in western part. Noted as growing most vigorously on *Datisca* on Santa Rita Creek, on *Thermopsis* in Rinconada district, and on *Sambucus* along Arroyo Grande. Other hosts observed include *Equisetum, Scirpus, Salix, Photinia, Rhamnus, Mentha, Stachys, Symphoricarpos,* and *Artemisia.*

3. **C. salina** Engelm. SALT-MARSH DODDER. Shores of Morro Bay and at Avila. Reported elsewhere as a parasite on a wide assortment of salt-marsh plants, but here, remarkably, nearly always on *Jaumea.* It was found once on *Frankenia.* Our plants are the typical form, not var. *major* Yuncker.

Polemoniaceae. PHLOX FAMILY

Leaves below inflorescence alternate, sometimes forming a basal rosette.
    Herbs; leaves (except sometimes the uppermost) not palmately parted.
        Flowers either loosely arranged or in bractless heads, or sometimes in a small cluster which is subtended by a single leafy bract.
            Calyx often enlarging in fruit but eventually ruptured by the growing capsule; stamens equally inserted; seeds usually many . . . . . . .1. *Gilia.*
            Calyx growing with the capsule, not usually ruptured by it; stamens unequally inserted; seeds 1 or 2 per locule . . . . .2. *Allophyllum.*
        Flowers in heads which are surrounded by bracts, and each flower usually subtended by a bract.

Calyx enlarging in fruit, the sinus between each 2 lobes extended into a projecting fold; bracts never spiny ....3. *Collomia.*
Calyx little enlarged in fruit, the sinuses without a projecting fold; bracts in most species spiny.
Corolla-lobes equal, though not always symmetrically arranged; leaves with acicular to filiform divisions, or entire.
Heads without tangled wool (bracts white-villous toward base in *N. intertexta* but the hairs hardly tangled)..4. *Navarretia.*
Heads with white tangled wool ....5..*Eriastrum.*
Corolla-lobes unequal, arranged in 2 lips; leaves with triangular lobes or teeth tipped with long bristles ....6. *Langloisia.*
Shrubs; leaves palmately parted nearly to base into spine-tipped divisions
7. *Leptodactylon.*
Leaves below inflorescence opposite.
Leaves entire, oblong or lanceolate to linear, the lowest often broader 8. *Phlox.*
Leaves palmately parted nearly to base into narrow segments, or sometimes entire but then acicular; lower leaves never broad and entire ....9. *Linanthus.*

1. **Gilia** R. & P.

Stems erect or ascending; leaves in a basal rosette or scattered along lower part of stem; flowers with greatly reduced bracts or none.
Flowers crowded in heads or very dense cymes which are globose to turbinate in outline; flowers in a head normally more than 6, often numerous.
Corolla-tube gradually expanded upward to base of lobes, the throat not clearly defined and not over 4 mm. in diameter ....1. *G. capitata.*
Corolla-tube above the slender base abruptly expanded into a throat usually more than 4 mm. wide ....2. *G. achilleaefolia.*
Flowers rather widely spaced or, if crowded, the clusters containing not more than 6 flowers and flat-topped.
Corolla 10 to 22 mm. long, showy.
Corolla normally with dark purple spots in throat .....3. *G. tricolor.*
Corolla without dark purple spots in throat (sometimes the throat dark purple throughout).
Basal leaves entirely glabrous, with divisions longer than width of rachis ....4. *G. tenuiflora.*
Basal leaves white-woolly to glabrate, or sometimes apparently glabrous from the first, with divisions about as long as width of rachis ....5. *G. latiflora.*
Corolla 3 to 10 mm. long.
Corolla without dark purple spots in throat.
Plant glandular only in inflorescence, or not at all.
Corolla 7 to 10 mm. long; capsule ovoid; leaves well developed except in drought-dwarfed plants, the basal ones usually over 3 cm. long.
Lower leaves with very narrow rachis; inflorescence without capitate glands; corolla pale violet or lavender to white ....6. *G. multicaulis.*
Lower leaves in vigorous plants with rather broad rachis (over 1 mm. wide); inflorescence bearing minute capitate glands; corolla normally deep violet, rarely white
7. *G. malior.*
Corolla 3 to 4 mm. long; capsule subglobose before dehiscence; leaves reduced, the basal ones not over 3 cm. long even in vigorous plants ....8. *G. minor.*

Plant glandular-pubescent throughout ..9. *G. austro-occidentalis.*
Corolla normally with dark purple spots in throat (or these present in
all but a very few plants in a stand) ....10. *G. clivorum.*
Stems widely spreading; leaves in a basal rosette and subtending small terminal and
intermediate clusters of flowers ....11. *G. jacens.*

1. **G. capitata** Sims var. **staminea** (Greene) Brand. Sandy or crumbling rocky soils,
occasional from Santa Lucia Range eastward to Palo Prieto Pass and Cottonwood
Pass. The plants vary in several details but are here regarded as forming essentially
a unit with plants occurring in sandy places northward through the inner South
Coast Ranges and, formerly, in the San Joaquin Valley. The usual flower-color is
sky-blue, occasionally darker. A white-flowered race occurs in the region of Price
Canyon in the coastal area. Careful comparison of living plants from various localities may, perhaps, provide evidence favoring a different classification for some of
the plants included here.

2. **G. achilleaefolia** Benth. Rocky slopes, often on serpentine, near coast. Dried
specimens are very unsatisfactorily distinguished from *G. capitata* var. *staminea,*
and for that reason I am not confident about listing localities for the interior. The
name *G. achilleaefolia* is here reserved for plants having a short corolla-tube
abruptly expanding into an ample throat. Such plants in the living state are readily separable from those with the corolla-tube gradually enlarged upward; the latter
are here included in *G. capitata.* My interpretation of *G. achilleaefolia* agrees with
that of Jepson (Fl. Cal. 3: 185) and is based on plate 1682 published in the "Botanical Register" (1835), which, however, does not represent the original publication
of the name.

3. **G. tricolor** Benth. var. **longipedicellata** Greenm. Along eastern border of
county: Carrizo Plain (*Weston* in 1926); Cottonwood Pass (*6890, Eastwood* and
Howell 2040). The last collection has some of the flowers rather crowded, varying
toward typical *G. tricolor.*

4. **G. tenuiflora** Benth. Common in sandy places from vicinity of Salinas River
eastward to Palo Prieto Canyon and La Panza district; upper Arroyo Grande to
Santa Maria. Plants having lighter-colored flowers than the common form of the
Salinas Valley have been found in the upper Arroyo Grande valley and along the
Santa Maria River. Some specimens of such plants have been identified by A. and
V. Grant as *G. splendens* Dougl., but I can find no reason for regarding them as
anything other than pale-flowered variants of *G. tenuiflora.*

5. **G. latiflora** (Gray) Gray. Cottonwood Pass, on gravelly slope (*6882*). A plant
noted as abundant on Caliente Mt. (*8213*) differs in having the basal leaves nearly
glabrous as in *G. tenuiflora.* Both collections in flower-size agree with subsp. *cuya-
mensis* A. & V. Grant, but neither conforms in all respects to the description of any
named subspecies. According to A. and V. Grant, subsp. *Davyi,* with much longer
corollas, also occurs in "southern San Luis Obispo Co."

6. **G. multicaulis** Benth. Uncommon, or at least seldom collected, mainly in sandy
places from San Luis Obispo southward. The name *G. multicaulis* has been subject
to different interpretations. It is used here in its traditional sense. The plants so
called are apparently the same as *G. angelensis* V. Grant. *G. peduncularis* Eastw.,
as I now understand it, is a name applied to plants of this species having the flowers on long pedicels, probably mainly as a result of environment, although genetic
factors may also tend to produce long pedicels.

7. **G. malior** Day & Grant. In crumbling shale or sand, Cottonwood Pass southward through Temblor Range to Cuyama Valley. I suspect that the plants said by Twisselmann to be the "most common gilia" in the Temblor Range and included by him under the name *G. minor* belong to this species. *Gilia malior* and *G. minor* are reported to differ in chromosome number and can usually be readily distinguished.

8. **G. minor** A. & V. Grant. Usually in sandy soils, common from vicinity of Atascadero and Pozo eastward. San Luis Obispo Co. plants which have been called *G. ochroleuca* subsp. *bizonata* seem exactly identical, but this statement does not apply to plants from other areas which have been included under that name. *Gilia minor* and *G. tenuiflora* often are found growing together with no indication of interbreeding.

9. **G. austro-occidentalis** A. & V. Grant. Sporadic in sandy soils, eastern and southern parts of county: Hughes Canyon, Red Hills (*Twisselmann 2143*); Bill Hill Ranch, foot of Caliente Mt. (*Twisselmann 3477*); 2 miles southwest of Nipomo (*7837*); Jack Lake (*10,538*).

10. **G. clivorum** (Jepson) V. Grant. Common and widespread in rocky or sandy soils except along coastline. Usually readily recognizable by the presence of dark purple spots in the corolla-throat. However, there are plants identified as *G. clivorum* which lack dark spots; these are not always clearly distinguishable from small-flowered plants of *G. multicaulis*. It is assumed that *G. clivorum* is the common species of this region, and accordingly specimens of doubtful identity are tentatively placed under this name.

11. **G. jacens** A. & V. Grant. Sandy hills and flats, west base of Temblor Range (near north end of Elkhorn Plain, *8135*). This collection differs from authentically determined material of *G. jacens* in having larger flowers and exserted stamens. The species, including the collection cited here, has a distinctive manner of growth, differing from that of any other *Gilia* in this region and suggesting *G. polycladon* (*Ipomopsis polycladon*). The plants reported by Twisselmann as *G. brecciarum* may belong here.

2. Allophyllum (Nutt.) A. & V. Grant

Herbage glandular throughout; stamens exserted .... 1. *A. glutinosum.*
Plant glandular only in inflorescence; stamens not exserted from corolla.
    Leaves mostly pinnately several-lobed except on dwarfed plants; flowers (on vigorous plants) in dense clusters of 4 or more .... 2. *A. gilioides.*
    Leaves mostly entire to 3-lobed even on vigorous plants; flowers loosely arranged, usually with a leaf subtending each pair .... 3. *A. violaceum.*

1. **A. glutinosum** (Benth.) A. & V. Grant. "Trout Creek" (*Condit* in 1908). There are two Trout Creeks: one entering the Salinas River near Santa Margarita, the other a confluent of the Huasna River. Probably the plant from Lopez Canyon, cited as *Gilia gilioides* by Jepson (Fl. Cal. 3: 197) also belongs here.

2. **A. gilioides** (Benth.) A. & V. Grant. Rarely collected in the county, but probably rather frequent in interior: Cottonwood Pass (*6876*); Navajo Creek (*Condit* in 1910, cited by A. & V. Grant, Aliso 3: 105).

3. **A. violaceum** (Heller) A. & V. Grant. East side of La Panza Range, in oak woods, granite soil (*7875*). Doubtfully distinct from *A. gilioides*, differing mainly in the more widely spaced flowers and the fewer lobes or divisions of the leaves.

3. **Collomia** Nutt.

Leaves entire; corolla pale orange fading to white after opening, 20 to 30 mm. long
1. *C. grandiflora.*
Most of leaves lobed; corolla pink or purplish, 10 to 15 mm. long 2. *C. heterophylla.*

1. **C. grandiflora** Dougl. About 1933, two greatly dwarfed plants were found in fruit in rocky soil along Town Creek in the Santa Lucia Mts. Seeds from these plants were planted at Modesto and grew into normal-sized *C. grandiflora.*

2. **C. heterophylla** Hook. Moist, usually shaded places in Santa Lucia Range: Santa Rita Creek; Old Creek; North Fork of San Simeon Creek.

4. **Navarretia** R. & P.

Corolla blue or violet to white.
  Heads not glandular.
      Main stem elongate, the branches, if any, never prostrate and rarely as long as main stem; stems white-puberulent; bracts and calyces markedly white-villous toward base . . . . .1. *N. intertexta.*
      Main stem very short, with longer prostrate branches below the central head; stems apparently quite glabrous; bracts and calyces sparsely villous toward base . . . . .2. *N. prostrata.*
  Heads markedly glandular.
      Upper leaves with narrow rachis (rarely over 2 mm. wide except in forms of *N. mitracarpa* where only upper part of rachis is broadened), with slender divisions or marginal spines.
          Stamens exserted from corolla at anthesis.
              Stems usually with spreading to prostrate branches from base; capsule 4-angled at apex . . . . .3. *N. mitracarpa.*
              Stem erect, usually simple or with ascending branches above; capsule not angled at apex . . . . .4. *N. pubescens.*
          Stamens not exserted from corolla.
              Plants 20 to 60 cm. tall; heads mostly over 25 mm. in diameter; corolla deep blue, not concealed by calyx . . . . . . .5. *N. squarrosa.*
              Plants 5 to 15 cm. tall; heads less than 25 mm. in diameter; corolla light blue, almost hidden by calyx . . . . .6. *N. mellita.*
      Upper leaves and bracts with broad rachis (usually over 3 mm. wide) and very stout marginal spines.
          Terminal spine of bracts continuous in direction with rachis or slightly bent downward, distinctly longer than the 2 adjacent spines; 2 subterminal spines not widely divergent; corolla deep violet
7. *N. atractyloides.*
          Terminal spine of bracts bent downward at a right angle or nearly so; terminal 3 spines subequal and widely divergent; corolla (in the geographic race found locally) lilac . . . . .8..*N. hamata.*
Corolla yellow with dark purple spots in throat . . . . .9. *N. nigellaeformis.*

1. **N. intertexta** (Benth.) Hook. Rare in Santa Lucia Mts., in soil which is wet during growing season: between Rocky Butte and Pine Mt. (*8004*).

2. **N. prostrata** (Gray) Greene. Bed of vernal pool, 1 mile northeast of Creston on road toward Shandon (*8322*).

3. **N. mitracarpa** Greene. Common in clay soils, less often in sand, from crest of Santa Lucia Range to our eastern border (no records from Carrizo Plain or Cuyama Valley). Plants with comparatively long calyces and consequently larger

flower-heads have been called subsp. *Jaredii* (Eastw.) Mason but constitute merely a tendency to vary in one feature, not a clearly defined entity. The "type sheet" of *N. Jaredii* Eastw., collected on Paso Robles Creek by Lorenzo Jared, bears four plants. The two upper represent exactly the prevailing form of *N. mitracarpa* as represented throughout its range, with slender, wide-spreading branches and flower-heads of average size for the species. The two lower plants on the sheet have simple erect stems and large heads. In aspect they resemble closely specimens of *N. pubescens* and are not mature enough to show the distinctive capsule characters of either *N. mitracarpa* or *N. pubescens*. From Miss Eastwood's original description of *N. Jaredii*, it is clear that she intended the name to apply to the two upper plants, because of the phrases "Slender annual, about 7 cm. high" and "heads of flowers about 15 mm. across" (Zoe 5: 89–90. 1900). Consequently, *N. Jaredii* is a synonym of *N. mitracarpa* as now interpreted. The two lower plants on the type sheet, being immature, can not with certainty be identified as *N. pubescens*, but it is suspected that they may belong to that species.

Var. **villosa** Hoover, n. var. Upper leaves, branches, and bracts conspicuously white-villous. Foliis superioribus, ramulis, et bracteis conspicue albo-villosis. Between Rocky Butte and Pine Mt., Santa Lucia Range, June 21, 1950, *Hoover 8006*. This collection represents an apparently local race which is distinctive enough in appearance to warrant a name.

4. **N. pubescens** (Benth.) H. & A. Salinas Valley near Paso Robles and San Miguel (Jepson, Fl. Cal. 3: 152). See also the note above concerning the plants mounted with the type of *N. Jaredii*.

5. **N. squarrosa** (Esch.) H. & A. SKUNKWEED. Occasional in winter-moist places in western part: Cambria; Santa Rita Creek; Los Osos Valley.

6. **N. mellita** Greene. Occasional in chaparral or wooded rocky areas, Santa Lucia Range. Found usually after the ground has been cleared, as by fire.

7. **N. atractyloides** (Benth.) H. & A. In sandy soils, frequent from coast to near western edge of Carrizo Plain. The name *N. atractyloides* is used here for a common species of the California Coast Ranges, ranging northward to Humboldt and Trinity Counties. The definition of this species and the following has been greatly confused. There is no question as to the existence of two distinct species in San Luis Obispo Co., but their correct names are difficult to establish. With little doubt, *N. hirsutissima* Brand is another name for the species here called *N. atractyloides*, and would be used for these plants if it could be definitely shown that the latter name belongs to the following species.

8. **N. hamata** Greene? Sandy soil in coastal area only. As determined by herbarium records, the species tentatively designated by this name extends northward only to Santa Cruz County. These plants differ noticeably from southern specimens of *N. hamata* (type locality, Guadalupe Mt., Baja California) in having smaller and paler corollas, and probably should be classified at least as a subspecies, or perhaps even as a distinct species. When this and *N. atractyloides* grow together, as they often do, "*N. hamata*" comes earlier into flower. Because of a combination of features which would seem unimportant if considered individually, the two look decidedly different. By using Mason's key in Abrams's "Illustrated Flora," the plants here called *N. hamata* are identified as *N. atractyloides*, and the drawing labelled as *N. atractyloides* also seems to represent what is here called *N. hamata*.

9. **N. nigellaeformis** Greene. Open hillsides, usually in heavy clay, rarely in sandy

soil: frequent in interior from near Salinas River (San Miguel to Templeton) eastward to Cholame and La Panza District. Farther north, this species often grows in
beds of vernal pools. *Navarretia ocellata* Eastw., based on a collection of L. Jared
from "Goodwin, San Luis Obispo Co.," is included here.

5. **Eriastrum** Wooton & Standley

Perennials with woody-based stems and a woody tap-root . . . . 1. *E. densifolium.*
  Corolla-lobes ⅔ as long as tube to nearly as long . . . . subsp. *densifolium.*
    Branches erect or ascending; leaves glabrate . . . . var. *densifolium.*
    Branches widely spreading from base; leaves white-woolly . . . . . var. *patens.*
  Corolla-lobes half as long as tube . . . . subsp. *elongatum.*
    Plants 2 to 5 dm. tall; leaves persistently white-woolly . . . . . . var. *elongatum.*
    Plants 1 to 2 dm. tall; leaves glabrate . . . . var. *austromontanum.*
Annuals with slender tap-root.
  Corolla blue or white.
    Corolla 12 to 20 mm. long.
      Corolla regular; stamens inserted in sinuses of corolla-lobes
                                                                                2. *E. pluriflorum.*
        Anthers about 2 mm. long . . . . var. *pluriflorum.*
        Anthers hardly over 1 mm. long . . . . var. *Sherman-Hoytiae.*
      "Corolla normally irregular; stamens inserted at base of throat" [4]
                                                                                3. *E. eremicum.*
    Corolla 6 to 11 mm. long.
      Corolla blue, rarely white, exceeding calyx; anthers exserted from
      corolla-throat.
        Stamens extending almost to tips of corolla-lobes . . 4. *E. filifolium.*
        Stamens much shorter than corolla-lobes.
          Stem with mostly ascending branches; corolla-lobes 3 to 4
          mm. long; anthers 1 mm. long . . . . 5. *E. Wilcoxii.*
          Stem with widely spreading branches; corolla-lobes 2 to 2.5
          mm. long; anthers 0.5 mm. long . . . . 6. *E. diffusum.*
      Corolla white or sometimes pale blue, barely equalling calyx; anthers
      hardly extending to base of corolla-lobes . . . . 7. *E. Hooveri.*
  Corolla yellow . . . . 8. *E. luteum.*

1. **E. densifolium** (Benth.) Mason var. **densifolium.** Plants which fit the description of typical *E. densifolium,* as interpreted by Mason (Madroño 8: 73. 1945), are
restricted to the region of Nipomo Mesa, southward into western Santa Barbara
County.

Var. **patens** Hoover, n. var. Stems several, spreading from a woody root, the flowering branches ascending; herbage persistently woolly; corolla sky-blue, sometimes
white; corolla-tube (in dried specimens) 10–12 mm. long including the short and
poorly defined throat, the lobes nearly or quite as long.

Herba lanata; caulibus pluribus, patentibus, ramis floriferis ascendentibus; corolla azurea, tubo 10–12 mm. longo, lobis tubo aequilongis vel proxime.

In sand on south side of Morro Bay: Los Osos, *Hoover 8973* (type); Haynes
Ranch, *Ingalls* in 1912; coast south of Hazard Canyon, *Hoover 7186.* Var. *patens* is
a well-marked local race, differing from all previously named subspecies in its
spreading habit. From var. *densifolium* it differs additionally in the persistently

---

4. The flowers of the specimen here included in *E. eremicum* are in such condition that
application of the key characters is impractical.

woolly herbage, the lighter color (on the average) of the flowers, and the very large corolla-lobes. From var. *elongatum* it differs in its notably larger, lighter-colored corollas with lobes much longer in proportion to the tube, and the tendency for the bracts to have more lobes.

Var. **elongatum** (Benth.) Hoover, n. comb. *Huegelia elongata* Benth., Bot. Reg. 19: under pl. 1622. 1833. Occasional in sandy places from Atascadero eastward; rare in Temblor Range.

Var. **austromontanum** (Craig) Hoover, n. comb. *Gilia densifolia* var. *austromontana* Craig, Bull. Torr. Club 61: 391. 1934. Caliente Mt. (*Richard Mason* in 1965). The plant is dwarfed, with glabrate leaves and remarkably small flower-heads.

2. **E. pluriflorum** (Heller) Mason var. **pluriflorum.** In sands, gravels, or calcareous clays and silts, common in eastern part from Cottonwood Pass to Cuyama Valley, extending westward to vicinity of Shell Creek between Shandon and Creston.

Var. **Sherman-Hoytiae** (Craig) Hoover, n. comb. *Gilia Sherman-Hoytiae* Craig, Bull. Torr. Club 61: 415. 1937. *E. pluriflorum* subsp. *Sherman-Hoytiae* Mason. Southern Temblor Range (*10,598*); Caliente Mt. (*L. Belden* in 1965). Plants transitional toward typical *E. pluriflorum* are frequent.

3. **E. eremicum** (Jepson) Mason. Carrizo Plain, 8 miles south of Soda Lake (*Eben McMillan 157* in 1952). The plant is not typical *E. eremicum* but in most respects seems more like that species than any other. It suggests a vigorous form of *E. diffusum*, with diffusely branched stems and numerous heads of flowers. The flowers, however, are too large for *E. diffusum*.

4. **E. filifolium** (Nutt.) Wooton & Standley. "Blochman's Ranch" near Santa Maria (*Eastwood 453*); perhaps not actually in San Luis Obispo Co.

5. **E. Wilcoxii** (Nelson) Mason. Sandy or gravelly soil in or near La Panza Range; on disintegrated shale among junipers, in a canyon near Mariannas Ranch in the Temblor Range; west edge of Carrizo Plain. There is some uncertainty about the identity of our plants, but H. L. Mason classified a collection from the La Panza Range as *E. Wilcoxii.* This interpretation is here followed.

6. **E. diffusum** (Gray) Mason. Caliente Mt., in sandy soil along ridge (*8271*). The plants correspond in every respect to Mason's interpretation of *E. diffusum,* but it may be questioned whether there really is in this area an additional species distinct from what is here called *E. Wilcoxii.*

7. **E. Hooveri** (Jepson) Mason. Sandy soil, Temblor Range near north end of Elkhorn Plain (*10,601*); eastern part of Cuyama Valley, near Cuyama River just north of Santa Barbara Co. line (*E. R. Chandler 2365*).

8. **E. luteum** (Benth.) Mason. In sandy or gravelly soil: rare in Santa Lucia Mts. west of Paso Robles; locally common east of Santa Margarita from near Salinas River to west base of La Panza Range.

6. **Langloisia** Greene

1. **L. Schottii** (Torr.) Greene. In sandy soil, foothills on either side of Carrizo Plain; Cuyama River.

7. **Leptodactylon** H. & A.

1. **L. californicum** H. & A. PRICKLY PHLOX. Rocky or sandy soil in areas of granite or sandstone: Rocky Canyon between Atascadero and Creston (apparently the northern limit for the species) to La Panza Range and southward; near coast from

Indian Knob Ridge southward. Plants on the Nipomo Dunes differ from the common form of the interior in having crowded leaves, more woolly branchlets and calyces, and paler corollas with broader lobes.

8. Phlox L.

1. **P. gracilis** (Hook.) Greene. In openly wooded areas and herbaceous plant communities, common throughout the interior. No coastal occurrences are at present known, but this fact may be due only to insufficient collecting.

9. Linanthus Benth.

Flowers widely spaced, on slender pedicels or some of them subsessile in forks of stem.
    Corolla-throat longer than the slender tube.
        Corolla 3 to 5 mm. long, barely exceeding the calyx ...... 1. *L. pygmaeus.*
        Corolla 6 to 16 mm. long, at least twice as long as calyx ....2. *L. liniflorus.*
          Corolla 10 to 16 mm. long ....var. *liniflorus.*
          Corolla 6 to 10 mm. long ....var. *pharnaceoides.*
    Corolla-throat shorter than the slender tube.
        Corolla 25 to 30 mm. long, the lobes longer than the tube 3. *L. dichotomus.*
        Corolla 15 to 20 mm. long, the lobes shorter than the tube 4. *L. Bigelovii.*
Flowers in dense clusters or heads, sessile or nearly so.
    Corolla-throat well developed, at least as long as the slender basal part of the tube; corolla-lobes at least as long as tube and throat combined.
        Dwarf plants mostly 2 to 10 cm. tall ....5. *L. Parryae.*
        Plants mostly 15 to 30 cm. tall ....6. *L. grandiflorus.*
    Corolla-throat shorter than the slender tube; corolla-lobes much shorter than tube and throat combined.
        Bracts with coarsely hispid-ciliate divisions; calyx membranous to base below sinuses ....7. *L. ciliatus.*
        Bracts with soft-ciliate to puberulent divisions; calyx membranous only in upper part below sinuses.
          Corolla-lobes 4 to 10 mm. long; plant rarely less than 10 cm. tall.
            Plant 20 to 30 cm. tall; corolla-lobes 6 to 10 mm. long, 3 to 5 mm. wide ....8. *L. androsaceus.*
            Plant 5 to 20 cm. tall; corolla-lobes 4 to 6 mm. long, 2 to 3.5 mm. wide ....9. *L. parviflorus.*
        Corolla-lobes 2 to 4 mm. long; plant not over 7 cm. tall 10. *L. bicolor.*

1. **L. pygmaeus** (Brand) J. T. Howell. Sandy soils in chaparral areas of the interior; uncommon or overlooked: Creston-La Panza road; Parkhill district; La Panza Range.

2. **L. liniflorus** (Benth.) Greene var. **liniflorus.** "San Luis Obispo Co.," according to Munz (Cal. Fl. 509). No local specimens have been seen of the large-flowered plants which are presumed to be typical *L. liniflorus* and which are almost entirely restricted to the San Francisco Bay region.

Var. **pharnaceoides** (Benth.) Hoover, n. comb. *L. pharnaceoides* Benth., Bot. Reg. 19: under pl. 1922. 1833. *Gilia liniflora* var. *pharnaceoides* Gray. Common throughout the interior (except perhaps Carrizo Plain and Cuyama Valley), mostly in sandy or gravelly soils.

3. **L. dichotomus** Benth. Evening Snow. Frequent in sandy areas of eastern part; not found nearer coast than upper Arroyo Grande.

4. **L. Bigelovii** (Gray) Greene. Between San Juan River and Carrizo Plain, in calcareous sandy soil; 10 miles south of Soda Lake (*Keck & Clausen 3150*); rare in Temblor Range.

5. **L. Parryae** (Gray) Greene. In barren sandy spots: frequent locally from hills east of Salinas River to western margin of Carrizo Plain.

6. **L. grandiflorus** (Benth.) Greene. "San Luis Obispo," *Blochman* in 1893; vicinity of Santa Maria; otherwise from Monterey Co. northward. Included here because of the probability that it occurs, or formerly occurred, near Nipomo.

7. **L. ciliatus** (Benth.) Greene. Mostly in shade in loose soil, La Panza Range and doubtless elsewhere in interior. Coastal records are lacking.

8. **L. androsaceus** (Benth.) Greene. Open north-facing hillslopes, Old Creek above Cayucos (*8460*). Reference to the current literature raises a doubt as to whether the name *L. androsaceus* actually belongs to these plants, since the name is applied to plants here called *L. parviflorus*. In any event, the plants in question, whatever their proper name may be, are entirely distinct, in their taller growth and markedly larger flowers, from the common species here called *L. parviflorus*.

9. **L. parviflorus** (Benth.) Greene. Frequent in rocky places near coast, often in areas of serpentine; very common in hilly country throughout the interior. Plants may be "keyed out" to certain subspecies named by H. L. Mason under *L. androsaceus,* but these do not seem natural groups on a geographical basis or otherwise. The prevailing confusion with *L. androsaceus* may be traced to the close similarity in appearance of dried specimens, but no such difficulty is encountered with fresh plants. *Linanthus parviflorus* is variable in flower-color, but not to a remarkable degree in other respects. A noteworthy local race, prevalent mainly in or near the La Panza Range, has a white to pale lilac corolla with yellow throat and a deep purple-red spot at the base of each lobe.

10. **L. bicolor** (Nutt.) Greene. Open hills and plains in interior, uncommon or overlooked: Choice Valley (*E. McMillan 92*); 11 miles east of Creston on La Panza road (*7780*); "occasional throughout Temblor Range," according to Twisselmann.

Hydrophyllaceae. WATERLEAF FAMILY

Herbs (including a few species with woody roots and woody-based stems).
  Style 1, 2-branched above base; flowers with pedicels or peduncles or sometimes sessile, in bractless helicoid cymes or, when solitary, nearly always extending beyond the subtending leaves.
    Corolla-lobes blue or violet to white (the tube or throat sometimes yellow); corolla deciduous after flowering.
      Calyx with projecting or deflexed appendages ("auricles" or "bracts") alternating with the lobes.
        Stem climbing by means of stiff retrorse hairs; flowers partly in raceme-like helicoid cymes which are only slightly coiled in bud

                    1. *Pholistoma.*
        Stem not climbing or with stiff retrorse hairs; flowers solitary and axillary . . . . 2. *Nemophila.*
      Calyx without appendages between the lobes.
        Lower leaves, or all, opposite; calyx-lobes widely spreading in fruit.
          Stem weak, usually reclining; leaves pinnately lobed, the lobes entire or rarely sparingly toothed . . . . 1. *Pholistoma.*
          Stem erect; leaves 2 to 3 times pinnately lobed or divided

                    3. *Eucrypta.*

Leaves alternate (in all San Luis Obispo Co. species); calyx-lobes
in fruit ascending or closely appressed to capsule . . . . 4. *Phacelia*.
Corolla yellow, fading to white in age, persistent and papery in fruit
5. *Emmenanthe*.
Styles 2, distinct to base; flowers sessile, shorter than the subtending leaves
6. *Lemmonia*.
Shrubs (stems woody for about 1 m. from base in *Turricula*, called by authors an "herb").
Herbage strongly glandular-hirsute throughout . . . . 7. *Turricula*.
Herbage tomentose at least on backs of leaves, not hirsute except sometimes on
juvenile shoots . . . . 8. *Eriodictyon*.

1. Pholistoma Lilja

Herbage rough-hairy; calyx with appendages alternating with the lobes; corolla violet, 12 to 25 mm. wide . . . . 1. *P. auritum*.
Herbage glabrous or very sparsely hairy; calyx without appendages; corolla white,
4 to 6 mm. wide . . . . 2. *P. membranaceum*.

1. **P. auritum** (Lindl.) Lilja. Fiesta Flower. Common in western part, usually under trees.

2. **P. membranaceum** (Benth.) Constance. Common in interior, from Salinas Valley to our eastern border, in rocky or sandy places, usually in shade.

2. Nemophila Nutt.

Corolla white or usually so, 3 to 6 mm. wide.
Appendages half as long as calyx-lobes; corolla-lobes usually with dark veins
and a purple spot at apex . . . . 1. *N. pedunculata*.
Appendages not more than ⅓ as long as calyx-lobes; corolla-lobes without dark
veins or a purple spot . . . . 2. *N. Fremontii*.
Corolla blue, 15 to 35 mm. wide . . . . 3. *N. Menziesii*.

1. **N. pedunculata** Dougl. Mostly in sandy places, widely scattered over all parts of the county. Apparently not plentiful, but this is explained by the plants being small and inconspicuous.

2. **N. Fremontii** Elmer. Shaded places, La Panza Range, rare or overlooked.

3. **N. Menziesii** H. & A. Baby Blue-Eyes. Sandy soils, common from Salinas Valley eastward; upper Arroyo Grande and Huasna district; Morro Bay (Jepson, Fl. Cal. 2: 229).

3. Eucrypta Nutt.

1. **E. chrysanthemifolia** (Benth.) Greene. Common in rocky places, or under trees and bushes in sandy soil, especially near coast, sparingly toward interior, and not reported from Temblor Range. Notably abundant and vigorous after fires.

4. Phacelia Juss.

Leaves pinnate with entire-margined leaflets, or some of them simple and entire;
perennials or sometimes biennial.
Plant less than 6 dm. tall; stems usually several from a branching caudex;
leaves all compound except the reduced upper ones.
Leaflets sparsely strigose; petioles sparsely hispid; calyx-lobes in fruit ovate
to lanceolate or elliptic, overlapping . . . . 1. *P. imbricata*.

Leaflets densely strigose; petioles canescent and densely hispid; calyx-lobes in fruit narrowly lanceolate or linear-elliptic, not overlapping   2. *P. egena.*
Plant usually at least 6 dm. tall; stems usually few or solitary; leaves mostly simple or some with 3 or 5 leaflets, the upper not greatly reduced

3. *P. nemoralis.*

Leaves either simple or, when compound, the divisions toothed, lobed, or further divided; annuals except *P. ramosissima.*

Leaves simple, entire or more often toothed or shallowly lobed (the lower ones in *P. malvaefolia* sometimes compound or deeply lobed).

Plant soft-hairy, without stiff bristles.

Leaves entire to simply toothed or shallowly lobed; corolla less than 15 mm. wide, medium violet to dull white.

Stems not over 10 cm. long, ascending or spreading; herbage not obviously glandular; flowers widely spaced; pedicels averaging about as long as calyces . . . . 15. *P. Douglasii* var *petrophila.*
Stems more than 10 cm. long, ascending to erect; herbage conspicuously glandular, at least in inflorescence; flowers crowded; pedicels much shorter than calyces.

Calyx-lobes obovate-spatulate; corolla dull white or pale bluish, with no yellow, its tube not extending beyond calyx

4. *P. grisea.*

Calyx-lobes narrowly oblanceolate-spatulate; corolla violet or lavender with yellow tube, the tube exserted from calyx

5. *P. suaveolens.*

Leaves (except the reduced upper ones) doubly toothed: i.e., with coarse teeth or shallow lobes which are in turn toothed; corolla 15 to 25 mm. wide, deep blue with white center (rarely all white)

6. *P. viscida.*

Plant with stiff bristles.

Corolla 6 to 8 mm. long; stamens and style conspicuously exserted

7. *P. malvaefolia.*

Corolla 4 to 5 mm. long; stamens and style included . . . . 8. *P. Rattanii.*

Leaves compound or deeply pinnately lobed (see also *P. malvaefolia*).

Leaves compound; leaflets well separated, deeply incised, lobed, or subdivided.

Perennials, with lower parts of stems often woody . . 9. *P. ramosissima.*
Annuals.

Calyx-lobes oblanceolate to narrowly linear, not visibly veined in fruit, variously hairy but not markedly ciliate.

Calyx-lobes ascending from the base, closely appressed to capsule; capsule with hairs all of one sort, not bristly.

Leaves extending up to inflorescence; branches of inflorescence each usually with a reduced leaf at base; stamens about equalling corolla or shortly exserted   10. *P. distans.*
Inflorescences of groups of 2 to 5 bractless helicoid cymes borne above the leaves; stamens about twice as long as corolla . . . . 11. *P. tanacetifolia.*

Calyx-lobes divergent at base, then curved upward, loosely enclosing capsule; capsule with scattered stiff bristles as well as a fine pubescence.

Vertical portion of corolla-scales shorter than transverse portion; bristles of capsule divergent . . . . . . 12. *P. eximia.*

Vertical portion of corolla-scales longer than transverse portion; bristles of capsule erect ......13. *P. cryptantha.*
Corolla about equalling calyx, dull white to pale blue ....var. *cryptantha.*
Corolla showy and surpassing calyx, sky-blue var. *heliophila.*
Calyx-lobes ovate or broadly lanceolate in fruit, venulose, ciliate 14. *P. ciliata.*
Leaves deeply pinnately lobed, or sometimes compound with leaflets decurrent on rachis; lobes few-toothed or sparingly lobulate to entire.
Pedicels longer than calyces ....15. *P. Douglasii.*
Pedicels shorter than calyces.
Corolla not surpassing calyx ....16. *P. affinis.*
Corolla surpassing calyx.
Stem erect; inflorescence usually many-branched, the longest branch much shorter than height of plant; corolla white or rarely lavender ....17. *P. brachyloba.*
Stems usually spreading at base or ascending; inflorescence simple or few-branched, the primary branch at least half the total height of plant; corolla blue ....18. *P. Fremontii.*

1. **P. imbricata** Greene. Rocky places in and near Santa Lucia Range; especially common around San Luis Obispo. Because of the close resemblance between this and *P. egena* in appearance and the lack of adequate collections, the detailed distribution of the two remains yet to be worked out.

2. **P. egena** (Brand) J. T. Howell. Mostly in rocky places, apparently common in Santa Lucia Range, occasional eastward to Temblor Range.

3. **P. nemoralis** Greene. Shady, often rather moist places: San Luis Range; west of Templeton and probably elsewhere in Santa Lucia Range; Cambria.

4. **P. grisea** Gray. Chaparral and wooded areas, most plentiful and vigorous after fire or clearing: Santa Lucia Mts. from Rinconada Mine northward; Black Mt. in La Panza Range.

5. **P. suaveolens** Greene. Black Mt., La Panza Range, abundant near summit in burned brushy area (*8244* in 1952).

6. **P. viscida** (Benth.) Torr. Usually in disintegrated shale, Santa Lucia Range, and near coast from San Luis Range southward. The report of *P. Parryi* in this county is based on a wrong identification of *P. viscida.*

7. **P. malvaefolia** Cham. var. **loasaefolia** (Benth.) Brand. Under trees and in loose rocky soils, near coast from San Luis Range northward.

8. **P. Rattanii** Gray. Loose soils in shady places: Calf Canyon; Santa Lucia Range from Paso Robles Creek northward.

9. **P. ramosissima** Dougl. Mainly on coastal sand-dunes; extending inland to upper Salinas Valley and upper Arroyo Grande, but specimens from these districts are not now available for study. The separation of our plants into varieties is of doubtful value. On Nipomo Mesa and the Nipomo Dunes, where the species is most abundant, plants have been found which correspond to the descriptions of var. *suffrutescens* Parry, var. *austrolitoralis* Munz, and var. *montereyensis* Munz.

10. **P. distans** Benth. Very common almost throughout the county, mostly in rocky or sandy soils, but largely absent from flat valleys or plains and from densely wooded areas. Near the coast the flowers are generally bright violet or blue; toward

the interior, commonly a dingy white. Blue or lavender flowers appear again in and near the Temblor Range, as in the San Joaquin Valley to the east.

11. **P. tanacetifolia** Benth. Mainly in disintegrating shale or in friable gypseous clay: east side of Santa Lucia Mts. near Paso Robles; commoner eastward from Cottonwood Pass and the Creston-Shandon road to Cuyama Valley.

12. **P. eximia** Eastw. *P. hispida* Gray, not Buckley. *P. cicutaria* Greene var. *hispida* (Gray) J. T. Howell. Rocky slopes, particularly where the surface is unstable or where the land has recently been cleared: summits of Santa Lucia Range northwest of Cuesta Pass, eastward to La Panza Range, and southward.

13. **P. cryptantha** Greene var. **cryptantha.** Southern part of Temblor Range and Elkhorn Hills; less plentiful locally than the following large-flowered variety.

Var. **heliophila** (Macbr.) Hoover, n. comb. *P. hispida* var. *heliophila* Macbr., Contr. Gray Herb. 49: 20. 1917. A few scattered colonies are known in the southern Temblor Range (*10,594*) and Elkhorn Hills (*10,464*). Both collections show a range in size of the corolla on different plants from barely equalling the calyx to large and showy. For this and other reasons, I believe that this variety is more closely related to *P. cryptantha* than to *P. vallis-mortae,* in which most botanists have placed it. The larger corollas are always sky-blue, so far as observed, while the small ones vary to dull white.

14. **P. ciliata** Benth. Clay soils and crumbling shales, very common in interior. Notably abundant from Cholame and Shandon to Cuyama Valley, where in good years it colors extensive areas purple. The only coastal record is one mile south of Nipomo (*8546*).

15. **P. Douglasii** (Benth.) Torr. var. **Douglasii.** Common in sandy areas throughout the interior, and near coast from Morro Bay southward.

Var. **petrophila** Jepson. *P. curvipes* Torr., probably. Rocky places, Cottonwood Pass and Temblor Range.

16. **P. affinis** Gray. Caliente Mt., summit ridge west of peak, in sandy soil (*8210*).

17. **P. brachyloba** (Benth.) Gray. Sandy or gravelly soils, usually appearing after fire, Salinas River eastward to La Panza Range. The corolla is usually white, but occasionally a few plants with light purple flowers are found.

18. **P. Fremontii** Torr. Sandy soils, crumbling shale, or sometimes in friable gypseous clay, from Shell Creek (south of Shandon) eastward to Cottonwood Pass, Temblor Range, and Cuyama Valley.

5. **Emmenanthe** Benth.

1. **E. penduliflora** Benth. Whispering Bells. Gravelly or rocky slopes, growing best after fires or clearing, in hilly areas throughout the county, but becoming scarce eastward.

6. **Lemmonia** Gray.

1. **L. californica** Gray. Sandy or gravelly soils, rare in Temblor Range; locally plentiful in 1952 on ridge west of Caliente Peak (*8209*).

7. **Turricula** Macbr.

1. **T. Parryi** (Gray) Macbr. Poodle-Dog Bush. Local on east slope of La Panza Range, in soil derived from granite.

8. **Eriodictyon** Benth.

Upper leaf-surfaces dark green, glutinous, glabrous.
    Leaves narrowly linear, with revolute margins, entire . . . . 1. *E. altissimum.*
    Leaves oblong or lanceolate to ovate, not revolute, usually serrate
                                              2. *E. californicum.*
Upper leaf-surfaces from green and thinly pubescent to white-tomentose, not glutinous.
    Calyx not glandular; corolla with open throat, the lobes gradually spreading
                                        3. *E. crassifolium.*
    Calyx glandular; corolla constricted at throat, the lobes abruptly spreading.
        Calyx-lobes dark-colored, thinly hairy on upper part, ciliate below; corolla
        5 to 8 mm. long . . . . .4. *E. Traskiae.*
        Calyx-lobes white-hairy throughout; corolla 3 to 4 mm. long
                                      5. *E. tomentosum.*

1. **E. altissimum** Wells. On sandstone, a local component of chaparral on Indian Knob Ridge between San Luis Obispo and Pismo Beach. A remarkable, highly localized endemic.

2. **E. californicum** (H. & A.) Torr. MOUNTAIN BALM. YERBA SANTA. Mountain north of Arroyo de la Cruz (*Susan Allison* in 1966).

3. **E. crassifolium** Benth. var. **denudatum** Abrams. Along Cuyama River at upper end of Cuyama Canyon (near Gypsum Canyon).

4. **E. Traskiae** Eastw. Locally plentiful on white shale, ridge southeast of Cuesta Pass.

5. **E. tomentosum** Benth. Dry hilly areas, east slope of Santa Lucia Range west of Paso Robles; Salinas River eastward to La Panza Range, where notably abundant.

Boraginaceae. BORAGE FAMILY

Corolla not yellow, usually white, sometimes blue or purplish.
    Leaves of fleshy texture; ovary not deeply 4-lobed, the fruit splitting into 4 nutlets; stigma sessile on top of ovary . . . . .1. *Heliotropium.*
    Leaves not fleshy; ovary deeply 4-lobed, the nutlets distinct from the first (some of them often abortive); style present, not connected with nutlets.
        Nutlets with barbed or hooked prickles.
            Erect to prostrate annuals with narrow leaves; corolla white, inconspicuous; nutlets flat, the margin with hooked prickles or bristles
                                        2. *Pectocarya.*
            Erect perennial with broad leaves; corolla blue; nutlets subglobose, prickly all over . . . . .3. *Cynoglossum.*
        Nutlets without prickles (or minute ones in *Allocarya:*—annuals with small flowers, erect nutlets, and the lower leaves opposite).
            Corolla blue, 15 to 20 mm. wide . . . . .4. *Borago.*
            Corolla white, not over 8 mm. wide.
                Basal leaves forming a rosette, distinctly larger than the cauline leaves . . . . .5. *Plagiobothrys.*
            Lowest leaves not forming a clearly defined rosette.
                Calyx circumscissile, the upper part falling away; flowers each in the axil of a bract, crowded in very short spikes
                                       6. *Greeneocharis.*
                Calyx not circumscissile; flowers in more or less elongate "spikes" (helicoid cymes), usually bractless or with bracts fewer than flowers.
                  Nutlets keeled on ventral side.

Lower leaves opposite; nutlets without a stipe
    7. *Allocarya.*
Leaves all alternate; nutlets attached by a stipe near
middle of ventral side ....8. *Echidiocarya.*
Nutlets not keeled on ventral side, with narrow ventral
groove ....9. *Cryptantha.*
Corolla orange or yellow ....10. *Amsinckia.*

1. **Heliotropium** L. Heliotrope

1. **H. curassavicum** L. var. **oculatum** (Heller) Johnston. Common in saline, alkaline, or moist ground in the valleys and along the seashore.

2. **Pectocarya** DC.

Nutlets divergent in pairs, those of a pair less widely separated than are the pairs.
    Stems widely spreading or nearly prostrate; nutlets elongate.
        Wing of nutlet bordered by triangular teeth tipped with a short hooked
        bristle ....1. *P. linearis.*
        Wing of nutlet not toothed, or with a very few broad teeth near base, with
        hooked bristles at apex and sometimes a few below ......2. *P. penicillata.*
            Nutlets all approximately alike, wing-margined ....var. *penicillata.*
            One nutlet of each pair with very narrow wing or none; nutlets near
            base of plant often all unmargined ....var. *heterocarpa.*
    Stem erect; nutlets orbicular to obovate ....3. *P. setosa.*
Nutlets divergent at equal angles from center ....4. *P. pusilla.*

1. **P. linearis** DC. var. **ferocula** Johnston. Dry sandy or gravelly places in eastern part; in 1948 noted as very abundant along San Juan River south of Shandon (*7464*).

2. **P. penicillata** (H. & A.) A. DC. var. **penicillata.** Common in sandy soils, Salinas Valley eastward to Temblor Range. There is no record for this species in our coastal region, although it may be there. Most specimens from the county have nutlets more or less heteromorphic, thus varying toward var. *heterocarpa.*

Var. *heterocarpa* Johnston. Common in eastern part.

3. **P. setosa** Gray. Sandy or gravelly soils, north of Pozo; more frequent eastward to Cottonwood Pass and Temblor Range.

4. **P. pusilla** (A. DC.) Gay. Local in seepage spot, between Rocky Butte and Pine Mt., Santa Lucia Range (*7899*).

3. **Cynoglossum** L.

1. **C. grande** Dougl. Hound's Tongue. In woods, summit between See Canyon and Coon Creek in San Luis Range; Santa Lucia Mts. from Lopez Canyon northward, where often locally abundant.

4. **Borago** L.

1. **B. officinalis** L. Borage. Escaped from cultivation: weedy roadside, Cambria, *J. T. Howell 40,817* in 1964

5. **Plagiobothrys** F. & M.

Calyx circumscissile, the lobes curved inward over fruit; nutlets in most flowers fewer than 4.
    Spikes solitary, usually with a few bracts; nutlets strongly arched in side view
    1. *P. arizonicus.*

Spikes mostly paired or in 3's, bractless or sometimes with a bract at base; nutlets not strongly arched ....2. *P. nothofulvus.*
Calyx not circumscissile, the lobes erect or somewhat spreading; nutlets in most flowers 4.
Nutlets with narrow transverse ridges separated by broader flat intervals, or in *P. infectivus* without clearly defined transverse ridges.
Calyx with reddish-brown hairs.
Herbage staining paper violet; midribs and margins of leaves not conspicuously dark; spikes bractless or with bracts only near base
3. *P. fulvus.*
Herbage staining paper reddish-purple; midribs and margins of leaves conspicuously dark on lower side; spikes bracteate throughout
4. *P. infectivus.*
Calyx white-hairy, or only slightly brownish.
Stems over 12 cm. long, or rarely less; calyx 4 to 6 mm. long in fruit
5. *P. canescens.*
Stems less than 12 cm. long, or rarely more; calyx 2.5 to 3 mm. long in fruit ....6. *P. myosotoides.*
Nutlets with low broad transverse ridges separated by narrow grooves.
Spikes bracteate; fruiting calyx 6 to 7 mm. long ....7. *P. shastensis.*
Spikes not bracteate or only at base; fruiting calyx 3 to 4 (rarely to 6) mm. long ....8. *P. tenellus.*

1. **P. arizonicus** (Gray) Greene. Hills near our eastern border, in sand or crumbling shale: Cottonwood Pass; according to Twisselmann, "common throughout the Temblor Range."

2. **P. nothofulvus** Gray. POPCORN FLOWER. Grassland or thinly wooded areas: occasional on coastal hills; abundant at many places in Santa Lucia Range and in upper Salinas Valley; rare or apparently absent east of La Panza Range.

3. **P. fulvus** (H. & A.) Johnston var. **campestris** (Greene) Johnston. Locally common in upper Salinas Valley from Santa Margarita to Pozo, and eastward to La Panza district.

4. **P. infectivus** Johnston. Hillside patches of clay soil which probably contain lime or gypsum: occasional from vicinity of Cammatti Creek eastward.

5. **P. canescens** Benth. In more or less sandy soils, common from Salinas Valley eastward; infrequent westward (San Luis Obispo) or at least seldom collected. Along our eastern border (Cottonwood Pass to Cuyama Valley) are plants which somewhat imitate *P. shastensis* in aspect. They have fewer and more erect branches than most specimens of *P. canescens* and are less conspicuously white-hairy. The characters of their nutlets, however, place them in *P. canescens*.

6. **P. myosotoides** (Lehm.) Brand. Rare in Santa Lucia Mts.: 9 miles northwest of Adelaida, *Eastwood & Howell 2370* in 1936.

7. **P. shastensis** Greene. Local in red gravelly clay, 18 miles east of Creston on La Panza road (*7589, 7618*).

8. **P. tenellus** (Nutt.) Gray. Common in sandy soils of the interior, but apparently absent from the driest localities. Perhaps present also near coast, but records are lacking.

6. **Greeneocharis** Guerke & Harms

1. **G. circumscissa** (H. & A.) Rydb. Occasional in dry sand in interior: north of Pozo; San Juan River basin; north end of Elkhorn Plain.

7. **Allocarya** Greene

Stems spreading or ascending; calyx symmetrical or nearly so.
    Scar of nutlet small, not deeply hollowed, basal, or ventral and extending from base upward.
        Scar of nutlet exactly basal ....1. *A. stipitata.*
        Scar of nutlet ventral or obliquely basal.
            Nutlets with linear scar.
                Ventral keel of nutlet lying in a groove ....2. *A. Chorisiana.*
                Ventral keel of nutlet not in a groove ....3. *A. undulata.*
            Nutlets with ovate or oblong scar.
                Nutlets with smooth transverse or oblique ridges.
                    Ventral keel of nutlet not in a groove; scar obliquely basal
                                            4. *A. bracteata.*
                    Ventral keel of nutlet in a groove; scar ventral but reaching base ....5. *A. californica.*
                Nutlets with rough-edged or dentate transverse ridges
                                      6. *A. trachycarpa.*
    Scar about ⅓ length of nutlet, deeply hollowed, situated on ventral side above base ....7. *A. acanthocarpa.*
        Nutlets bearing minutely barbed spines ....var. *acanthocarpa.*
        Spines of nutlets few and reduced, or absent ....var. *oligochaeta.*
Stems prostrate; calyx-lobes all directed toward upper side.
    Fruiting calyx 4 to 6 mm. long; nutlets attached at base, the scar circular, not bordered by a ridge ....8. *A. leptoclada.*
    Fruiting calyx 6 to 10 mm. long; nutlets attached on ventral side near base, the scar ovate or deltoid, bordered by a ridge ....9. *A. humistrata.*

1. **A. stipitata** Greene var. **micrantha** (Piper) Macbr. Drying beds of vernal pools near Creston and perhaps elsewhere. A record from Cholame Valley (Jepson, Fl. Cal. 3: 360) is perhaps based on an incorrect identification.

2. **A. Chorisiana** (Cham.) Greene var. **Hickmanii** (Greene) Jepson. San Simeon, *K. Brandegee* (Jepson, Fl. Cal. 3: 359).

3. **A. undulata** Piper. In the bed of a vernal pool, 4 miles southeast of Santa Margarita (*8757*).

4. **A. bracteata** Howell. Frequent in Salinas River basin, in places which are moist during growing season: 7 miles east of Adelaida; near Estrella; between Santa Margarita and Pozo.

5. **A. californica** (F. & M.) Greene. Moist depressions in coastal region: Los Osos Valley (*9030*); possibly at Cambria. The identity of our plants is uncertain, largely because *A. californica* and *A. trachycarpa* are difficult to tell apart when mature fruit is lacking.

6. **A. trachycarpa** (Gray) Greene. Moist clay soils in widely scattered localities: San Miguel; 4 miles southeast of Santa Margarita; 3 miles east of Pozo; Laguna near San Luis Obispo.

7. **A. acanthocarpa** Piper var. **acanthocarpa.** Frequent in clay soils in interior: west base of La Panza Range near Pozo; Carrizo Plain and Temblor Range.

Var. **oligochaeta** (Piper) Jepson. At scattered localities within the range of the species: Carrizo Plain, *Condit.*

8. **A. leptoclada** Greene. Frequent in valleys of eastern part in clay, usually alkali soils, Cholame Valley to Carrizo Plain; probably also in Cuyama Valley.

9. **A. humistrata** Greene. Rare in low places flooded during rains: 18 miles east of Creston on La Panza road (*7592, 7620*).

8. **Echidiocarya** Gray

1. **E. californica** Gray var. **fulvescens** (Johnston) Hoover, n. comb. *Plagiobothrys californicus* (Gray) Greene var. *fulvescens* Johnston, Contr. Gray Herb. 68: 74. 1923. Infrequent in sandy soils, often appearing after fire: bluffs near Ragged Point; Oak Park district near Arroyo Grande; Navajo Creek, La Panza Range; Huasna district.

9. **Cryptantha** Lehm.

Nutlets tuberculate, granulate, papillate, or muricate.
    Nutlets, or at least most of them, bordered by a white erose wing.
        Corolla 4 to 7 mm. wide ....1. *C. oxygona.*
        Corolla 1 to 2 mm. wide ....2. *C. pterocarya.*
    Nutlets not wing-margined.
        Nutlets normally 4 (sometimes a few flowers with not all of nutlets developing).
            Calyx in fruit 1 to 1.5 mm. long, with fine hooked bristles; one of nutlets smooth and slightly larger than others ....3. *C. micromeres.*
            Calyx in fruit 3 to 8 mm. long, the bristles not hooked; nutlets alike.
                Calyx narrowed above a ventricose base, 4 to 8 mm. long; nutlets ovate-lanceolate.
                    Calyx mostly 4 to 6 mm. long, with short appressed hairs and longer bristles; leaves narrowly linear even on vigorous plants; lower part of stem strigose as well as spreading-bristly
                                        4. *C. intermedia.*
                    Corolla 3 to 7 mm. wide ....var. *intermedia.*
                    Corolla 1 to 3 mm. wide ....var. *rigida.*
                  Calyx mostly 6 to 8 mm. long, with copious long hairs on lower part; leaves on vigorous plants linear-oblong; lower part of stem hispid, with few or no appressed hairs
                                        5. *C. barbigera.*
                Calyx broadly ovoid or subglobose, 3 to 4 mm. long; nutlets triangular-ovate ....6. *C. muricata.*
                Corolla 4 to 7 mm. wide ....var. *muricata.*
                Corolla 1 to 4 mm. wide ....var. *Jonesii.*
        Nutlet solitary ....7. *C. corollata.*
    Nutlets smooth.
        Hairs on calyx straight, those on lower part not deflexed.
            Hairs on upper part of calyx-lobes ascending or spreading.
                Leaves mostly oblong or oblanceolate; spikes bracteate
                                        8. *C. leiocarpa.*
                Leaves mostly linear to lanceolate; spikes bractless except sometimes at base.
                    Fruiting calyx 2 to 5 mm. long; style more than half as long as nutlets, usually equalling them; nutlets 1 to (usually) 4
                                    9. *C. Clevelandii.*
                    Corolla 1 to 2 mm. wide ....var. *Clevelandii.*
                    Corolla 2 to 5 mm. wide ....var. *hispidissima.*
                Fruiting calyx 1 to 2 mm. long; style less than half as long as nutlets; nutlet solitary ....10. *C. microstachys.*

Hairs on upper part of calyx-lobes retrorse ....11. *C. nemaclada.*
Hairs on calyx curved, those on lower part deflexed ....12. *C. flaccida.*

1. **C. oxygona** (Gray) Greene. On crumbling shale or sandy slopes and in sandy washes: Cottonwood Pass, Temblor Range, and Cuyama Valley. Reaches its maximum abundance on north-facing slopes in the southern Temblor Range. In this area *C. oxygona* is not clearly differentiated from *C. pterocarya* and probably should be made a variety of it.

2. **C. pterocarya** (Torr.) Greene. Occasional in Temblor Range and washes issuing from it. Small-flowered plants are scarce here as compared with *C. oxygona.*

3. **C. micromeres** (Gray) Greene. Sandy soil in wooded or chaparral areas, generally appearing after fire: Oak Park district near Arroyo Grande (*6765*).

4. **C. intermedia** (Gray) Greene var. **intermedia.** Common in sandy soils, except very near the sea, inland to Palo Prieto Canyon and western edge of Carrizo Plain.

Var. **rigida** (Johnston) Brand. Common in eastern part, in sands, gravels, and crumbling shale; intergrading with typical *C. intermedia* where their ranges come together. Also very similar to narrow-leaved plants of *C. barbigera,* from which it differs in its strigose rather than spreading-hirsute stems. The two species, when growing together at Cottonwood Pass, looked much alike but were distinct.

5. **C. barbigera** (Gray) Greene. Gravelly or sandy soil, occasional in eastern part: Cottonwood Pass; just west of San Juan River, La Panza district; north end of Elkhorn Plain. Also Santa Maria River bed, *Eastwood* in 1906; an occurrence which would seem improbable except that many other species of the desert interior are likewise carried down toward the coast.

6. **C. muricata** (H. & A.) Nels. & Macbr. var. **muricata.** Rocky places, notably plentiful after fire or clearing: Common in Santa Lucia Range, eastward to La Panza Range and Cuyama Canyon; rare in Red Hills near Shandon.

Var. **Jonesii** (Gray) Johnston. Associated with typical *C. muricata* in Santa Lucia Mts., and intergrading with it; rare on Caliente Mt.

7. **C. corollata** Johnston. On rocky or gravelly slopes: Santa Lucia Range (Lopez Canyon); La Panza Range; Suey Creek (*Eastwood 122*); notably abundant on north side of lower Cuyama Valley.

8. **C. leiocarpa** (F. & M.) Greene. Occasional on coastal dunes: Point Sierra Nevada; north end of Morro Bay; Oceano Beach; Oso Flaco Lake.

9. **C. Clevelandii** Greene var. **Clevelandii.** As distinguished from var. *hispidissima* by its very small corollas, rare and sporadic: coastal bluffs near Ragged Point; Cottonwood Pass; Tassajera Peak and southeast of Cuesta Pass in Santa Lucia Mts. The variety *hispidissima* is variable, however, and many plants are intermediate.

Var. **hispidissima** (Greene) Johnston. Very common in sandy or rocky soils, often on serpentine, from coast eastward to La Panza Range. Especially vigorous and abundant after fires.

10. **C. microstachys** Greene. Gravelly or sandy soils in chaparral or open woods, growing most commonly among *Adenostoma* and thriving after fire, from coastal hills eastward to La Panza Range.

11. **C. nemaclada** Greene. Barren clay or rocky slopes in interior: Paso Robles (Jepson, Fl. Cal. 3: 352); Cottonwood Pass to Temblor Range and Caliente Mt., where locally abundant.

12. **C. flaccida** (Dougl.) Greene. Rare in the county, and known only in eastern part, in sandy soils: hills west of Carrizo Plain; Temblor Range.

10. **Amsinckia** Lehm. Fiddleneck

Calyx-lobes mostly 2 wide and 1 narrow, or 1 wide and 3 narrow.
  Leaves glaucous, ciliate, pustulate but scarcely hairy on surfaces, nutlets smooth

1. *A. vernicosa.*

  Corolla 8 to 12 mm. long, 2 to 8 mm. wide; ventral groove of nutlet not forked at base . . . . var. *vernicosa.*
  Corolla 12 to 18 mm. long, 8 to 14 mm. wide; ventral groove of nutlet forked at base . . . . var. *furcata.*
  Leaves not glaucous, hairy with mostly pustulate-based hairs; nutlets tuberculate or tessellate.
    Corolla 6 to 14 mm. wide.
      Flowers heterostylic (some plants with short style and stamens attached high in corolla-throat, others with long style and stamens attached low in corolla-tube); corolla 15 to 20 mm. long, 10 to 14 mm. wide

2. *A. Douglasiana.*

      Flowers homostylic (plants approximately uniform with regard to length of style and attachment of stamens); corolla 12 to 16 mm. long, 6 to 9 mm. wide . . . . .3. *A. tessellata* var. *gloriosa.*
    Corolla 2 to 6 mm. wide.
      Calyx and upper stem densely bristly and densely to sparsely appressed-hairy; stem (except under drought conditions) branching throughout

3. *A. tessellata* var. *tessellata.*

      Calyx densely covered with soft appressed brown hairs, the bristles few and short; stem simple or with short strictly ascending branches above

3. *A. tessellata* var. *elegans.*

Calyx-lobes 5, all narrow (partly fused toward base in *A. spectabilis*).
  Leaves mostly irregularly dentate, each tooth ending in a hair, the leaves otherwise sparsely hirsute; nutlets dark brown, 1.5 to 2 mm. long . . .4. *A. spectabilis.*
    Some of calyx-lobes usually partly fused; corolla 8 to 12 mm. long

var. *spectabilis.*

    Calyx-lobes distinct to base or nearly so; corolla 12 to 16 mm. long

var. *microcarpa.*

  Leaves entire or obscurely denticulate, moderately to densely hirsute or strigose; nutlets whitish, grayish, or light brown, 2 to 3.5 mm. long.
    Corolla 7 to 11 mm. long; stem hirsute but not strigose or sparsely so.
      Corolla with open throat, the stamens exserted from the tube

5. *A. intermedia.*

      Corolla constricted at throat, which is closed by hairy swellings; stamens included in corolla-tube . . . . .6. *A. lycopsoides.*
    Corolla 5 to 7 mm. long; stem usually finely strigose as well as hirsute

7. *A. Menziesii.*

1. A. vernicosa H. & A. var. **vernicosa.** Frequent in crumbling shale or white clay, usually on steep slopes and often where the soil is rich in gypsum, from 9 miles east of Creston (*8099*) eastward. On east slope of Santa Lucia Range in Monterey Co., and therefore to be sought west of Paso Robles. The plants reported by Twisselmann as var. *furcata* are actually of the typical form of the species. A local race with larger corollas (5 to 8 mm. wide) is frequent in the vicinity of Cholame. Such plants vary toward var. *furcata* in flower-size but lack the forked groove of the nutlet, which is the distinctive feature of that variety.

Var. **furcata** (Suksd.) Hoover. The type is labelled as collected on the "white hills" on the northern side of Cuyama Valley. Because no recent observer has been able to confirm this record, the possibility of an error is to be considered. This very showy plant, if present, should be easy to find. Otherwise it occurs in western Fresno and Kings Counties, some distance to the north of the recorded type locality.

2. **A. Douglasiana** A. DC. As defined by Ray and Chisaki (Am. Journ. Bot. 44: 529–536), this species is relatively rare, occurring as dense but small colonies at widely scattered places: near Fernandez Creek between La Panza and Creston; La Panza district. It is somewhat more widespread in Monterey Co. The large, brightly colored flowers form splashes of color on the hillsides which greatly enhance the character of the dry country where they are found. *Amsinckia Lemmonii* Macbr., based on a collection from Cholame, is included here by Ray and Chisaki; however, all collections seen from the vicinity of Cholame seem rather to belong to the species which Ray and Chisaki called *A. gloriosa.*

3. **A. tessellata** Gray var. **tessellata.** From Salinas Valley eastward, growing in sandy soils, friable clays, or crumbling shale. The small-flowered densely bristly plants ("typical" *A. tessellata*) are widespread but not so numerous in this area as the larger-flowered ones called *A. gloriosa* in recent references. A large proportion of the plants are intermediate.

Var. **gloriosa** (Eastw. ex Suksd.) Hoover, n. comb. *A. gloriosa* Eastw. ex Suksd., Werdenda 1: 103. 1931; type locality White Hills, Cuyama. Very abundant in the same areas where var. *tessellata* occurs, although apparently absent from the southern Temblor Range and hills bordering Cuyama Valley. One of the very few native plants which in this region can be properly regarded as a weed. Although Ray and Chisaki maintained *A. gloriosa* as a species, they stated that it and *A. tessellata* have the same chromosome number and are interfertile (Am. Journ. Bot. 44: 537, 539). The prevalence of intermediates prevents the segregation of San Luis Obispo Co. plants into distinct groups. I suspect that the correct name for var. *gloriosa* may be *A. tessellata* var. *Lemmonii* (Macbr.) Jepson, because of the prevalence of such plants around Cholame, the type locality of *A. Lemmonii*; but Ray and Chisaki have listed that name as a synonym of *A. Douglasiana.* No large-flowered plants could be found in the white hills on the north side of Cuyama Valley in 1967, although small-flowered *A. tessellata* was abundant. There is thus reason to suspect, as with *A. furcata*, that the type collection is erroneously labelled.

Var. **elegans** (Suksd.) Hoover, n. comb. *A. elegans* Suksd., Werdenda 1: 103. 1931. *A. Douglasiana* var. *elegans* Jepson & Hoover in Jepson, Fl. Cal. 3: 321. 1943. Locally abundant on Carrizo Plain (*9757*), and found at scattered localities northward to Contra Costa Co. Ray and Chisaki separated *A. gloriosa* and *A. tessellata* because of a difference in size of the corolla, along with associated differences in flower structure. If the whole plant be considered, rather than the flowers only, a somewhat different classification is suggested. Although in flower-size var. *elegans* matches typical *A. tessellata*, the stems have a few strictly ascending branches or are branched only at the top. The calyx is not conspicuously bristly but is densely covered with appressed rusty-brown hairs. As long as the chromosome number of var. *elegans* remains unknown, the possibility should be kept in mind that it may be a diploid, like *A. Douglasiana.* There is, however, apparent intergradation with the tetraploid *A. tessellata.*

4. **A. spectabilis** F. & M. var. **spectabilis.** Sandy soils near the sea, at least from Toro Creek northward.

Var. **microcarpa** (Greene) Jepson & Hoover. Plentiful on south side of Morro

Bay, and from Pismo Beach southward, extending inland to sandy hills on south side of San Luis Valley.

5. **A. intermedia** F. & M. FIDDLENECK. Grasslands and open woods from coast to vicinity of La Panza Range. Probably very common; although, in the absence of extensive collections, it can not now be determined how many of the plants so often seen and assumed to be this species may instead be *A. lycopsoides.* "Fiddleneck" is the usual English name for any *Amsinckia* in the San Joaquin Valley and else-where. Locally *A. tessellata* and its varieties have frequently been called "fireweed," but that name has long been established in the English language as the designation for *Epilobium angustifolium.*

6. **A. lycopsoides** Lehm. From Salinas Valley eastward to Carrizo Plain, often locally plentiful. Readily confused with *A. intermedia* and *A. Menziesii,* so perhaps more widely distributed.

7. **A. Menziesii** (Lehm.) Nels. & Macbr. At least from summits of Santa Lucia Range eastward to Temblor Range, often abundant.

Verbenaceae. VERBENA FAMILY

Stems erect to spreading, not rooting at nodes; calyx 5-toothed .... 1. *Verbena.*
Stems creeping, rooting at nodes; calyx 2-lobed .... 2. *Phyla.*

1. Verbena L.

Bracts shorter than calyx or barely equalling it.
    Leaves soft-pubescent; nutlets not papillate on inner surface 1. *V. lasiostachys.*
        Calyx-teeth about 1 mm. long .... var. *lasiostachys.*
        Calyx-teeth nearly obsolete .... var. *Abramsii.*
    Leaves scabrous; nutlets densely papillate on inner surface ...... 2. *V. robusta.*
Bracts longer than calyx .... 3. *V. bracteata.*

1. **V. lasiostachys** Link var. **lasiostachys.** Common in dry to moist places from coast inland to La Panza Range; also summit of Palo Prieto Canyon (*Twisselmann 4887*), grading toward var. *Abramsii.* This species is evidently the one which Twis-selmann reported under the name *V. menthaefolia* as occurring in the Choice Val-ley Hills. Sometimes the flowers are white instead of the usual violet color. Some of the specimens have been identified by H. N. Moldenke as var. *septentrionalis,* which is described as having a shorter calyx, but measurements of the calyx on such speci-mens show no difference from those plants which Moldenke identified simply as *V. lasiostachys.*

Var. **Abramsii** (Moldenke) Jepson. Hills at summit of Palo Prieto Canyon, in a vernal pool bed (*Twisselmann 7422*).

2. **V. robusta** Greene. Frequent along streams and around springs in coastal area, often in areas of serpentine; rare on east slope of Santa Lucia Range and not ob-served farther inland. No specimens have been seen to verify Twisselmann's report of this species as "occasional in the Choice Valley hills." Probably the report was based on plants of *V. lasiostachys* or its var. *Abramsii.*

3. **V. bracteata** Lag. & Rodr. Low places in interior, where subject to occasional flooding: Owen's Lake at summit of Palo Prieto Pass; Grant Lake; Salinas River near Templeton; Atascadero Lake; between Santa Margarita and Atascadero.

2. Phyla Lour.

1. **P. nodiflora** (L.) Greene. Presumably escaped from cultivation in low valleys: Paso Robles (*Chester Dudley* in 1926); Laguna near San Luis Obispo; Nipomo

Creek. The Dudley collection is identified by H. N. Moldenke as var. *canescens* (H.B.K.) Moldenke.

Labiatae. MINT FAMILY

Corolla-lobes unequal, arranged in 2 lips.
  Flowers in reduced cymes or solitary in the axils of the upper leaves or bracts (sometimes forming a "panicle" when bracts are very small); pedicels nearly always present though sometimes very short (*Melissa* and to some degree *Satureja mimuloides* have flowers in whorls but with evident pedicels).
    All or most of axils bearing few-flowered cymes.
      Stamens 4, extending far beyond longer corolla-lip; leaves not white-felted . . . . .1. *Trichostema.*
      Stamens 2, not extending beyond longer corolla-lip; leaves covered with a close white felt . . . . .9. *Salvia.*
    Flowers solitary in the axils, or sometimes a few of the axils bearing one or few additional flowers (*Melissa* has more than one flower in most axils).
      Herbs; corolla with 2 lobes in upper and 3 in lower lip.
        Calyx 2-lipped with entire lips . . . . .2. *Scutellaria.*
        Calyx-teeth 5.
          Calyx-teeth unequal, forming 2 lips . . . . .3. *Melissa.*
          Calyx-teeth equal or slightly unequal . . . . .4. *Satureja.*
      Shrubs; corolla with 4 short lobes and 1 long lower lobe 5. *Lepechinia.*
  Flowers in whorls; pedicels absent or nearly so (see *Melissa* above).
    Perennials (our species), with rhizomes, or with stems clustered from a woody base, or a few shrubby.
      Flower-whorls crowded to form a single dense terminal "spike," the bracts (except the lowermost) hidden by the flowers.
        Calyx-tube with about 12 to 15 longitudinal veins; calyx-teeth purple . . . . .6. *Agastache.*
        Calyx-tube with 5 evident veins; calyx-teeth green
                               8. *Stachys pycnantha.*
      Flower-whorls separated, with evident bracts.
        Calyx-teeth 10, hooked . . . . .7. *Marrubium.*
        Calyx-teeth 5, or fused to appear fewer, not hooked.
          Calyx not 2-lipped, with 5 subequal teeth . . . . .8. *Stachys.*
          Calyx 2-lipped . . . . .9. *Salvia.*
    Annuals (our species) with slender tap-root.
      Bracts not conspicuously veined, not translucent or with marginal spines.
        Leaves pinnatifid or bipinnatifid; bracts spine-tipped; corolla blue or violet . . . . .9. *Salvia.*
        Leaves entire to crenate or serrate; bracts not spine-tipped; corolla rose-red or purple.
          Bracts about as broad as long, crenate . . . . .10. *Lamium.*
          Bracts much longer than broad, more or less widened upward, entire . . . . .11. *Pogogyne.*
      Bracts conspicuously veined, translucent, bearing spines on the margin
                                12. *Acanthomintha.*
Corolla-lobes approximately equal, the corolla from 2-lipped to almost regular.
  Flowers in heads terminating the branches, surrounded by bracts
                                13. *Monardella.*
  Flowers in whorls in the upper leaf-axils or in interrupted terminal spikes
                                14. *Mentha.*

1. **Trichostema** L. Blue-Curls

Shrubs; leaves linear ....1. *T. lanatum.*
Annual herbs; leaves lanceolate or broader.
    Leaves narrowly elliptic or lanceolate ....2.. *T. lanceolatum.*
    Middle leaves ovate-acute or broadly elliptic ....3. *T. ovatum.*

1. **T. lanatum** Benth. Woolly Blue-Curls. Rocky hills from summits of Santa Lucia Range eastward to La Panza Range and Cuyama Canyon. Common, but not occurring in dense stands or large numbers.

2. **T. lanceolatum** Benth. Turpentine Weed. Vinegar Weed. Very common and often exceedingly plentiful in uncultivated ground back from the coast. On the west side of Carrizo Plain in September, 1965, this species grew in such abundance that the flowers colored whole hillsides blue. An important honey plant.

3. **T. ovatum** Curran. East edge of Cuyama Valley near Kern Co. line (*6491*), on dry hill with *Ephedra*. Otherwise restricted to the southern San Joaquin Valley.

2. **Scutellaria** L. Skullcap

1. **S. tuberosa** Benth. In leaf-mold or in loose rocky soil, and especially in recently burned places, in Santa Lucia Range and near coast. Probably all our plants of this species belong to subsp. *australis* Epling.

3. **Melissa** L.

1. **M. officinalis** L. Garden Balm. Occasionally established in stream-beds: Santa Rosa Creek; Santa Rita Creek; Coon Creek in San Luis Range. This plant is closely related to *Satureja,* and there seems no adequate reason for classifying it as a distinct genus.

4. **Satureja** L. Savory

Stems trailing; leaves glabrous; corolla white or partly purple, 8 to 12 mm. long
                                    1. *S. Douglasii.*
Stems erect or ascending; leaves soft-hairy; corolla orange or scarlet, 3 to 4 cm. long
                                    2. *S. mimuloides.*

1. **S. Douglasii** (Benth.) Briq. Yerba Buena. Common near coast, mostly under trees, sometimes in moist open places; less common inland about to Salinas River.

2. **S. mimuloides** (Benth.) Briq. Rare on moist banks in Santa Lucia Mts.: Lopez Canyon.

5. **Lepechinia** Willd.

1. **L. calycina** (Benth.) Epling. Pitcher Sage. At scattered localities, a minor component of chaparral or in open woods: San Luis Range; Santa Lucia Mts. and eastward to La Panza Range.

6. **Agastache** Clayt.

1. **A. urticifolia** (Benth.) Ktze. Stream-bottoms and shaded brushy hillsides: upper Coon Creek in San Luis Range; several places in Santa Lucia Range.

7. **Marrubium** L.

1. **M. vulgare** L. Horehound. Well established in all parts of county, often around rocks or at the sites of abandoned dwellings.

8. Stachys L. Hedge-Nettle

Plant with soft to bristly nearly straight hairs (sometimes the leaves white-silky but the hairs not tangled).

    Fruiting calyx funnelform-campanulate, 9 to 13 mm. long; corolla 25 to 30 mm. long from base to end of upper lip, 15 to 19 mm. wide from top to end of lower lip ....1. *S. Chamissonis.*

    Fruiting calyx short-campanulate or short-tubular, 6 to 9 mm. long; corolla 10 to 17 mm. long from base to end of upper lip, 7 to 14 mm. wide from top to end of lower lip.

        Flower-whorls, except sometimes the upper ones, distinct; corolla-tube obviously extending beyond calyx.

            Leaf-blades rounded or truncate to cordate at base.

                Corolla-tube with a transverse ring of hairs inside near base, without a constriction ....2. *S. bullata.*

                Corolla-tube constricted near base, with an oblique ring of hairs inside ....3. *S. ajugoides* var. *quercetorum.*

            Leaf-blades cuneate at base ....3. *S. ajugoides* var. *ajugoides.*

        Flower-whorls, except sometimes the lowermost, crowded to form an ovoid or cylindric "spike"; corolla-tube slightly if at all exserted from calyx

                        4. *S. pycnantha.*

Plant covered with tangled white wool, the older parts often partly glabrate

                        5. *S. albens.*

    1. **S. Chamissonis** Benth. Occasional in wet, usually shaded places near coast: Baywood Park; See Canyon; Oak Park district; region of Dune Lakes.

    2. **S. bullata** Benth. Moist, or more frequently dry soil, generally in woods, sometimes in sunny places, common in coastal region. Introduced at Twisselmann Ranch (reported by Twisselmann as *S. rigida* var. *quercetorum*). Otherwise there are no records at hand of the occurrence of this species east of the main divide of the Santa Lucia Range.

    3. **S. ajugoides** Benth. var. **ajugoides.** Occasional in wet places: coast north of Cambria; Atascadero; Laguna near San Luis Obispo; marsh at Edna. A plant found near the San Luis Obispo Laguna (*6324*) has the aspect of a possible hybrid with *S. pycnantha.*

    Var. **quercetorum** (Heller) Jepson & Hoover. The extreme form with cordate leaf-bases is unknown here. However, some of the plants along the northern coast have leaf-blades with rounded or truncate bases and thus represent intergrades between var. *ajugoides* and var. *quercetorum*. Apart from the obscure feature of having an oblique instead of transverse ring of hairs in the corolla-tube, these plants are not obviously distinct from *S. bullata*. The measurements of the corolla are not consistently smaller, nor are other differences readily apparent.

    4. **S. pycnantha** Benth. Frequent in moist places, often in areas of serpentine, from coast inland to near Santa Margarita; also at Cottonwood Pass (*7273*).

    5. **S. albens** Gray. Moist places in interior: Frequent between Salinas River and La Panza Range; Las Yeguas Creek and Horseshoe Canyon in Temblor Range.

9. Salvia L. Sage

Annual herbs; leaves pinnatifid or bipinnatifid.

    Leaves white-woolly, with spine-tipped teeth ....1. *S. carduacea.*

    Leaves not woolly, without spines ....2. *S. columbariae.*

Evergreen perennial herbs with somewhat woody base, or definitely shrubby; leaves crenate or crenulate.

    Herbaceous, or the woody stems, if any, creeping on the surface of the ground.

        Clump-forming plants, the stems a little woody at base or in young plants not at all so; leaf-blades truncate to sagittate at base; corolla red

                                                          3. *S. spathacea.*

        Plants with creeping woody stems bearing erect tufts of leaves and erect peduncles; leaf-blades gradually narrowed into the petiole; corolla violet or blue ....4. *S. sonomensis.*

    Shrubs, with well-developed woody stems.

        Flowers in whorls.

            Leaf-blades bright green on upper surface, cuneate at base; bracts acuminate, more or less white-villous but not felty .....5. *S. mellifera.*

            Leaf-blades gray-green on upper surface, broadly cuneate to truncate or subcordate at base; bracts obtuse to acute, white-felty

                                              6. *S. leucophylla.*

        Flowers in a modified compound cyme or "panicle," the axils of the bracts bearing condensed cymes ....7. *S. apiana.*

1. **S. carduacea** Benth. Thistle Sage. Very common in sandy soils in the interior; in coastal region found only near Nipomo. South of Shandon, considerable areas are in some years colored blue by the flowers of this species.

2. **S. columbariae** Benth. Chia. Common in hilly regions throughout the county, usually on gravelly or sandy slopes. The seeds, reported to be highly nutritious, were eaten by California Indians and are even yet in great demand by health-food enthusiasts.

3. **S. spathacea** Greene. Hummingbird Sage. Common in sheltered places near coast, usually under trees, inland to east side of Santa Lucia Range and Cuyama Canyon.

4. **S. sonomensis** Greene. Creeping Sage. Locally abundant on the ridge northwest of Cuesta Pass in Santa Lucia Range, and south of the Pozo-La Panza road in La Panza Range. A single plant on Cuesta Ridge which was obviously a hybrid between *S. sonomensis* and *S. mellifera* was destroyed by the United States Forest Service in the course of a "fuel-break" project. Such an interspecific hybrid, if it could be found again, would be of exceptional interest.

5. **S. mellifera** Greene. Black Sage. Very common in sandy or rocky soils, except on level valleys or plains, from coast inland to La Panza Range and upper end of Cuyama Canyon; Red Hills near Shandon; sparingly in higher parts of Temblor Range. In some places an important component of chaparral; in others forming extensive pure stands. The corolla is most commonly white or lavender-tinged, but near the coast some shrubs have bright blue or lilac flowers. Such forms of the species seem desirable for ornamental planting.

6. **S. leucophylla** Greene. Dry hills, Suey Creek and Huasna River watershed, eastward along Cuyama River to west end of Caliente Mt.

7. **S. apiana** Jepson. White Sage. In 1946 two shrubs were found in the Oak Park district north of Arroyo Grande (*6197*). These may have been planted by some unknown person some years earlier and in any case have since disappeared. As the species is present in the Santa Inez Valley of Santa Barbara Co., it may reasonably be sought in the Cuyama River basin.

10. **Lamium** L.

1. **L. amplexicaule** L. Sometimes appearing as a weed in cultivated ground: Paso Robles; Twisselmann Ranch; Santa Margarita.

11. **Pogogyne** Benth.

Bracts and calyx-lobes hirsute-ciliate (sometimes the hairs quite short); corolla 10 to 20 mm. long; fertile stamens 4 . . . . .1. *P. Douglasii.*
Bracts and calyces sparsely short-hirsute to glabrous; corolla 3 to 5 mm. long; fertile stamens 2 . . . . .2. *P. serpylloides.*

1. **P. Douglasii** Benth. In clay soils from head of Perfumo Canyon, San Luis Range, and upper Chorro Creek northward along the coast; upper Salinas Valley at Santa Margarita (*R. W. Summers*), where not recently seen and perhaps exterminated.

Plants found on the coast north of Arroyo de la Cruz differ from other specimens of *P. Douglasii* in having the hairs on the margins of the bracts and calyx-lobes sparse and short. In this manner, such plants approach *P. nudiuscula* Gray, a local endemic of San Diego which, as far as can be learned, has been exterminated in that area. It may be deduced that *P. nudiuscula* is a relict species which in this region has lost its distinctness through hybridization with *P. Douglasii.*

2. **P. serpylloides** (Torr.) Gray. Moist places in Santa Lucia Range, apparently rare locally: near summit on Atascadero-Morro Bay road; Cambria; 9 miles northwest of Adelaida.

12. **Acanthomintha** Gray

1. **A. obovata** Jepson. Open clay hillsides in the interior, apparently rare: La Panza district (*7273*); 2½ miles west of Painted Rocks (*Johannsen 1179*).

13. **Monardella** Benth.

Leaf-margins not undulate.
    Perennials, forming clumps and usually rooting from slender rhizomes.
        Calyx 9 to 12 mm. long; corolla 15 to 20 mm. long, its lobes about half as long as its tube . . . . .1. *M. Palmeri.*
        Calyx 7 to 9 mm. long; corolla 12 to 15 mm. long, its lobes nearly as long as its tube . . . . .2. *M. villosa.*
            Herbage more or less white-woolly with branched hairs var. *obispoensis.*
            Herbage sparsely puberulent, seemingly glabrous . . . . . .var. *subglabra.*
    Annuals with slender tap-root, the stem erect at base and branching above.
        Bracts acute, the intervals between the veins green or purple, not scarious
                3. *M. lanceolata.*
        Bracts acuminate, the intervals between the veins straw-colored and scarious or purplish toward apex . . . . .4. *M. Breweri.*
Leaf-margins undulate.
    Stems, at least the upper parts, densely covered with white curly hairs; leaves oblanceolate to round-oblong, short-pubescent or woolly, sparsely glandular-dotted . . . . .5. *M. crispa.*
    Stems sparsely pubescent to subglabrous; leaves linear to narrowly oblanceolate, sparsely long-hairy or nearly glabrous, moderately to densely glandular-dotted . . . . .6. *M. undulata.*

Annual with slender tap-root ....var. *undulata.*
Perennial or biennial, the tap-root and lower part of stem becoming woody
var. *frutescens.*

1. **M. Palmeri** Gray. Areas of serpentine rock: Atascadero-Morro road; ridge northwest of Cuesta Pass; Steiner Creek near San Luis Obispo; Reservoir Canyon; Rinconada Mine. Plants are not scarce, but they seldom bloom except after the ground has been cleared.

2. **M. villosa** Benth. var. **obispoensis** Hoover. Common on brushy or wooded hills in western part, most frequently in rocky soil, extending inland sparingly as far as La Panza Range. The fresh flowers are not usually "white" as originally described. The specimen which was the basis for the report of var. *franciscana* in San Luis Obispo Co. (Jepson, Fl. Cal. 3: 435) is actually var. *obispoensis.*

Var. **subglabra** Hoover. Slopes of hills near Cambria, locally rare.

3. **M. lanceolata** Gray. Summit between head of Salinas River and Stoney Creek, on gravelly slopes (7957); Tajea Spring, head of San Juan River (*Graham 572*).

4. **M. Breweri** Gray. Occasional in dry sand in interior, from Atascadero to Caliente Mt. and Huasna River.

5. **M. crispa** Elmer. Coastal sands from Oceano southward, growing most vigorously on shifting dunes but also found on stabilized sand. An occasional plant shows a combination of features which suggests hybridization with *M. undulata* var. *frutescens.* Such intermediate plants are, however, scarce as compared with individuals which are definitely referable to one species or the other.

6. **M. undulata** Benth. var. **undulata.** Sandy soil near coast: south side of Morro Bay to Hazard Canyon; from Price Canyon southward.

Var. **frutescens** Hoover. In stabilized sand from Oceano to north edge of Santa Maria Valley.

14. **Mentha** L. MINT

Flower-whorls subtended by bracts which are much smaller and narrower than the foliage leaves; upper bracts greatly reduced or apparently absent completely concealed by the flowers.
Herbage glabrous.
Leaves sessile or with petioles not over 2 mm. long; inflorescence becoming elongate, not over 7 mm. in diameter (excluding the lowermost bracts)
1. *M. spicata.*
Leaves with well-developed petioles; inflorescence short and dense, about 10 to 15 mm. in diameter ....2. *M. piperita.*
Herbage white-woolly, the upper leaf-surfaces tending to be glabrate
3. *M. rotundifolia.*
Flower-whorls, or at least the lower ones, subtended by bracts which are exactly like the foliage leaves; upper bracts all clearly visible, even if sometimes shorter than the flowers.
Bracts gradually reduced upward, the upper shorter than the flowers; corolla bright purple ....4. *M. Pulegium.*
Bracts little reduced upward, even the uppermost longer than the flowers; corolla light purple to white, or the lobes white with purple mid-vein
5. *M. arvensis.*
Leaf-blades broadly rounded at base ....var. *arvensis.*
Leaf-blades narrowed toward base ....var. *villosa.*

1. **M. spicata** L. Spearmint. Occasionally established in stream-beds and moist places: Las Tablas Creek; Trout Creek near Santa Margarita.

2. **M. piperita** L. Peppermint. San Luis Obispo, *Condit,* according to Jepson (Fl. Cal. 3: 448).

3. **M. rotundifolia** (L.) Hudson. Apple Mint. Rarely established in wet places: Mariannas Spring, Temblor Range (*Twisselmann 743*); Santa Rita Creek west of Templeton (*9525*).

4. **M. Pulegium** L. Pennyroyal. Well established in drying stream-bed, Santa Rita Creek (*9544*).

5. **M. arvensis** L. var. **arvensis** f. **lanata** (Piper) S. R. Stewart. Low moist ground near Laguna, San Luis Obispo.

Var. **villosa** (Benth.) Stewart f. **brevipilosa** Stewart. Santa Margarita Creek and Trout Creek near Santa Margarita; upper Arroyo Grande.

Solanaceae. Nightshade Family

Fruit a berry.
    Thorny shrubs with small fleshy leaves . . . . .1. *Lycium.*
    Plants herbaceous or, if shrubs, not thorny; leaves not fleshy.
        Calyx inflated, completely enclosing the berry . . . . .2. *Physalis.*
        Calyx not inflated, covering only the base of the berry.
            Corolla urn-shaped . . . . .3. *Salpichroa.*
            Corolla open, bowl-shaped or rotate . . . . .4. *Solanum.*
Fruit a capsule.
    Capsule bearing stout spines; plants not glandular-hairy . . . . .5. *Datura.*
    Capsule spineless; plants glandular-hairy except in *Nicotiana glauca.*
        Flowers solitary in the leaf-axils; stems widely spreading or prostrate
                                                        6. *Petunia.*
        Flowers in open terminal cymes; stems erect . . . . .7. *Nicotiana.*

1. Lycium L.

1. **L. Andersonii** Gray. Near north end of Elkhorn Plain, west base of Temblor Range, in sandy washes and on rocky hills.

2. Physalis L.

1. **P. ixocarpa** Brot. Tomatillo. Salinas River bed, Paso Robles (*Rodin 6195* in 1957).

3. Salpichroa Miers

1. **S. rhomboidea** (Gill. & Hook.) Miers. Lily-of-the-Valley Vine. Weed in cultivated ground and on creek-banks at San Luis Obispo.

4. Solanum L. Nightshade

Corolla deeply lobed, usually white, sometimes varying to purple.
    Annuals; corolla 3 to 6 mm. in diameter.
        Herbage glabrous or nearly so; berry purple-black when ripe
                                                1. *S. nodiflorum.*
        Herbage glandular-hairy; berry yellow . . . . .2. *S. sarachoides.*
    Perennials, often becoming somewhat woody below; corolla 8 to 15 mm. in
        diameter . . . . .3. *S. Douglasii.*
Corolla 5-angled but only shallowly lobed, violet or blue.
    Plant without spines.

Herbage with close fine pubescence consisting mostly or entirely of branched hairs, not glandular ....4. *S. umbelliferum.*
    Stems grayish with moderately dense pubescence ..var. *umbelliferum.*
    Stems densely white-tomentose ....var. *incanum.*
Herbage with all or nearly all simple hairs, often glandular, sometimes nearly glabrous ....5. *S. Xanti.*
    Herbage only slightly glandular, or not at all so ....var. *Xanti.*
    Herbage markedly glandular-hairy.
        Plants more or less shrubby, with woody stems above the base; leaf-margins not undulate to moderately so ....var. *intermedium.*
        Plants herbaceous, with spreading stems from a woody root; leaf-margins undulate ....var. *obispoense.*
Plant with spines ....6. *S. elaeagnifolium.*

1. **S. nodiflorum** Jacq. Black Nightshade. Weed in moist places: Twisselmann Ranch; San Luis Obispo; Salinas River near Santa Margarita. Probably widely spread, although no other specimens are at hand.

2. **S. sarachoides** Sendt. Hairy Nightshade. Occasional as a weed in cultivated ground, as at San Luis Obispo; Rancho Montana de Oro.

3. **S. Douglasii** Dunal. Common in canyons, hollows among sand-dunes, or along streams in western part. There is no record of it east of Santa Lucia Range.

4. **S. umbelliferum** Esch. var. **umbelliferum.** Paso Robles (*Eastwood 13,843*); Santa Maria River bed (*Eastwood 318*). Specimens from San Francisco, presumably the type locality, do not differ significantly from these two collections. Some plants within the area of var. *incanum* are intermediate toward var. *umbelliferum.*

Var. **incanum** Torr. *S. californicum* Dunal. Common in interior, from east base of Santa Lucia Mts. to Palo Prieto Canyon, thence south to Cuyama Valley and in Temblor Range. Absent, or nearly so, from the area where *S. Xanti* var. *obispoense* occurs. In view of the apparent close relationship between *S. umbelliferum* var. *incanum* and *S. Xanti* var. *obispoense,* the absence of hybrids is notable.

5. **S. Xanti** Gray var. **Xanti.** From coast inland to (rarely) east slope of Santa Lucia Range. Most of the plants probably represent var. *intermedium.* The species is highly variable, and varieties can be distinguished only rather arbitrarily. Some coastal specimens have a few of the hairs branched, suggesting the influence of *S. umbelliferum,* and some have the stems densely long-hairy as in the Catalina Island plant called *S. Wallacei.* I am including in *S. umbelliferum* only those plants which have the stems densely covered with hairs that are all, or mostly, branched. There may possibly be a distinct species on Catalina Island, but I believe that the San Luis Obispo Co. plants which have been reported as *S. Wallacei,* at least, are not distinct from *S. Xanti.*

Var. **intermedium** Parish. Especially plentiful in coastal region from Morro Bay southward. Because the variety is not sharply defined, the boundaries of its distribution are rather indefinite.

Var. **obispoense** (Eastw.) Wiggins. Sandy or rocky soil, upper Salinas Valley from Atascadero to Pozo, east to La Panza Range and south to Huasna district.

6. **S. elaeagnifolium** Cav. Along railway at Paso Robles.

5. **Datura L.**

Perennial; herbage finely white-hairy; corolla 15 to 20 cm. long ...1. *D. meteloides.*
Annual; herbage green; corolla 6 to 8 cm. long ....2. *D. Stramonium.*

1. **D. meteloides** DC. Tolguacha. Jimson Weed. Common in valleys of the interior, most frequently in cultivated ground: Salinas Valley and upper Arroyo Grande eastward.

2. **D. Stramonium** L. var. **Tatula** (L.) Torr. Purple Thorn-Apple. Occasional in sandy cleared or cultivated ground near coast; perhaps not permanently established.

6. Petunia Juss.

1. **P. parviflora** Juss. Low places where subject to seasonal flooding: Salinas River at San Miguel and Paso Robles; Grant Lake; Arroyo Grande; Santa Maria River.

7. Nicotiana L. Tobacco

Shrubs; herbage glabrous, glaucous; corolla yellow ....1. *N. glauca.*
Annual herbs; herbage glandular-hairy; corolla white.
    Middle and upper leaves narrowed below into a petiole, or narrow throughout; corolla about 1 cm. wide ....2. *N. acuminata.*
    Middle and upper leaves broadest near base, sessile; corolla 2 to 5 cm. wide
                                     3. *N. Bigelovii.*
        Corolla 3 to 5 cm. wide ....var. *Bigelovii.*
        Corolla 2 to 3 cm. wide ....var. *Wallacei.*

1. **N. glauca** Graham. Tree Tobacco. Along dry stream-beds and rocky banks, in both eastern and western parts but usually as widely scattered plants.

2. **N. acuminata** (Graham) Hook. var. **multiflora** (Phil.) Reiche. Occasional in stream-beds: Las Tablas Creek (*7375*); abundant in bed of Salinas River in 1969; near Atascadero; Calf Canyon.

3. **N. Bigelovii** (Torr.) Wats. var. **Bigelovii**. Usually in dry sandy or silty soil, near Atascadero, Caliente Mt. (*8270*), and Cuyama River; collected in Temblor Range in Kern Co.

Var. **Wallacei** Gray. Upper Salinas Valley: abundant in Salinas River bed at Paso Robles, mixed with *N. acuminata*; Santa Margarita (*6373*); Calf Canyon (*9597*); site of Eldorado School north of Pozo (*6984*).

Scrophulariaceae. Figwort Family

Leaves opposite throughout, or sometimes in whorls of 3 at a few of the nodes.
    Calyx 5-lobed; corolla 5-lobed, from strongly 2-lipped to nearly regular.
        Calyx divided to below the middle, usually nearly to base.
            Middle lobe of lower corolla-lip folded into a "keel" which encloses the stamens and style ....1. *Collinsia.*
            Middle lobe of lower corolla-lip not folded or enclosing the stamens and style.
                Corolla brownish or dark red, its tube subglobose, its upper lobes longer than the lower ....2. *Scrophularia.*
                Corolla variously colored but not brownish or dark red, its tube not subglobose, its upper lobes shorter than the lower or at most equalling them ....3. *Penstemon.*
        Calyx tubular or cup-shaped, its lobes relatively short.
            Shrubs ....4. *Diplacus.*
            Herbs, the stems not at all woody even in the perennial species.
                Calyx sharply 5-angled and usually more or less folded or winged on the angles; corolla usually obviously 2-lipped, rarely almost regular ....5. *Mimulus.*
                Calyx not angled; corolla nearly regular ....6. *Mimetanthe.*

Calyx 4-lobed; corolla 4-lobed, not 2-lipped, the uppermost lobe often broader than the others but of equal length .... 7. *Veronica.*
Leaves mainly alternate (or a few lower ones opposite), or partly in a basal rosette.
    Corolla not 2-lipped, with equal or subequal lobes.
        Calyx and corolla 4-lobed; slender plants mostly less than 2 dm. tall

7. *Veronica.*

        Calyx and corolla 5-lobed; coarse plants mostly over 6 dm. tall

8. *Verbascum.*

    Corolla 2-lipped.
        Upper corolla-lip not forming a galea, spreading upward at an angle from the tube, obviously 2-lobed.
            Corolla with a short blunt sac at base on lower side .. 9. *Antirrhinum.*
            Corolla with a slender spur at base on lower side.
                Flowers in a terminal raceme, the bracts smaller than the foliage leaves; pedicels shorter than flowers .... 10. *Linaria.*
                Flowers partly in the axils of the lower foliage leaves as well as of reduced bracts upward; pedicels longer than flowers

11. *Kickxia.*

        Upper corolla-lip forming a galea (i.e., continuous in direction with the tube, not 2-lobed).
            Calyx tubular or urceolate-tubular; lower corolla-lip very different from the upper, either saccate or much reduced.
                Annuals; lower corolla-lip slightly to strongly saccate.

12. *Orthocarpus.*

                Perennials or annuals; lower corolla-lip very short, not saccate.
                    Leaves entire to sparingly laciniate-pinnatifid; bracts toothed at apex, or pinnatifid, or entire, not serrate .... 13. *Castilleja.*
                    Leaves finely divided ("fern-like"); bracts serrate

14. *Pedicularis.*

        Calyx one-sided, or appearing to consist of 2 distinct sepals because of a sepal-like bractlet opposite the true calyx; lower corolla-lip quite similar to the upper, of equal length .... 15. *Cordylanthus.*

1. **Collinsia** Nutt.

Flowers mostly in whorls of 3 or more; pedicels much shorter than flowers.
    Upper leaf-blades triangular, widened to a truncate or broadly rounded base (this often obscured by folding under of the margins); calyx-lobes sharply acute or slightly acuminate .... 1. *C. heterophylla.*
    Upper leaf-blades oblong or oblong-lanceolate, a little widened toward base or not at all so; calyx-lobes barely acute to obtuse .... 2. *C. bartsiaefolia.*
        Flowers 17 to 22 mm. long .... var. *bartsiaefolia.*
        Flowers 10 to 15 mm. long .... var. *Davidsonii.*
Flowers solitary in the upper axils; pedicels longer than flowers ... 3. *C. sparsiflora.*

1. **C. heterophylla** Buist ex Graham. Common throughout the hilly parts of the county, usually in partial or complete shade, on north slopes or in woods. Plants included in this species vary in many respects. The flowers in the coastal region vary from white to light purple. Inland, the corolla usually has a bright purple lower lip and a white or purple-tinged upper lip. At some places in the Santa Lucia Mts., a few of the plants have both upper and lower lips bright purple.

2. **C. bartsiaefolia** Benth. var. **bartsiaefolia.** Sandy soil, northern part of Nipomo Mesa.

Var. **Davidsonii** (Parish) Newsom. Common in sandy places in interior, usually in sunny exposures and frequently on south-facing slopes, mostly from Salinas River to La Panza district, sparingly to Temblor Range. The typical form of var. *David-sonii* consists of low-growing plants not usually more than 15 cm. tall. Such plants are common in the interior of San Luis Obispo Co. In the same region, and often in the same localities, are also found taller plants which look like typical *C. bartsiae-folia* except for their relatively small flowers. The nature of their occurrence indicates that the taller plants represent a response to better environmental conditions. Because smaller flowers are a constant feature of all plants of interior San Luis Obispo Co., as distinguished from those of Nipomo Mesa, flower-size is here taken as the basis for separating the two varieties.

3. **C. sparsiflora** F. & M. var. **solitaria** (Kell.) Newsom. 18 miles east of Creston on La Panza Road (*7591*), in open oak woodland.

2. Scrophularia L. Figwort

Inflorescence obscurely puberulent or apparently glabrous; corolla very dark red, nearly black ....1. *S. atrata.*
Inflorescence glandular-pubescent or glandular-puberulent; corolla brownish red
2. *S. californica.*

1. **S. atrata** Pennell. Rocky places or sandy woods: Mallagh's Landing just east of Avila (*9018*) and hills bordering San Luis Valley on the south, to western Santa Barbara Co.

2. **S. californica** Cham. California Figwort. Common in coastal region, mostly under trees and on north-facing canyon-sides. Not reported east of Santa Lucia Mts. Largely absent from the restricted area where *S. atrata* grows, and less likely than that species to be found in rocky places. R. J. Shaw (Aliso 5: 172) has commented upon the distinctness of *S. atrata* from *S. californica* where growing adjacent to it.

3. Penstemon Mitch.

Herbaceous, or woody only at the base; corolla slightly to moderately 2-lipped, not deeply cleft down the sides.
    Leaves dentate; corolla sky-blue ....1. *P. Grinnellii.*
    Leaves all entire; corolla not blue.
        Corolla red, rarely yellow, nearly regular, with very short lobes
2. *P. centranthifolius.*
        Corolla violet, obviously 2-lipped ....3. *P. heterophyllus.*
Shrubs or woody vines; corolla cleft half its length at the sides, thus strongly 2-lipped.
    Leaves narrowed at base; corolla white or cream with dark purple lines
4. *P. breviflorus.*
    Leaves broadly rounded or subcordate at base; corolla scarlet, rarely orange or yellow ....5. *P. cordifolius.*

1. **P. Grinnellii** Eastw. subsp. **scrophularioides** (Jones) Munz. "Pass on road from Paso Robles to Bakersfield," *Mary H. Fleisher* in 1927. This locality would presumably be Polonio Pass. Not reported by Twisselmann from Temblor Range or observed on Caliente Mt. (where to be expected), but found in Kern Co. on road to Mt. Abel.

2. **P. centranthifolius** Benth. Scarlet Bugler. Sandy or rocky places in hills of interior, from west of Paso Robles to Temblor Range; also on Nipomo Mesa.

Widely scattered but seldom plentiful. Highly desirable as a garden flower in suitable soil, as may be seen particularly at the Ian McMillan Ranch in the Shandon Hills.

3. **P. heterophyllus** Lindl. var. **heterophyllus.** Frequent, but mostly as scattered plants, from Santa Lucia Mts. eastward to La Panza Range.

Var. **australis** M. & J. Huasna River (*7228*); Cuyama Canyon, and probably overlooked elsewhere. A trivial variety, having exactly the appearance of var. *heterophyllus* and differing only in being minutely puberulent throughout instead of only toward the base. Many plants of var. *heterophyllus* vary more or less toward var. *australis.*

4. **P. breviflorus** Lindl. Rocky places in Santa Lucia Range west of Paso Robles.

5. **P. cordifolius** Benth. CLIMBING PENSTEMON. In woods or among tall chaparral, from Morro Bay-Atascadero road (probably the northern limit for the species) southward; extending inland to La Panza Range and Cuyama Canyon.

4. **Diplacus** Nutt. BUSH MONKEY-FLOWER

Corolla-lobes not bifid, at most retuse.
    Corolla 5 to 6.5 cm. long; calyx typically hairy (but freely hybridizing with *D. aurantiacus* and showing various combinations of features) . . . . 1. *D. longiflorus.*
    Corolla 4 to 5 cm. long; calyx glabrous . . . . 2. *D. aurantiacus.*
Corolla-lobes deeply bifid . . . . 3. *D. fasciculatus.*

1. **D. longiflorus** Nutt. Rocky hills from crest of Santa Lucia Range near Cuesta Pass eastward to hills on west side of Carrizo Plain; upper Arroyo Grande to Cuyama Canyon; east side of Temblor Range in Kern Co. The flowers in some localities are yellow, elsewhere decidedly orange, and in Calf Canyon have been observed to have a pinkish tone. As to var. *calycinus* (Eastw.) Jepson, I believe either that it is not a distinguishable entity or that it does not occur in San Luis Obispo Co. In either case, no basis can be found for separating our plants into two groups. The Santa Lucia Range forms a natural boundary separating *D. longiflorus* from *D. aurantiacus.* As would be expected, hybrid-like intermediates which are not entirely typical of either species are plentiful along the summit of that range.

2. **D. aurantiacus** (Curtis) Jepson. STICKY MONKEY-FLOWER. Common in coastal area in woods, on sand-flats, and mostly on north slopes of hills, extending inland to west slope of Santa Lucia Mts. The most distinctive form of this species has relatively long pedicels. Plants with short pedicels (a feature of *D. longiflorus* and perhaps indicating hybridization), found from Morro Bay southward, represent *D. lompocensis* McMinn, which is not here regarded as distinct from *D. aurantiacus.* Short pedicels are also frequently found far beyond our limits to the north.

3. **D. fasciculatus** (Pennell) McMinn. Rocky places in Santa Lucia Mts. from Rocky Butte and Paso Robles-Adelaida road northward. All forms of *Diplacus* make very fine ornamental plants in cultivation when properly treated. This species is an especially attractive dwarf shrub.

5. **Mimulus** L. MONKEY-FLOWER

Pedicels relatively slender, at least half as long as flowers, often much longer.
    Calyx-lobes equal or nearly so; corolla with open throat, the lower lip not forming a palate.
        Perennial; corolla scarlet, strongly irregular, its lower lip reflexed
                                             1. *M. cardinalis.*

Annuals; corolla not scarlet, regular or slightly irregular.
  Pedicels mostly shorter than subtending leaves; corolla yellow or white.
    Leaf-blades petioled; corolla yellow.
      Stems 2 to 3 mm. thick below; herbage moist with copious glandular hairs . . . . .2. *M. floribundus.*
      Stems scarcely 1 mm. thick, except at the swollen lower nodes; herbage sparsely if at all glandular-hairy . .3. *M. geniculatus.*
    Leaf-blades, except the lowermost, sessile; corolla white
                                   4. *M. latidens.*
  Pedicels longer than subtending leaves; corolla rose-red or red-purple, or sometimes the lower lip pale yellow in *M. androsaceus.*
    Corolla 15 to 20 mm. long, more than twice as long as calyx
                                5. *M. Palmeri.*
    Corolla 7 to 10 mm. long, less than twice as long as calyx
                         6. *M. androsaceus.*
Calyx with upper lobe longest, the lower curved up toward it in fruit; corolla with lower lip forming swellings or a "palate" which partly close the throat
                                7. *M. guttatus.*
  Corolla 10 to 45 mm. long.
    Usually annual; stems erect or ascending . . . . .var. *guttatus.*
    Perennial; main stems widely spreading or reclining, rooting at nodes
                                var. *grandis.*
  Corolla 5 to 10 mm. long . . . . .var. *micranthus.*
Pedicels stout, much shorter than flowers, or virtually absent.
  Lower corolla-lobes at least as long as upper.
    Corolla 15 to 25 mm. long; capsule barely equalling calyx 8. *M. Fremontii.*
    Corolla 8 to 10 mm. long; capsule exserted from calyx . . . . .9. *M. Rattanii.*
  Lower corolla-lobes obsolete or nearly so . . . . .10. *M. Douglasii.*

1. **M. cardinalis** Dougl. Scarlet Monkey-Flower. Moist places, usually along streams: Cottonwood Pass (*7274*); Cambria; Cuesta Grade; Poly Canyon; Pettitt (Sycamore) Canyon in San Luis Range; Lopez Canyon.

2. **M. floribundus** Lindl. Sandy moist places in interior, rare locally: Navajo Creek, La Panza Range (*8187*).

3. **M. geniculatus** Greene. Plants which seem to represent an extremely dwarfed form of this species were collected from damp crevices in sandstone caves, foot of Caliente Mt. (presumably on north side), *Hardham 1781.*

4. **M. latidens** (Gray) Greene. Occasional in drying beds of rain-pools: Palo Prieto summit; Templeton-Creston road; 18 miles east of Creston.

5. **M. Palmeri** Gray. Sand-filled crevices in sandstone outcrops, 1 mile northwest of Bee Rock, *Hardham 5252.*

6. **M. androsaceus** Curran. Navajo Creek district, east side of La Panza Range (*8186*), scattered in burned area, scarce except in small seepage spot.

7. **M. guttatus** Fischer var. **guttatus.** Common Monkey Flower. Common where soil is permanently or temporarily wet. Highly variable, particularly in size. The varieties listed below are probably of little significance, because their distinguishing features might as easily be produced by the environment as by genetic factors. Although there may be some justification for separating from this species *M. nasutus* Greene as originally described, most specimens which botanists have called by that name do not differ in the slightest detail from plants which the same authorities have identified as *M. guttatus.* I have, anyhow, seen no San Luis Obispo Co. plants showing the sort of calyx described for *M. nasutus.*

Var. **grandis** Greene. In permanently wet places near the coast grow very robust plants which may be considered as this variety. However, a quite similar plant was collected at the summit of Palo Prieto Canyon *(Twisselmann 1889)*.

Var. **micranthus** (Heller) Grant. Frequent in interior: upper Cammatta Creek *(Twisselmann 2058)*. Dwarfed plants with equally small flowers are sometimes found in rapidly drying soil adjacent to normal-sized *M. guttatus* growing where the soil remains wet longer.

8. **M. Fremontii** Gray. Fairly common in sandy places in interior: Franklin Creek west of Adelaida *(Hardham 6901)* eastward to La Panza district; sparingly in Temblor Range. A collection from 7 miles east of La Panza *(Eastwood & Howell 2340)* shows considerable variation in size of the corolla. The smaller-flowered plants of this collection seem identical with Monterey Co. plants which contemporary botanists identify as *M. subsecundus*.

(*Mimulus subsecundus* Gray is an ambiguous name which probably does not apply to a distinguishable species. The type was collected on Pine Mt. in the Santa Lucia Range by Dr. Edward Palmer. The only similar plants found there in recent years belong to *M. Rattanii*. There is thus reason to suspect that *M. subsecundus* may be a synonym of *M. Rattanii*. However, the name has been used in recent years for Monterey Co. plants which in flower-size are intermediate between *M. Rattanii* and *M. Fremontii*. I believe such plants to be a small-flowered variant of *M. Fremontii*, as mentioned under that species).

9. **M. Rattanii** Gray. Dry brushy slopes, especially following fires: Pine Mt. above San Simeon *(8017, Hardham 6839)*.

10. **M. Douglasii** (Benth.) Gray. Serpentine grassland and gravelly outcrops, Cypress Mt., *Hardham 5292*.

6. **Mimethanthe** Greene

1. **M. pilosa** (Benth.) Greene. Sandy stream-beds in interior: Paso Robles; Calf Canyon; La Panza Range.

7. **Veronica** L.

Flowers solitary in the axils of alternate leaves.
    Leaves linear to narrowly oblong, mostly entire . . . . .1. *V. peregrina.*
    Leaves ovate, crenate . . . . .2. *V. arvensis.*
Flowers in racemes which are borne in the axils of opposite leaves
                           3. *V. Anagallis-aquatica.*

1. **V. peregrina** L. var. **xalapensis** (H.B.K.) St. John and Warren. Moist places, often in beds of vernal pools: Palo Prieto summit; near Estrella and Creston; Laguna near San Luis Obispo.

2. **V. arvensis** L. Occasional as a weed in lawns, not well established; also in bed of Salinas River at Paso Robles.

3. **V. Anagallis-aquatica** L. Frequent in water or wet soil: Maxey Ranch near Cholame; upper Salinas Valley; Laguna near San Luis Obispo; Oceano. Some of the plants growing in Rinconada Creek are *V. comosa* Richt., according to the key in Munz's Flora, but the nature of their occurrence suggests them to be merely dwarfed plants of *V. Anagallis-aquatica.*

8. **Verbascum** L. Mullein

Herbage below inflorescence glabrous (or perhaps sometimes "stellate-pubescent" as described) . . . . .1. *V. virgatum.*
Herbage densely tomentose . . . . .2. *V. Thapsus.*

1. **V. virgatum** Stokes. Dry soil near Atascadero and Santa Margarita; apparently long established but the plants few and not spreading beyond this small area.

2. **V. Thapsus** L. WOOLLY MULLEIN. Locally established in gravelly stream-beds on east side of Santa Lucia Mts. (Santa Rita Creek, *9536*).

9. **Antirrhinum** L. SNAPDRAGON

Flowers sessile or with stout pedicels less than 3 mm. long; corolla white to rose-pink.
    Leaf-blades linear to narrowly lanceolate; corolla with throat partly closed by a palate formed by swellings of the lower lip.
        Perennial, without twining branches; corolla 17 to 20 mm. long, rose-pink to white ....1. *A. multiflorum.*
        Annual, with twining branches when vigorous; corolla 10 to 12 mm. long, white ....2. *A. Coulterianum.*
    Leaf-blades mostly ovate; corolla with open throat, the lower lip without obvious swellings ....3. *A. ovatum.*
Flowers on slender pedicels at least 3 mm. long, usually much longer; corolla violet.
    Leaf-blades ovate; pedicels about twice the length of the flowers or less, with little or no tendency to twine ....4. *A. Nuttallianum.*
    Leaf-blades, except some of the lower ones, lanceolate, usually narrowly so; pedicels several times as long as flowers, twining when support is available
                                                5. *A. Kelloggii.*

1. **A. multiflorum** Pennell. *A. glandulosum* Lindl., not Lejeune. Rocky hills south of San Luis Valley and from Santa Lucia Mts. eastward to La Panza Range; widely scattered and never abundant.

2. **A. Coulterianum** Benth. Blochman's Ranch near Santa Maria, *Eastwood 486* in 1906. This locality may actually have been in San Luis Obispo Co.

3. **A. ovatum** Eastw. Open hillsides in gypsum-rich clay, or in washes, west base of Cottonwood Pass (*7734*) to Caliente Mt. On the western part of the Caliente Mt. ridge, the species in 1952 was locally abundant. A second center of abundance is near Pinole Spring north of Carrizo Plain. Even at these places plants are scarce except in years of unusually good rainfall. Other occurrences, consisting of only one or a few plants, probably result from seeds occasionally washed down from the hills above.

4. **A. Nuttallianum** Benth. Rare among volcanic rocks near coast: White's Point in Morro Bay State Park; ridge west of Cerro Romauldo; Mallagh's Landing east of Avila.

5. **A. Kelloggii** Greene. *A. strictum* (H. & A.) Gray, not Sibth. & Smith. *Asarina stricta* Pennell. *Neogaerrhinum strictum* Rothm. In woods or chaparral, especially vigorous following fire: near San Luis Obispo and Arroyo Grande to Cuyama Canyon and La Panza Range. The separation of this species from the genus *Antirrhinum* is not convincing, in view of its apparent close relationship to *A. Nuttallianum,* which has slender pedicels that are often longer than the flowers and sometimes seem to show a slight tendency to twine just under the flower.

10. **Linaria** Mill. TOAD-FLAX

Perennial; upper leaves ovate, broadly rounded at base or subcordate; corolla yellow
                                                1. *L. dalmatica.*
Annual; upper leaves linear.
    Corolla usually purple, sometimes varying to rose, orange, or pale yellowish; spur nearly straight ....2. *L. pinifolia.*
    Corolla violet; spur strongly curved (an occasional plant has greatly reduced spurless corollas) ....3. *L. texana.*

1. **L. dalmatica** (L.) Mill. Upper Toro Creek east of Cayucos, *W. H. Frey*. Perhaps not permanently established.

2. **L. pinifolia** (Poir.) Thell. Seeds of what seems to be this species are often included in "wildflower mixtures" and come up spontaneously in sandy fields near the coast. Probably not permanently established but often taken for a native plant.

3. **L. texana** Scheele. Sandy soils, mainly near coast; also Salinas Valley near Paso Robles and Atascadero.

11. Kickxia Dum.

1. **K. spuria** (L.) Dum. Well established in wet clay, Trout Creek east of Santa Margarita (*9571*).

12. Orthocarpus Nutt.

Bracts, at least the upper ones, with tips colored somewhat like the corolla, white or cream to pink, rose-red, or purple-red.
  Leaves oblong, obtuse or abruptly acute, entire or with short obtuse lobes
                                                     1. *O. castillejoides.*
  Leaves lanceolate or narrower, acuminate, entire or with narrow acuminate divisions.
    Galea straight, densely pubescent but the hairs evident only under magnification.
      Corolla moderately to strongly saccate, 2 to 5 mm. wide at widest point.
        Style longer than stamens, the stigma usually visibly exserted
                                                     2. *O. densiflorus.*
          Calyx 8 to 17 mm. long; corolla-sacs gradually narrowed into the tube.
            Spikes rose-purple . . . . var. *densiflorus.*
            Spikes white to pale cream-yellow. . . . . . . var. *obispoensis.*
          Calyx 5 to 8 mm. long; corolla-sacs abruptly narrowed into the tube . . . . var. *gracilis.*
        Style normally shorter than stamens, the stigma not exserted
                                                     3. *O. brevistylus.*
      Corolla only slightly saccate, hardly 2 mm. wide at widest point
                                                     4. *O. attenuatus.*
    Galea hooked at tip, visibly velvety without magnification
                                                     5. *O. purpurascens.*
      Upper bracts mostly not over 5 mm. wide, cleft into lobes over 2 mm. long.
        Sacs of corolla not tipped with orange.
          Sacs colored like the rest of corolla, rose-pink to purple-red
                                                     var. *purpurascens.*
          Sacs bright yellow to white or pale pink, conspicuous in the spikes . . . . var. *pallidus.*
        Sacs of corolla orange-tipped, the corolla otherwise bright rose-red
                                                     var. *ornatus.*
      Upper bracts mostly 5 to 7 mm. wide, cleft at apex with teeth 2 mm. long or less . . . . var. *latifolius.*
Bracts uniformly green or purplish to the tips.
  Upper flowers about equalling bracts or usually exceeding them; corolla yellow or white, often with dark purple galea, well exserted from calyx.

Flowers about twice as long as bracts; corolla 13 to 22 mm. long, its sacs 3 to 4 mm. deep.

Herbage green; galea yellow . . . . .6. *O. faucibarbatus.*

Herbage more or less purple-tinged; galea dark purple   7. *O. erianthus.*

Corolla yellow . . . . .var. *erianthus.*

Corolla white, sometimes turning pink in age . . . . .var. *roseus.*

Flowers about as long as bracts or only a little exceeding them; corolla 9 to 11 mm. long, its sacs 1.5 to 2 mm. deep . . . . .8. *O. micranthus.*

Upper flowers all shorter than bracts; corolla dull red, hardly exceeding the calyx . . . . .9. *O. pusillus.*

1. **O. castillejoides** Benth. Moist places on northern coast: Piedras Blancas (*Eastwood & Howell 5986*); hilltop south of Arroyo de la Cruz (*7947*).

2. **O. densiflorus** Benth. var. **densiflorus.** OWL'S CLOVER. Mainly in interior: grasslands in Santa Lucia Range, and notably abundant in upper Salinas Valley, eastward to Carrizo Plain. In some places, as near Santa Margarita, considerable areas are colored a rosy red when this species is in flower. Formerly a few plants were found near San Luis Obispo, scattered among colonies of var. *obispoensis*; these seem to have been exterminated.

Var. **obispoensis** Keck. Common in grasslands in coastal area, extending northward nearly to San Carpoforo Creek, and southward at least to San Luis Valley.

Var. **gracilis** (Benth.) Keck. Meadow-like openings in Santa Lucia Mts.: between Rocky Butte and Pine Mt. (*7898*); Atascadero road (*Eastwood 14,378,* intergrading to var. *densiflorus*). Farther inland, occasional plants are to some degree intermediate toward this variety, especially when growing under dry conditions.

3. **O. brevistylus** Hoover. Open hills or plains in eastern part: San Juan River to Carrizo Plain; north side of lower Cuyama Valley (*8119,* the type collection). A few obvious hybrids between this and *O. purpurascens* were found where the two grew together on Carrizo Plain. Drought-dwarfed plants, which resemble *O. attenuatus* in appearance, are readily distinguished by the pink color in the corollas and in the tips of the bracts and calyx-lobes.

4. **O. attenuatus** Gray. At scattered localities from Santa Lucia Mts. eastward to Temblor Range; apparently common but inconspicuous as compared with *O. purpurascens* or *O. densiflorus.*

5. **O. purpurascens** Benth. var. **purpurascens.** PINK PAINTBRUSH. OWL'S CLOVER. Mostly in sandy soils, throughout the county, often abundant and coloring the landscape. Near the coast, a considerable proportion of the plants are intermediate between the varieties *purpurascens* and *pallidus.*

Var. **pallidus** Keck. In sand, south side of Morro Bay and region of Nipomo Dunes.

Var. **ornatus** Jepson. Many of the plants found on dry sand-flats near Jack Lake showed the distinctive coloring of this variety, which has previously been reported as occurring only in the Mohave Desert.

Var. **latifolius** Wats. Coast from at least San Simeon northward; intergrading with var. *pallidus* southward.

6. **O. faucibarbatus** Gray. In low meadow near northwest end of Laguna, San Luis Obispo (*7920* in 1950); a few plants also seen in Perfumo Canyon. The species was locally plentiful in that one year but otherwise has not been noticed. The plants were the typical form with yellow flowers, not subsp. *albidus* Keck.

7. **O. erianthus** Benth. var. **erianthus.** BUTTER-AND-EGGS. Common in grasslands

near coast, especially around San Luis Valley and Morro Bay. Replaced northward by the following variety.

Var. **roseus** Gray. Coastal fields and bluffs from Cambria northward.

8. **O. micranthus** (Gray) Greene. Occasional in hard gravelly clay soils from Atascadero eastward to La Panza district. Although most authors have classified this species as a variety of *O. erianthus,* there is no evidence that the two are in the slightest degree interfertile. Usually the herbage and the galea contain a purple pigment, as in *O. erianthus,* but a colony found 18 miles east of Creston *(8442)* resembled *O. faucibarbatus* in being completely devoid of purple pigment.

9. **O. pusillus** Benth. Moist places near coast from San Luis Obispo northward, uncommon or overlooked.

13. **Castilleja** Mutis. INDIAN PAINTBRUSH.

Perennials; bracts, or some of them, usually cleft, lobed, or toothed.
    Plants without woolly hairs.
        Calyx not more deeply cleft on lower than on upper side; lower corolla-lip hidden within calyx.
            Stems pubescent to glabrate but not glandular below inflorescence.
                Leaves essentially glabrous, all or mostly entire; stems usually erect from base ....1. *C. miniata.*
                Leaves pubescent and often a little scabrous, mostly lobed (except the lower ones); stems usually decumbent at base ....2. *C. affinis.*
            Stems glandular-pubescent below inflorescence ......3. *C. Applegatei.*
        Calyx deeply cleft on lower side, the lower corolla-lip thus usually exposed
                                     4. *C. Jepsonii.*
    Plants with some woolly hairs, at least on the bracts or calyces.
        Spikes bright-colored, red or sometimes varying to orange or yellow.
            Plants sparsely woolly, mainly in the inflorescence ....5. *C. mollis.*
            Plants closely white-woolly throughout ....6. *C. foliolosa.*
        Spikes not bright-colored, green or slightly purple-tinged 7. *C. plagiotoma.*
Annual; bracts entire, mostly narrowly linear ....8. *C. stenantha.*

1. **C. miniata** Dougl. Highly localized in moist hollows among the sand-dunes at Oceano. A remarkable occurrence for a species found mostly in the higher mountains of western America.

2. **C. affinis** H. & A. INDIAN PAINTBRUSH. In woods, among or near shrubs, or on rocky slopes in western part. Not observed east of Santa Lucia Range. The local representation of this species has been named var. *contentiosa* (Macbr.) Bacigalupi. Usually the spikes are bright red. Plants growing on coastal bluffs and brushy north-facing slopes north of Cambria have spikes varying to orange, salmon-pink, bright yellow, and creamy white; and in addition the leaves tend to be on the average shorter, broader, and more obtuse at the apex. Such plants correspond to *C. Wightii* Elmer. Although these coastal plants often look quite different from those found in woods and sheltered canyons, no way has been found to separate the plants into two coherent groups. Those features which might be used as a basis for separating *C. Wightii* are not correlated but vary independently. While it is useful to make note of the variation, the fact that these plants form what seems to be a single freely interbreeding group makes it impractical to draw a line separation.

3. **C. Applegatei** Fernald. *C. Roseana* Eastw. Interior mountains, apparently rare: near Hi Mt. Spring Camp *(8793).*

4. **C. Jepsonii** Bac. & Heckard. Among *Artemisia californica* and other shrubs,

frequent in hills of eastern part from Cottonwood Pass to Caliente Mt., extending westward to hills between Shandon and Creston.

5. **C. mollis** Pennell. Sand-dunes, usually among shrubs, between Pismo Beach and Oceano Beach (*8033*), where now probably exterminated, and southward.

6. **C. foliolosa** H. & A. WOOLLY INDIAN PAINTBRUSH. Mostly in rocky soil, sometimes in hard-packed sand, from coast eastward to La Panza Range and Red Hills near Shandon. Near the coast the spikes are commonly yellow or orange; inland, always flaming scarlet.

7. **C. plagiotoma** Gray. Usually among shrubs, Temblor Range to Caliente Mt. and Chalk Mt., Cuyama Valley. Partially parasitic usually on *Eriogonum fasciculatum* var. *polifolium,* sometimes on *Stenotopsis linearifolius* or *Eastwoodia elegans.* The plants are widely scattered but never plentiful in any one place.

8. **C. stenantha** Gray. Moist soil along streams: Paso Robles; between Atascadero and Morro Bay (*Winblad* in 1937); Rinconada Creek; Stoney Creek (*7959*); Cuyama Ranch.

14. Pedicularis L.

1. **P. densiflora** Benth. INDIAN WARRIOR. In woods or chaparral: common in Santa Lucia Range and on coastal hills; at scattered localities eastward to La Panza Range. Partially parasitic on roots of woody plants.

15. Cordylanthus Nutt. BIRD'S BEAK

Flowers in a spike; bracts mostly 3-toothed at apex . . . . .1. *C. maritimus.*
Flowers in small heads, or some of them solitary; bracts deeply parted into 3 narrow segments . . . . .2. *C. rigidus.*
    Bracts with hairs less than twice as long as the width of the segments, often with spreading hairs short and sparse . . . . .var. *rigidus.*
    Bracts with conspicuous spreading hairs longer than twice the width of the very narrow segments . . . . .var. *filifolius.*

1. **C. maritimus** Nutt. Haynes Ranch, *Ingalls* in 1912. From other plant records, it is evident that this locality was on the shore of Morro Bay, where the species has not recently been seen. It is restricted to salt-marshes.

2. **C. rigidus** (Benth.) Jepson var. **rigidus.** BIRD'S BEAK. Common in the hills, usually in wooded areas, from coast to Palo Prieto Canyon, La Panza district, and Temblor Range. Notably variable.

Var. **filifolius** (Nutt.) Macbr. Occasional in Santa Lucia Mts.: between Rocky Butte and Pine Mt. (*8401*). The differences which authors have described between the corolla of this plant and that of typical *C. rigidus* are not visible in available specimens.

Martyniaceae. UNICORN-PLANT FAMILY

1. Proboscidea Keller

1. **P. louisianica** (Mill.) Thell. UNICORN PLANT. Clarence Still Ranch, Annette Hills (*Allen Still* in 1955), "small colony in plowed adobe ground."

Orobanchaceae. BROOMRAPE FAMILY

1. Orobanche L.

Flowers on peduncles several times the length of the flowers.
    Flowers one or few; peduncles from a stem nearly always less than 5 cm. long; corolla violet or white . . . . .1. *O. uniflora.*

Flowers several; peduncles from a stem more than 5 cm. long; corolla yellowish
or red-tinged . . . . .2. *O. fasciculata.*
Flowers sessile or on pedicels which bear 2 small bracts and are usually shorter than
the flowers.
Flowers less than 15 mm. long, sessile or on pedicels 5 mm. long or less
                                                                          3. *O. bulbosa.*
Flowers over 15 mm. long, on pedicels over 5 mm. long.
Corolla 30 mm. long or more; anthers densely woolly . . . . . .4. *O. Grayana.*
Corolla 18 to 25 mm. long; anthers sparsely long-hairy . . .5. *O. californica.*

1. **O. uniflora** L. var. **purpurea** (Heller) Achey. Moist grassy places: Stoney Creek,
parasitic on *Saxifraga* and *Dodecatheon*; Anderson Canyon near upper San Juan
River.

2. **O. fasciculata** Nutt. Poly Canyon, San Luis Obispo *(9368)*, on dry rocky slope,
parasitic on *Artemisia californica*; 5 miles east of Pozo.

3. **O. bulbosa** Beck. Usually, if not always, parasitic on *Adenostoma*: Santa Lucia
Mts. (Tassajera Peak); eastward to La Panza Range.

4. **O. Grayana** Beck var. **violacea** (Eastw.) Munz. Coastal bluffs from Cambria
northward, probably parasitic on *Grindelia*.

5. **O. californica** C. & S. Atascadero, apparently parasitic on *Quercus agrifolia*
(Jack Huffman in 1965); 2 miles southeast of Huerhuero School; Yaro Creek north
of Pozo; dunes south of Oso Flaco Lake, apparently parasitic on *Convolvulus Soldanella*. Plants of this species at different localities, although having similar flowers,
differ widely in manner of growth, probably in response to different hosts. The
plant from Atascadero is remarkably tall and stout, and flowers late in the fall. It
may be var. *claremontensis* Munz.

Plantaginaceae. Plantain Family

1. **Plantago** L. Plantain

Leaves pinnately lobed . . . . .1. *P. Coronopus.*
Leaves entire to dentate.
Perennials with stout caudex or root-crown.
Leaves succulent, narrowly linear to oblong . . . . .2. *P. maritima.*
Leaves much elongate, tapering to a narrow point; scapes erect or
nearly so . . . . .var. *juncoides.*
Leaves broadly linear to oblong, obtuse or moderately tapering to-
ward apex; scapes spreading, curved upward . . . . .var. *californica.*
Leaves not succulent, ovate to elliptic or lanceolate.
Spike very dense, rounded or truncate at base, rarely more than 1/10
total height of plant . . . . .3. *P. lanceolata.*
Spike with flowers less crowded toward base, nearly always more than
1/6, and often over half, total height of plant.
Leaf-blades ovate to broadly elliptic, narrowed at base into a
clearly defined petiole . . . . .4. *P. major.*
Leaf-blades oblanceolate to narrowly elliptic, gradually narrowed
into a poorly defined petiole . . . . .5. *P. hirtella.*
Annuals with slender tap-root.
Leaves with silky hairs; corolla-lobes spreading.
Bracts similar to calyx-lobes, equalling them in length . .6. *P. insularis.*
Bracts unlike calyx-lobes, about half as long . . . . .7. *P. erecta.*
Leaves glabrous; corolla-lobes erect .8. *P. Bigelovii.*

1. **P. Coronopus** L. Locally established at San Simeon Creek Beach.

2. **P. maritima** L. var. **juncoides** (Lam.) Gray. SEASIDE PLANTAIN. South of Hazard Canyon, in seepage spot at edge of rocky beach (*7371*).

Var. **californica** (Fernald) Pilger. Moist places along the shore, or on bluffs reached by salt spray: south of Hazard Canyon, where associated with var. *juncoides*; common from Cambria northward.

3. **P. lanceolata** L. ENGLISH PLANTAIN. Common in western part; particularly abundant in open pastures around San Simeon; scattered in irrigated places in interior.

4. **P. major** L. Moist places, especially in lawns and along streams near coast; uncommon in interior.

5. **P. hirtella** H.B.K. var. **Galeottiana** (Dec.) Pilger. Common in wet places near coast.

6. **P. insularis** Eastw. Gravelly or sandy hills on east side of Carrizo Plain.

7. **P. erecta** Morris. Very common throughout the county, absent only from densely brushy or wooded areas. Highly variable, especially in size.

8. **P. Bigelovii** Gray. Coastal bluffs from Morro Bay northward; alkaline soil on Carrizo Plain. A collection from near Morro (*Eastwood & Howell 2235*) has been identified by I. J. Bassett as subsp. *californica* (Greene) Bassett, which scarcely seems distinguishable.

Rubiaceae.[5] MADDER FAMILY

Leaves (in our species) nearly always in whorls of 4 or more; herbs (one species with slender fragile woody stems).
    Flowers lilac or pink, in small heads . . . . .1. *Sherardia.*
    Flowers green, white, or yellowish, not in heads . . . . .2. *Galium.*
Leaves opposite; shrubs, small trees, or woody vines.
    Leaves pinnate; flowers in large flat-topped cymes . . . . .3. *Sambucus.*
    Leaves simple; flowers not in cymes.
        Flowers in axillary and terminal clusters . . . . .4. *Symphoricarpos.*
        Flowers in pairs, or whorled in terminal spikes . . . . .5. *Lonicera.*

1. Sherardia L.

1. **S. arvensis** L. Lopez Canyon, abundant on open flats (*8282*); also a scarce lawn weed at San Luis Obispo.

2. Galium L. BEDSTRAW. CLEAVERS

Annuals with very slender root.
    Fruits with bristles.
        Inflorescence an open panicle with reduced bracts . . . . . . .1. *G. parisiense.*
        Flowers few in axillary cymes, or paired or solitary in the axils.
            Leaves in whorls of 6 to 8.
                Fruit at maturity 4 to 5 mm. long . . . . .2. *G. Aparine.*
                Fruit at maturity 1.5 to 3 mm. long . . . . .3. *G. spurium.*
            Leaves opposite or in whorls of 4 . . . . .4. *G. biflorum.*
        Fruits tuberculate, not bristly . . . . .5. *G. tricornutum.*
Perennials, usually with somewhat woody roots (except *G. trifidum*).
    Fruits glabrous or nearly so.

---

5. The customary separation of the *Caprifoliaceae* as a family, however sound its theoretical basis might be, can not be supported by features which the actual plants show.

> Stems soft, herbaceous to the base ....6. *G. trifidum.*
> Stems rigid in lower part, often more or less woody at least at base.
>> Leaves 6 to 8 in a whorl ....7. *G. Hardhamiae.*
>> Leaves 4 in a whorl.
>>> Plants normally forming a close tuft, with ascending rigid spine-tipped leaves ....8. *G. Andrewsii.*
>>> Plants not usually forming a close tuft, with spreading leaves not spine-tipped.
>>>> Stems woody, if at all, only at base ....9. *G. californicum.*
>>>> Woody stems long and slender, extensively sprawling or reclining on other plants ....10. *G. Nuttallii.*
> Fruits with bristly hairs ....11. *G. angustifolium.*
>> Internodes glabrous ....var. *angustifolium.*
>> Internodes pubescent ....var. *siccatum.*

1. **G. parisiense** L. Garden weed at Twisselmann Ranch.

2. **G. Aparine** L. Widely distributed and apparently common except in the most arid localities, but see note under *G. spurium.*

3. **G. spurium** L. var. **echinospermum** (Wallr.) Hay. In loose soil under trees, apparently not common but perhaps not noticed because of its resemblance to *G. Aparine,* from which it may actually not be distinct.

4. **G. biflorum** Hochst.? Calf Canyon, on steep slope under scrub-oaks (*8344*). The plants resemble certain Old-World specimens which European botanists have identified as *G. spurium* var. *tenerum,* but are surely distinct from Californian specimens which are called *G. spurium.* There is some degree of similarity to the native American species *G. bifolium* Wats.

5. **G. tricornutum** Dandy. Occasional in cultivated ground: Palo Prieto Canyon; Paso Robles airport; See Canyon.

6. **G. trifidum** L. var. **subbiflorum** Wieg. Common in marshy spots near coast.

7. **G. Hardhamiae** Dempster. Santa Lucia Range from Cypress Mt. northward into southern Monterey Co.; usually, if not always, in serpentine areas where *Cupressus Sargentii* grows.

8. **G. Andrewsii** Gray. Stony or sandy soil, usually among chaparral, from coast to La Panza Range; higher parts of Temblor Range. The commonest native species locally.

9. **G. californicum** H. & A. Common in shaded places near coast, extending inland at least to Pilitas Creek and Pozo-Arroyo Grande summit. Notably variable in aspect: plants growing in sunny places have small and more crowded leaves.

10. **G. Nuttallii** Gray var. **ovalifolium** Dempster. Common on coastal hills; inland at least to Hi Mt. Some of the plants may fit the descriptions of one or more other varieties of the species, but collections for study are lacking.

11. **G. angustifolium** Nutt. Rocky hills from Santa Lucia Range eastward. Plants at American Canyon Campground in La Panza Range represent var. *siccatum* (Wight) Hilend & Howell, an inconsequential form having approximately the same geographic distribution as typical *G. angustifolium.*

3. **Sambucus** L. ELDERBERRY.

1. **S. mexicana** Presl ex DC. BLUE ELDERBERRY. Common along streams, or occasional individuals on dry hillsides, mostly near coast but extending sporadically far inland.

4. **Symphoricarpos** Duhamel

Erect shrubs; corolla 5 to 7 mm. long; berries 8 to 12 mm. in diameter . . 1. *S. albus.*
Low spreading or trailing shrubs; corolla 3 to 5 mm. long; berries 4 to 6 mm. in
diameter . . . . .2. *S. mollis.*

1. **S. albus** (L.) Blake var. **laevigatus** (Fernald) Blake. SNOWBERRY. Along streams
and in wooded canyon-bottoms in western part. Available specimens have the
branchlets minutely hairy when young, as in *S. mollis.* They are here called *S. albus*
because the shrubs are tall and the corollas larger than in *S. mollis.*

2. **S. mollis** Nutt. CREEPING SNOWBERRY. Common on wooded hillsides in western
part. No records are at hand from east of Santa Lucia Range, though the species
probably extends farther inland.

5. **Lonicera** L. HONEYSUCKLE

Flowers in pairs on axillary peduncles.
    Flowers subtended by showy red bracts; corolla nearly regular 1. *L. Ledebourii.*
    Flowers subtended by green leaf-like bracts; corolla strongly irregular
                                           2. *L. japonica.*
Flowers terminal, sessile, in whorls.
    At least one pair of upper leaves connate.
        Corolla rose-red or pink, 15 to 18 mm. long . . . . .3. *L. hispidula.*
        Corolla pale yellow, 10 to 12 mm. long . . . . .4. *L. interrupta.*
    Leaves all distinct . . . . .5. *L. denudata.*

1. **L. Ledebourii** Esch. TWINBERRY. Frequent in marshy places near coast, usually among trees and often climbing on them. An excellent ornamental shrub for
cultivation in suitables places.

2. **L. japonica** Thunb. JAPANESE HONEYSUCKLE. Bank of Santa Rosa Creek, Cambria (*J. T. Howell 40,793*); to be expected as escape from cultivation elsewhere on
stream-banks near coast.

3. **L. hispidula** (Lindl.) Dougl. ex T. & G. var. **vacillans** Gray. Climbing on shrubs
and trees along streams or on hillsides near coast and in Santa Lucia Range.

4. **L. interrupta** Benth. CHAPARRAL HONEYSUCKLE. Dry hillsides, east side of Santa
Lucia Range, not common, extending to Salinas River: 5½ miles southeast of Santa
Margarita (*9900*).

5. **L. denudata** (Rehder) Davidson & Moxley var. **Johnstonii** (Keck) Hoover, n.
comb. *L. subspicata* H. & A. var. *Johnstonii* Keck, Bull. S. Cal. Acad. 25: 67. 1926.
Common on dry hills among trees or in sparse chaparral, from Santa Lucia Mts. to
Palo Prieto Canyon and La Panza Range; higher parts of Temblor Range; probably on Caliente Mt. Distinct from *L. subspicata* H. & A., a well marked endemic of
the Santa Inez and Santa Monica Mts. with which authors have merged it, but
hardly distinguishable from the more southern *L. denudata.*

Valerianaceae. VALERIAN FAMILY

Annuals with a single stem from base; corolla less than 1 cm. long; stamens 3
                                                  1. *Plectritis.*
Perennials with several stems from base; corolla more than 1 cm. long; stamen 1
                                           2. *Centranthus.*

1. Plectritis DC.

Fruit sharply keeled, without a groove ....1. *P. congesta.*
Fruit with a low broad longitudinal ridge, which usually has a groove down the middle.
    Spur of corolla stout, shorter than the ovary or fruit ....2. *P. macrocera.*
    Spur of corolla slender, at least as long as the ovary or fruit ....3. *P. ciliosa.*

1. **P. congesta** (Lindl.) DC. Usually shaded places near coast, often in somewhat moist soil. The genus *Plectritis* is particularly difficult, because of the close resemblance in general appearance among the species, and because each species shows a comparable range of variation within itself. An attempt has been made to group our plants according to the classification of Dempster (Brittonia 10: 14–28. 1958).

2. **P. macrocera** T. & G. Frequent in grassland and open woods of interior, from east base of Santa Lucia Range eastward.

3. **P. ciliosa** (Greene) Jepson. Having about the same distribution in the county as *P. macrocera.* Not enough is known about the comparative ecology of the two, or their relative abundance or scarcity. Herbarium specimens, at least, are often difficult to differentiate.

2. Centranthus DC.

1. **C. ruber** (L.) DC. Jupiter's Beard. Occasionally escaping from cultivation, usually on rocky banks: Cambria; San Luis Obispo.

Dipsacaceae. Teasel Family

Spikes or heads with rigid spine-like bracts ....1. *Dipsacus.*
Heads with soft bracts ....2. *Scabiosa.*

1. Dipsacus L.

1. **D. fullonum** L. Fuller's Teasel. Well established in low places, usually in moist soil, near coast, especially near San Luis Obispo; also in upper Salinas Valley near Santa Margarita.

2. Scabiosa L.

1. **S. atropurpurea** L. Occasionally spreading from gardens: San Simeon Creek; Cuesta Grade; vacant lots at San Luis Obispo.

Cucurbitaceae. Gourd Family

Flowers yellow, solitary, at least 4 cm. long; fruit a pepo (gourd) ......1. *Cucurbita.*
Flowers green to white, the staminate in racemes or small panicles, less than 2 cm. in diameter; fruit a spiny capsule ....2. *Marah.*

1. Cucurbita L.

Leaf-blades triangular-ovate, very shallowly if at all lobed ....1. *C. foetidissima.*
Leaf-blades palmately lobed half way to base ....2. *C. palmata.*

1. **C. foetidissima** H.B.K. Calabazilla. Sandy soil, Santa Maria Valley (*7321*). A single plant seen in a vacant lot at San Luis Obispo doubtless represented a casual introduction which did not persist. Solitary plants were also noticed between Paso Robles and Estrella and in the pass at the southeast end of Carrizo Plain.

2. **C. palmata** Wats. Coyote Melon. Sandy soils, Temblor Range, Carrizo Plain, and Cuyama Valley.

2. **Marah** Kell.

Corolla pale green; rarely white, but only in plants of the interior which have fruits
with reduced spines ....1. *M. fabaceus.*
    Fruit densely covered with spines 5 to 25 mm. long ....var. *fabaceus.*
    Fruit with sparse spines mostly not over 5 mm. long ....var. *agrestis.*
Corolla white; coastal only ....2. *M. guadalupensis.*

1. **M. fabaceus** (Naud.) Greene var. **fabaceus.** WILD CUCUMBER. Common among
shrubs, less often in open fields, especially near coast, extending inland apparently
to Palo Prieto Canyon (*Twisselmann 889*). Most specimens from the interior lack
fruits but on geographical grounds would be expected to belong to the following
variety.

Var. **agrestis** (Greene) Stocking. Probably the common form of the species throughout the interior. The following specimens have well-developed fruits and definitely
are this variety: 5 miles east of Creston (*9824*); Eldorado School north of Pozo (*Wall
in 1933*); 7 miles east of La Panza (*Eastwood & Howell 2353*). A plant with more
deeply lobed leaves and white flowers, found in the Temblor Range east of Soda
Lake (*10,259*), is *M. inermis* (Congdon) Dunn, according to Abrams's "Illustrated
Flora." This form needs more study but is provisionally included in var. *agrestis.*
Some coastal specimens from near Cambria and Cayucos are perhaps intermediate
between the varieties.

2. **M. guadalupensis** (Wats.) Greene. A plant with white flowers, provisionally
identified as this species, was found a short distance south of Cayucos (*6640*) associated with *M. fabaceus,* which has greenish flowers. The ovules in the white-flowered
plant were no more than four, a feature which excludes this plant from *M. macrocarpa,* the occurrence of which would for geographic reasons be more probable.
White-flowered plants have also been seen in the upper valley of Arroyo Grande,
but no fruiting specimens have been collected.

Campanulaceae. BELLFLOWER FAMILY

Calyx-lobes lanceolate to ovate; capsule oblong-cylindric or cup-shaped.
    Leaves entire or obscurely serrate; calyx-lobes lanceolate; capsule oblong-cylin-
    dric .... 1. *Specularia.*
    Leaves conspicuously dentate; calyx-lobes broadly ovate; capsule broadly cup-
    shaped ....2. *Heterocodon.*
Calyx-lobes linear; capsule linear-clavate to linear ....3. *Githopsis.*

1. **Specularia** Heist.

1. **S. biflora** (R. & P.) F. & M. Widely scattered in burned or recently disturbed
areas, the plants always very few: Navajo Creek region in La Panza Range; Pettitt
(Sycamore) Canyon in San Luis Range; Tar Spring Creek.

2. **Heterocodon** Nutt.

1. **H. rariflorum** Nutt. Usually in moist places, sometimes in dry soil following
fires, from coast to La Panza Range; rare or more probably overlooked.

3. **Githopsis** Nutt.

1. **G. specularioides** Nutt. Apparently rare locally: Navajo Campground (*8199*)
and upper Navajo Creek (*8184*) in La Panza Range, following fire. A collection

from Suey Creek (*Eastwood 433*) represents *G. diffusa* Gray, which is not here accepted as a distinct species. The plants in the La Panza Range are variable with respect to those characters which have been regarded as separating *G. diffusa* from *G. specularioides*. All our plants are small-flowered.

Lobeliaceae. LOBELIA FAMILY

Ovary short and thick, abruptly narrowed into a very slender pedicel; corolla white or pale purplish; growing in dry places.
    Flowers in racemes, with elongate pedicels . . . . .1. *Nemacladus.*
    Flowers in head-like clusters, with very short pedicels . . . . .2. *Parishella.*
    Flowers sessile, with a long pedicel-like ovary; corolla blue; growing in drying beds of pools . . . . .3. *Downingia.*

1. **Nemacladus** Nutt.

Axis of raceme flexuous or nearly straight; seeds subglobose . . . .1. *N. ramosissimus.*
Axis of raceme slightly to obviously zigzag; seeds slightly longer than thick.
    Corolla-tube scarcely longer than calyx . . . . .2. *N. gracilis.*
    Corolla-tube distinctly longer than calyx . . . . .3. *N. secundiflorus.*

1. **N. ramosissimus** Nutt. Plants from sandy places in the interior (Salinas Valley to La Panza Range and Stoney Creek) key out to this species in current references, although, according to G. T. Robbins (Aliso 4: 139–147), *N. secundiflorus* occupies this area. Plants from the Oak Park district near the coast, although immature, seem to be the same. In view of the similarity in flower structure, inflorescence, and type of branching shown by the plants, the case for the existence of more than one species in the county is singularly unconvincing. I suspect that ultimately all the *Nemacladus* in San Luis Obispo Co. will be included in a single species, for which the name *N. ramosissimus* may be used. This conclusion is reinforced by the fact that use of the key provided by Robbins often leads to an identification of a plant as something other than the name to which Robbins assigned it.

2. **N. gracilis** Eastw. The only local collection assigned to this name by G. T. Robbins is *Twisselmann 2000* from San Juan River Valley 8.6 miles south of Shandon. However, a collection from Santa Margarita (*Eastwood* in 1902) shows to an equal or greater degree the features which, according to Robbins, distinguish *N. gracilis* from *N. secundiflorus*.

3. **N. secundiflorus** Robbins. Specimens included here by Robbins have been collected in the area between the Santa Lucia and La Panza Ranges, in Palo Prieto Canyon, and in the upper Arroyo Grande Valley. The type locality is East Fork of Huerhuero Creek 3 miles north of Creston.

2. **Parishella** Gray

1. **P. californica** Gray. Caliente Mt., mainly on western part, in decomposed sandstone, loam, or especially barren gypseous clay. It was noted in 1952 that each corolla contained a conspicuous drop of nectar.

3. **Downingia** Torr.

1. **D. cuspidata** Greene. Beds of vernal pools from near Paso Robles (east of Salinas River) to near Creston. The plants found locally are of the relatively large-flowered form named *D. immaculata* by Munz and Johnston.

Compositae. SUNFLOWER FAMILY

KEY TO TRIBES

Some or all of the flowers in a head with regular 5-lobed (rarely 4-lobed) corollas disk-flowers).

   Pistillate flowers, if any, on the outer margin of the heads, and the disk-flowers perfect (except in a few dioecious species); stamens with anthers united into a tube.

   Involucral bracts not in a single series, unequal in length, usually imbricated, but sometimes in 2 distinct series.

   Receptacle without bracts or dense long hairs.

   Style-branches clavate; ray-flowers none; disk-corollas white to greenish; bracts not scarious .9. *Eupatorieae.*
   Style-branches not clavate; ray-flowers either present or absent; disk-corollas normally yellow except in some species with scarious **bracts.**

   Bracts not scarious.

   Pappus of bristles or flattened scales, or none, sometimes of soft hairs but then not pure white and the plant not a shrub with scale-like leaves.

   Pappus of scales which are flattened at least at base although often awn-tipped; or, if of bristles, the leaves opposite; or pappus none . . . . . .4. *Helenieae.*
   Pappus normally present, of bristles or hairs (except *Gutierrezia* and one variety of *Lessingia* which have alternate leaves) . . . . .5. *Astereae.*
   Pappus of copious white hairs; shrub with leaves on flowering shoots reduced to green scales
   Lepidospartum in 7. *Senecioneae.*
   Bracts scarious or scarious-margined.

   Pappus of fine hairs, often very fragile; leaves entire or toothed.

   Leaves toothed; plants without woolly pubescence
   Baccharis in 5. *Astereae.*
   Leaves strictly entire; plants with woolly or glandular-woolly pubescence . . . . .8. *Inuleae.*
   Pappus absent or forming a short crown or ring on summit of achene; leaves toothed to finely divided
   6. *Anthemideae.*
   Receptacle with bracts or dense long hairs.

   Receptacle with scale-like bracts; involucral bracts neither spine-tipped nor with a callous border.

   Involucral bracts not scarious; heads over 3 cm. in diameter; bracts of receptacle easily visible . . . . .1. *Heliantheae.*
   Involucral bracts scarious or scarious-margined; heads less than 3 cm. in diameter; bracts of receptable inconspicuous
   6. *Anthemideae.*
   Receptacle densely long-hairy; involucral bracts tipped either with a spine or with a callous border which usually is fringed
   10. *Cynareae.*
   Involucral bracts essentially in a single series and of equal length, except sometimes for a few very short bracts at base of involucre (or, if described as being in 2 series, the series not clearly differentiated and about equal), not imbricated, sometimes the bracts united in a lobed cup or disk; also

included in this section are some *Inuleae* without a true involucre but with receptacular bracts enclosing or subtending the marginal pistillate flowers and simulating an involucre.

 Bracts present on receptacle, often forming a circle or fused to form a cup between ray-flowers and disk-flowers.

  Marginal pistillate flowers with ray-corollas .... 3. *Madieae.*

  Marginal pistillate flowers with reduced tubular corollas.

   Pubescence not woolly; our species perennial with extensively creeping rhizomes .... *Iva* in 2. *Ambrosieae.*

   Pubescence woolly; annuals .... 8. *Inuleae.*

 Receptable without bracts.

  Pappus present, except in a few species having rays.

   Pappus of flattened scales, at least at base, although often awn-tipped .... 4. *Helenieae.*

   Pappus of hairs or bristles, not visibly flattened.

    Pappus-hairs rather few, not pure white .... 5. *Astereae.*

    Pappus-hairs copious, white .... 7. *Senecioneae.*

  Pappus absent, or a rudimentary crown; rays absent

               6. *Anthemideae.*

Pistillate flowers not in heads, enclosed in an involucre which in fruit becomes indurate and nut-like (the involucre often spiny and then called a bur); staminate flowers in heads; stamens with anthers distinct or nearly so 2. *Ambrosieae.*

None of the flowers in a head with regular corollas.

 Corollas 2-lipped, bright rose-purple in our species; plant without latex

               11. *Mutisieae.*

 Corollas strap-shaped or ligulate (as in ray-flowers), variously colored; plants with latex (this not obvious in some smaller species or when plants are old)

               12. *Cichorieae.*

KEYS TO GENERA

Tribe 1. **Heliantheae.** SUNFLOWER TRIBE

Rays normally present, showy, yellow.

 Involucral bracts imbricated, not in distinct series (sometimes a few additional leaf-like bracts at base).

  Achenes 4-angled to moderately compressed, not ciliate; herbs (one species woody-based).

   Flowering stems bearing usually one large head and sometimes a few smaller additional heads; ray-achenes fertile .... 1. *Wyethia.*

   Flowering stems branching (except in starved plants) and usually bearing many heads of about equal size; ray-flowers not maturing achenes

             2. *Helianthus.*

  Achenes strongly flattened, villous-ciliate; shrubs .... 3. *Encelia.*

 Involucral bracts in 2 distinct series, the outer markedly different from the inner .... 4. *Coreopsis.*

Rays (in our species) absent or very inconspicuous.

 Herb; leaves opposite, compound or deeply parted .... 5. *Bidens.*

 Shrub; leaves alternate, simple, narrow .... 6. *Eastwoodia.*

Tribe 2. **Ambrosieae.** RAGWEED TRIBE [6]

Outer flowers in heads pistillate, fertile, the inner perfect, not developing achenes; leaves all simple, entire (in our species) .... 7. *Iva.*

---

6. The *Ambrosieae* would seem to differ enough from other *Compositae* to rank as a separate family, as some authors have classified them, except that the genus *Iva* is transitional.

Flowers in heads staminate; pistillate flowers not in heads, enclosed in an involucre which becomes indurate in fruit; leaves, or some of them, toothed, lobed, or compound (many of them filiform and entire in *Hymenoclea*).
    Fruiting involucre without spines, though sometimes with a few small pointed tubercles.
        Shrub with very narrow leaves or leaf-divisions; pistillate involucre with spreading scarious wings in fruit ....8. *Hymenoclea.*
        Herb with pinnatifid to bipinnatifid leaves; pistillate involucre without wings ....9. *Ambrosia.*
    Fruiting involucre bearing spines (a bur).
        Spines of bur straight ....10. *Franseria.*
        Spines of bur hooked at apex ....11. *Xanthium.*

Tribe 3. Madieae. TARWEED TRIBE

Upper leaves or bracts not tipped with tack-shaped or saucer-shaped glands.
    Leaves with appressed silky hairs (see also *Hemizonia luzulaefolia* with white corollas).
        Rays very short, erect, red (or at first yellow); disk-flowers developing black achenes with conspicuous white pappus ....12. *Achyrachaena.*
        Rays spreading, yellow; disk-flowers sterile; pappus none ..13. *Lagophylla.*
    Leaves variously hairy to subglabrous, but not with appressed silky hairs (except in *Hemizonia luzulaefolia*).
        Involucral bracts laterally compressed, somewhat keeled, so that involucre is longitudinally ribbed ....14. *Madia.*
        Involucral bracts not laterally compressed, rounded or flattened on back.
            Involucral bracts completely enclosing ray-achenes, with flattened tips
15. *Layia.*
            Involucral bracts only partly enclosing ray-achenes, with tips not flattened ....16. *Hemizonia.*
Upper leaves or bracts tipped with tack-shaped or saucer-shaped glands.
    Upper leaves tipped with a saucer-shaped gland ....17. *Holocarpha.*
    Upper leaves tipped with one or more tack-shaped glands.
        Ray-achenes without pappus; disk-achenes not prominently ribbed, bearing 11 or fewer pappus-paleae which are not plumose ....18. *Calycadenia.*
        Pappus when present plumose on both ray-achenes and disk-achenes, of 15 or more very narrow paleae (very short in our species), or pappus absent; disk-achenes prominently ribbed ....19. *Blepharizonia.*

Tribe 4. Helenieae. SNEEZEWEED TRIBE

Rays normally present (in a few species absent in occasional individual plants, but not in most plants in a stand).
    Involucral bracts unequal, imbricated, in about 3 series.
        Leaves alternate, broad, not fleshy; in sheltered canyons ....20. *Venegasia.*
        Leaves opposite, narrow, fleshy; coastal, mainly in salt-marshes 21. *Jaumea.*
    Involucral bracts equal or subequal, not imbricated, in 1 series or 2 poorly differentiated series.
        Leaves alternate.
            Involucral bracts reflexed; disk dome-shaped to globose 22. *Helenium.*
            Involucral bracts not reflexed; disk flat or only slightly convex.
                Involucral bracts numerous, narrow; rays in our species numerous, reddish ....23. *Hulsea.*
                Involucral bracts and rays 12 or fewer; rays yellow.

Leaves sparingly dentate to entire; ray-corollas with a minute
projecting lobe opposite the limb . . . . .24. *Monolopia.*
Leaves mostly lobed, from shallowly 3-lobed at apex to twice
pinnately lobed; ray-corollas without a minute lobe opposite
the limb.
 Involucral bracts not scarious-margined; pappus present;
 achenes not ribbed or obscurely so, not papillate
                   25. *Eriophyllum.*
 Involucral bracts with scarious brownish margin; pappus
 none; achenes ribbed, papillate . . . . . .26. *Blennosperma.*
Leaves opposite . . . . .27. *Lasthenia.*
Rays absent.
 Herbage not glandular or inconspicuously so, at least not viscid.
 Leaves mostly lobed (or the reduced upper ones entire).
 Dwarf tufted plant, or with short rigid spreading branches from base;
 leaves 3-lobed at apex . . . . .25. *Eriophyllum.*
 Plants with erect or ascending stems; leaves pinnately lobed, or the
 reduced upper ones entire (sometimes all fleshy-filiform and entire in
 C. lanosa) . . . . .28. *Chaenactis.*
 Leaves entire or dentate (see also *Chaenactis lanosa* above).
 Herbage woolly; stems mostly decumbent; achenes ciliate
                   29. *Eatonella.*
 Herbage not woolly; stem erect; achenes not ciliate . .30. *Rigiopappus.*
Herbage glandular-viscid . . . . .31. *Amblyopappus.*

<h3 style="text-align:center">Tribe 5. Astereae. Aster Tribe</h3>

Plants not dioecious; disk-flowers perfect; ray-flowers either present or absent.
Rays yellow or none, or (in *Conyza*) so inconspicuous as to seem absent (in
Lessingia the marginal corollas may be outwardly enlarged but are 5-lobed and
do not bear true rays).
 Pappus of scales.
 Small shrub; rays present . . . . .32. *Gutierrezia.*
 Annual herb; rays absent . . . . .44. *Lessingia tenuis* var. *Jaredii.*
 Pappus of hairs, bristles, or awns (sometimes flattened and slightly widened
 downward).
 Disk in bud covered with white gum; pappus early deciduous
                   33. *Grindelia.*
 Disk not covered with gum; pappus persistent.
 Shrubs or perennial herbs, mostly with woody base or at least
 from a woody root (in *Solidago* not at all woody but with rhi-
 zomes).
 Plants with woody root at least; stems usually woody at least
 near base; rhizomes none (or in *Chrysopsis* sometimes slightly
 developed).
 Rays present, yellow.
 Herbs; leaves without glandular dots or resin-pits
                 34. *Chrysopsis.*
 Shrubs; leaves with minute glandular dots or resin-
 pits.
  Involucral bracts in about 2 or 3 series; heads 2
  cm. in diameter or more . . . . . . .35. *Stenotopsis.*
  Involucral bracts in about 3 to 5 series; heads 1
  cm. in diameter or less . . . . .38. *Ericameria.*
 Rays absent.

Shrubs (although often dwarfed and the woody stems slender).

Involucral bracts not in vertical ranks.

Tips of involucral bracts curved outward or recurved; corollas turning brownish

36. *Hazardia.*

Tips of involucral bracts erect; corollas permanently yellow.

Leaves, or some of them, usually dentate, sometimes all entire; corolla-tube abruptly enlarged upward into the throat ....37. *Isocoma.*

Leaves all strictly entire; corolla-tube gradually enlarged upward, without a clearly defined throat . .38. *Ericameria.*

Involucral bracts in clearly defined vertical ranks ....39. *Chrysothamnus.*

Herbs, the stems in winter dying down to the woody root-crown and replaced by new shoots.

Stems loosely clustered, normally over 3 dm. tall; leaves lanceolate to elliptic-ovate; heads usually many ....34..*Chrysopsis.*

Stems tufted, usually not over 15 cm. tall; leaves linear to oblanceolate; heads not usually more than 5 ....47. *Erigeron.*

Plants entirely herbaceous, with rhizomes and fibrous roots

40. *Solidago.*

Annual herbs (in some species sometimes persisting through winter but probably never sending up second-year shoots).

Plants hirsute, strigose, sparingly pubescent, or subglabrous; corolla-lobes of disk-flowers very short in proportion to tube.

Coarse plants, hirsute or strigose.

Rays present ....41. *Heterotheca.*

Rays absent or apparently so ....42. *Conyza.*

Slender delicate plants, inconspicuously hairy or subglabrous ....43. *Chaetopappa.*

Plants woolly at least when young (upper parts often glabrate and glandular); corolla-lobes at least half as long as tube, those of the marginal corollas often spreading outward

44. *Lessingia.*

Rays present, white to lilac or violet-purple (absent in a rare individual plant occurring among the normal forms).

Involucral bracts imbricated, unequal; rays mostly fewer than 30.

Herbage neither woolly nor glandular ....45. *Aster.*

Herbage woolly at least below, often glandular above

46. *Corethrogyne.*

Involucral bracts not imbricated or scarcely so, approximately equal; rays mostly more than 30 (except in forms of *E. foliosus*) ....47. *Erigeron.*

Plants dioecious; flowers in a head either all pistillate or all staminate; ray-flowers absent ....48. *Baccharis.*

Tribe 6. **Anthemideae.** MAYWEED TRIBE

Rays present.

Heads solitary at ends of spreading branches.

Leaves finely divided; receptacle bearing bracts on upper part
49. *Anthemis.*
Leaves shallowly pinnatifid, the lobes toothed; receptacle without bracts
50. *Chrysanthemum.*
Heads in a large flat-topped cyme-like cluster at the top of a simple stem
51. *Achillea.*
Rays none.
Disk cone-shaped or shaped like a high dome . . . . .52. *Matricaria.*
Disk flat, or heads very small.
Heads terminating the branches, or sessile in the forks.
Heads sessile; style persistent on achene as a stout spine; achenes laterally winged . . . . .53. *Soliva.*
Heads peduncled; style deciduous; achenes without spine or wings
54. *Cotula.*
Heads in narrow paniculate inflorescences . . . . .55. *Artemisia.*

Tribe 7. **Senecioneae.** Groundsel Tribe

Herbs, shrubs, or vines; leaves not reduced to scales; involucral bracts essentially equal, except for a few very small ones at base of involucre in some species.
Leaves opposite . . . . .56. *Arnica.*
Leaves alternate.
Receptacle flat; annuals or perennials . . . . .57. *Senecio.*
Receptacle conical; small annual with showy yellow rays . . .58. *Crocidium.*
Shrub; leaves dimorphic, those on the flowering branches reduced to small green scales; involucral bracts unequal, imbricated . . . . .59. *Lepidospartum.*

Tribe 8. **Inuleae.** Everlasting Tribe

Involucral bracts imbricated, scarious or coriaceous.
Shrub with many slender erect or arching stems from base . . . . . . .60. *Pluchea.*
Herbs (sometimes the lower part of stem slightly woody and persisting through winter).
Plants not dioecious; rhizomes absent; leaves variously woolly or glandular
61. *Gnaphalium.*
Plants dioecious, bearing all pistillate or all staminate flowers, spreading by rhizomes; leaves dark green and glabrate on upper surface, white-tomentose on lower surface . . . . .62. *Anaphalis.*
Involucral bracts in a single series, or without a true involucre, the marginal pistillate flowers enclosed or subtended by receptacular bracts.
Heads without a 5-lobed involucre-like structure in the center (or, if this structure is present, the outer flowers or bracts conceal it completely).
Leaves alternate, except some of those subtending the heads.
Involucre present, of scarious bracts.
Scarious bracts imbricated . . . . .63. *Stylocline gnaphalioides.*
Scarious bracts in a single series, not imbricated . . . .64. *Micropus.*
Involucre absent or apparently so . . . . .65. *Filago.*
Leaves opposite, or those around the heads whorled or apparently so
66. *Psilocarphus.*
Heads with an evident 5-lobed involucre-like structure, which encloses the few, early-deciduous central sterile flowers.
Bracts of central involucre-like structure with hooked tips
63. *Stylocline filaginea.*
Bracts of central involucre-like structure without hooked tips . . . .67. *Evax.*

Tribe 9. **Eupatorieae.** Eupatory Tribe

Involucral bracts equal ....68. *Eupatorium.*
Involucral bracts unequal, imbricated ....69. *Brickellia.*

Tribe 10. **Cynareae.** Thistle Tribe

Leaves white-mottled ....70. *Silybum.*
Leaves not white-mottled.
    Leaves with marginal spines (in *Cynara,* with very large heads, the spines on the leaves are much reduced and weak).
        Achenes basally attached to receptacle.
            Pappus plumose.
                Involucral bracts ovate-acute; corollas violet-purple ...71. *Cynara.*
                Involucral bracts narrow; corollas variously colored but not violet-purple ....72. *Cirsium.*
            Pappus not plumose ....73. *Carduus.*
        Achenes obliquely or laterally attached.
            Leaves surrounding heads spreading ....74. *Carthamus.*
            Leaves surrounding heads curved upward, the tips erect ....75. *Cnicus.*
    Leaves without marginal spines; heads less than 4 cm. in diameter (cf. *Cynara* above) ....76. *Centaurea.*

Tribe 11. **Mutisieae.** Mutisia Tribe

Only one genus represented ....77. *Perezia.*

Tribe 12. **Cichorieae.** Chicory Tribe

Stems either leafy or more or less branching above base, the branches subtended by leaves or reduced bracts, except in drought-dwarfed plants, and even then the stem usually with a few small scale-like leaves.
    Achenes truncate, not beaked (sometimes constricted just below the summit).
        Pappus of short scales, without awns or bristles; corollas blue (or on an occasional plant white) ....78. *Cichorium.*
        Pappus of awns or slender bristles, or of scales tapering into awns or bristles (except sometimes on marginal achenes of head); corollas not blue.
            Pappus plumose ....79. *Stephanomeria.*
            Pappus not plumose.
                Pappus of slender bristles or soft fine hairs.
                    Pappus clear white, readily deciduous; herbage variously hairy or glabrous, but not villous.
                        Achenes not flattened; plants without prickles.
                          80. *Malacothrix.*
                      Achenes flattened; leaf-margins more or less prickly.
                          81. *Sonchus.*
                    Pappus sordid-white, persistent; herbage villous.
                          82. *Hieracium.*
                Pappus of scales tapering into awns, or on marginal achenes of short truncate scales.
                  Annual; outer achenes enclosed by involucral bracts, with pappus of short truncate scales ....83. *Hedypnois.*
                  Perennial; outer achenes not enclosed by involucral bracts, with pappus similar to that of inner ones ....84. *Scorzonella.*
    Achenes with a slender beak, except sometimes those on the margin of the head.

Pappus plumose.
   Herbage without pustulate-based hairs; involucral bracts not in 2 distinct series.
      Leaves well-distributed up the stem.
         Leaves pinnately lobed and toothed; corollas white.
                              85. *Rafinesquia.*
         Leaves entire, grass-like; corollas purple or yellow.
                              86. *Tragopogon.*
      Leaves forming a basal rosette, those above the base reduced to minute bracts ....87. *Hypochoeris.*
   Herbage with bristly hairs which are usually pustulate-based; involucral bracts in 2 distinct series, the outer ovate ....88..*Picris.*
Pappus not plumose, of soft white hairs ....89. *Lactuca.*
Leaves all in a basal rosette or at least confined to lower part of stem, the heads all on scapes or on long leafless and bractless peduncles (cf. *Hypochoeris,* in which minute cauline leaves are present).
   Achenes not beaked, although sometimes narrowed toward the summit.
      Involucral bracts tapering to apex, without a conspicuous median dark stripe; pappus not plumose.
         Pappus of nearly always 5 paleae (sometimes very short) tapering into awns ....90. *Microseris.*
         Pappus of soft white hairs, readily deciduous, with 2 slender bristles persistent ....80. *Malacothrix.*
      Outer involucral bracts rounded at apex, with a conspicuous median dark stripe; pappus plumose ....91. *Anisocoma.*
   Achenes beaked.
      Achenes with about 10 ribs, not spinulose at summit ....92. *Agoseris.*
      Achenes with 4 or 5 ribs, spinulose at summit ....93. *Taraxacum.*

1. **Wyethia** Nutt.

Plant not hirsute; outer involucral bracts leaf-like, ovate, not ciliate, about equalling the rays or even surpassing them.
   Plant covered with minute glands, otherwise nearly or quite glabrous.
                               1. *W. glabra.*
   Plant woolly or in age glabrate, the wool persistent on the involucre.
                         2. *W. helenioides.*
Plant more or less hirsute; involucral bracts linear to lanceolate, ciliate, well surpassed by the rays ....3. *W. angustifolia.*

   1. **W. glabra** Gray. MULE-EARS. Openly wooded hills, Paso Robles west to Adelaida district.

   2. **W. helenioides** (DC.) Nutt. Wooded hills, vicinity of Atascadero and Templeton.

   3. **W. angustifolia** (DC.) Nutt. "Near San Luis Obispo," *J. E. Roadhouse* in 1905. As it has not been possible to confirm this record, it must be concluded that the species has since been exterminated, or else the label is erroneous.

2. **Helianthus** L. SUNFLOWER

Annuals with single stem; leaves serrulate or crenulate to entire; disk-corollas purple-brown or dark purplish-red.
   Involucral bracts ovate-acuminate or orbicular and cuspidate ....1. *H. annuus.*
   Involucral bracts lanceolate-acuminate ....2. *H. Bolanderi.*

Perennial with stems in a clump (sometimes flowering the first year and then with a single stem); larger leaves usually coarsely serrate, the smaller entire; disk-corollas yellow ....3. *H. gracilentus.*

1. **H. annuus** L. SUNFLOWER. Occasional in flat valleys, along streams, or in cultivated ground. In 1966 it was fairly plentiful in the Cholame-Shandon area but scarce elsewhere.

2. **H. Bolanderi** Gray. Frequent in upper Salinas Valley: Creston; Atascadero; Santa Margarita; Calf Canyon. Plants from Rinconada Mine (*6398*) represent the reduced form called *H. exilis.*

3. **H. gracilentus** Gray. Gravelly or stony slopes: ridge northwest of Cuesta Pass; La Panza Range, where many seedlings appeared in 1952 following a fire the preceding year; "burnt area east of Arroyo Grande" (*A. S. Albert* in 1933).

3. Encelia Adans.

Leaves densely covered with fine velvety white hairs; disk-corollas yellow.
1. E. actoni.

Leaves green, sparsely strigose or apparently glabrous; disk-corollas dark purplish.
2. E. californica.

1. **E. actoni** Elmer. Eastern part of Cuyama Valley, in gravelly stony washes (*7931*).

2. **E. californica** Nutt. Sandy soil, north edge of Santa Maria Valley, the northern limit for the species, and in bed of Santa Maria River.

4. Coreopsis L.

Shrub with a stout trunk, bearing at the summit leafy branches terminated by the flower-heads ....1. *C. gigantea.*
Annual herbs.
    Outer involucral bracts linear, narrower than the inner.
        Disk-achenes without pappus, not ciliate ....2. *C. Douglasii.*
        Disk-achenes with pappus, densely appressed-ciliate ....3. *C. Bigelovii.*
    Outer involucral bracts ovate, at least as wide as the inner ..4. *C. calliopsidea.*

1. **C. gigantea** (Kell.) Hall. GIANT COREOPSIS. Sand-dunes from Oso Flaco Lake southward.

2. **C. Douglasii** (DC.) Hall. Rocky, gravelly, or sandy soil, the known localities rather few but the plants often locally abundant: vicinity of San Luis Obispo, where it grows on serpentine, eastward to San Juan River south of Shandon, where it grows in coarse sandy soil.

3. **C. Bigelovii** (Gray) Hall. Gravelly or sandy soil in eastern part from Cottonwood Pass to Cuyama Valley, extending west to hills between Creston and Shandon. Plentiful in years of good rainfall.

4. **C. calliopsidea** (DC.) Gray. Open hills and plains, usually in gypseous clay, sometimes in hard-packed sandy soil, from Cholame Valley to Cuyama Valley, throughout Temblor Range, and westward to near Creston. Extensive tracts of country are colored deep yellow by this species in a good year.

5. Bidens L.

1. **B. pilosa** L. BEGGAR-TICKS. Weed in cultivated ground and city lots at San Luis Obispo and probably elsewhere.

6. Eastwoodia Bdg.

1. **E. elegans** Bdg. Sandstone or shale hills on either side of Carrizo Plain, southward to Cuyama Valley.

7. Iva L.

1. **I. axillaris** Pursh. POVERTY WEED. Mostly in interior valleys, sometimes in the hills: near Shandon to northern part of Temblor Range; spreading along roads and railways elsewhere.

8. Hymenoclea T. & G.

1. **H. Salsola** T. & G. WHITE BURROBUSH. Sandy soil, southern part of Temblor Range to Cuyama Valley. This shrub is called "cheesebush" by some people, who state that it has a cheese-like odor.

9. Ambrosia L. RAGWEED

1. **A. psilostachya** DC. WESTERN RAGWEED. Common in hard soil in low valleys and canyons: San Luis Obispo, Tassajera Creek, Santa Margarita, Nipomo Creek, and undoubtedly throughout western part.

10. Franseria Cav.

Annual with ascending branches ....1. *F. acanthicarpa.*
Perennial with prostrate or decumbent stems from a long fleshy tap-root.
                                                2. *F. Chamissonis.*
    Leaves shallowly pinnately lobed ....var. *Chamissonis.*
    Leaves twice or thrice pinnatifid ....var. *bipinnatisecta.*

1. **F. acanthicarpa** (Hook.) Cov. BURWEED. Sandy soil from Salinas Valley eastward and in Cuyama Valley, extending downstream to near Santa Maria. Probably native in sandy river-beds, but also spreading as a roadside weed.

2. **F. Chamissonis** Less. var. **Chamissonis.** Only near Oso Flaco Lake, where intergrading with the following variety.

Var. **bipinnatisecta** Less. BEACH-BUR. Very common on beaches and adjacent dunes all along our coast.

11. Xanthium L.

Leaves light green, alike on both surfaces, without axillary spines. 1. *X. strumarium.*
Leaves dark green on upper surface, finely white-hairy on the lower, the axils bearing 3-branched yellow spines ....2. *X. spinosum.*

1. **X. strumarium** L. var. **canadense** (Mill.) T. & G. COCKLEBUR. Common in irrigated land or where subject to seasonal flooding, both coastal and interior regions.

2. **X. spinosum** L. SPINY CLOTBUR. SPANISH THISTLE. Common weed in agricultural areas: San Luis Obispo; Temblor Range, according to Twisselmann; probably widespread through the county, though collections are lacking.

12. Achyrachaena Schauer

1. **A. mollis** Schauer. BLOW-WIVES. Clay soils, common on open hills and plains of the interior. There are no records from our coastal region, but with little doubt the species occurs there also.

13. **Lagophylla** Nutt.

1. **L. ramosissima** Nutt. Widely scattered and common in all parts of the county except the most arid localities.

14. **Madia** Mol.

Perennial (or biennial?), with previous year's dried leaves persistent on a caudex; disk-achenes with a pappus of short scales .... 1. *M. madioides.*
Annuals; pappus absent.
    Rays showy, 8 to 15 mm. long.
        Involucral bracts short-pubescent, with short tips above the ray-achenes; anthers yellow .... 2. *M. radiata.*
        Involucral bracts hirsute, with elongate tips above the ray-achenes; anthers purple-black .... 3. *M. elegans.*
            Plants mostly 3 to 5 dm. tall, flowering in spring and early summer; leaves on lower part of stem rather few, not usually crowded.
                                  var. *elegans.*
            Plants mostly 5 to 10 dm. tall, flowering in summer and fall; leaves on lower part of stem many, usually crowded .... var. *densifolia.*
    Rays inconspicuous, 1 to 5 mm. long.
        Plant usually more than 3 dm. tall (if shorter, the stem stout and densely leafy); involucre 5 to 12 mm. long; disk-flowers several.
            Plant moderately glandular, mild-scented; stem with long ascending branches from near middle; heads solitary and terminal, or a few racemosely arranged along the branches .... 4. *M. gracilis.*
            Plant heavily glandular, strong-scented; stem simple or with short branches above; heads subspicately arranged or crowded at ends of branches.
                Heads mostly racemose or subspicate; subtending leaves narrowly lanceolate, shorter than heads .... 5. *M. sativa...*
                Heads mostly in dense terminal clusters; subtending leaves deltoid-lanceolate, equalling or exceeding heads .... 6. *M. capitata.*
        Plant rarely as much as 3 dm. tall; stem slender; involucre 3 to 4 mm. long; disk-flower 1 .... 7. *M. exigua.*

1. **M. madioides** (Nutt.) Greene. Occasional in woods, Santa Lucia Range: upper Lopez Canyon (*8803*); North Fork San Simeon Creek to Rocky Butte (*7671*).

2. **M. radiata** Kell. Open hillsides in gypseous clay throughout eastern part, extending west to Cammatti Creek. In its mode of occurrence, this plant is representative of several species of similar ecology, which form dense patches, but these comparatively few and widely separated.

3. **M. elegans** D. Don var. **elegans.** Under this name are included relatively small plants blooming from April to June, with large flower-heads. These plants probably correspond to subsp. *vernalis* Keck; but, aside from the early flowering season, no adequate basis has been found for separating that subspecies from typical *M. elegans.* Such plants are apparently rare here: upper Arroyo Grande (*7706*); Caliente Mt. (*8223*).

Var. **densifolia** (Greene) Jepson. Canyon bottoms and valleys, fairly common in upper Salinas River basin from Santa Rita Creek to Calf Canyon and Rinconada Creek. Fall-blooming plants are included in this variety, perhaps somewhat arbitrarily, as some of them are in appearance hardly different from typical *M. elegans.*

4. **M. gracilis** (Smith) Keck. Hills and mountains of western part: near Rocky

Butte Fire Lookout (*9059*); ridge southeast of Cuesta Pass (*7924*); Suey Creek (*Eastwood 403*). Probably common.

5. **M. sativa** Molina. Coast Tarweed. Common near coast in fields and by roadsides, in clays or sandy loams, extending inland to upper Salinas Valley from Templeton southward.

6. **M. capitata** Nutt. Coastal fields: Point Sierra Nevada; Morro; Los Osos Valley; near Avila. Doubtfully distinct from *M. sativa* and seemingly less common, but many of the plants assumed to be *M. sativa* may prove to belong here. *Madia capitata* has been described as "honey-scented," as distinguished from *M. sativa*. No difference in odor has been noticed in the plants of our coastal region, but further observation on this point is desirable.

7. **M. exigua** (Smith) Gray. In chaparral or open woods, especially after fire: Oak Park district; La Panza Range. Probably common and more widespread, although there is a lack of definite records.

15. **Layia H. & A.**

Tips of involucral bracts glabrous or puberulent; pappus of acuminate paleae without woolly hairs.

    Stem not purple-dotted; leaves glabrous or sparsely hairy except for the often minutely ciliate margin; involucral bracts flattened on back to base, the lower part not sharply differentiated from the upper ....1. *L. Munzii.*

    Stem usually purple-dotted; leaves hispidulous; lower portion of involucral bracts (enclosing ray-achenes) bowed out, sharply differentiated from the tips.

                    2. *L. Jonesii.*

Tips of involucral bracts visibly hairy, though often less so than lower part; pappus of bristles or awns (the flattened awns of *L. glandulosa,* called "paleae" by some authors, have tangled woolly hairs near base on inner side).

    Rays showy, 10 to 25 mm. long, or as short as 6 mm. in starved plants.

        Anthers black; rays usually yellow with white tips, sometimes all yellow.

            Stems not purple-spotted; heads erect in bud; involucral bracts with tips as long as basal portion enclosing ray-achenes ...3. *L. platyglossa.*

                Involucral bracts oblanceolate obtuse. .... var. *platyglossa.*

                Involucral bracts linear to narrowly lanceolate, acute var. *breviseta.*

            Stems purple-spotted; heads nodding in bud; involucral bracts with tips much shorter than basal portion ....4. *L. gaillardioides.*

        Anthers yellow; rays of uniform color, white or cream-color to deep yellow.

            Stems persistently white-pubescent, though often sparsely so, or rarely glabrate, glandular only near the heads; lower leaves (at least) pinnately lobed or coarsely dentate except in plants dwarfed by drought or sterile soil; pappus persistent.

                Lobed or coarsely dentate leaves confined to base or lower part of stem; stems nearly always purplish, though sometimes only on upper part or only on lower part, and usually dark purple; pappus of 10 to 12 flattened but very narrow awns, with tangled woolly hairs near base on inner side ....5. *L. glandulosa.*

                    Rays pure white, sometimes turning pink or rose-purple in age ....var. *glandulosa.*

                Rays cream-color (in intermediate plants) to deep yellow

                              var. *lutea.*

            Lobed leaves extending well up the stem except in dwarfed plants; stems not purplish (very rarely slightly purple-tinged on

some branches); pappus of 10 to 18 slender bristles, plumose at
base with very fine straight hairs . . . . .6. *L. pentachaeta.*
Rays yellow . . . . .var. *Hansenii.*
Rays white . . . . .var. *albida.*
Stems glabrate, with minute black glands scattered nearly to base;
largest leaves dentate to entire, not lobed even on the most vigorous
plants; pappus deciduous, of soft slender bristles . . .7. *L. heterotricha.*
Rays inconspicuous, 2 to 4 mm. long .8. *L. hieracioides.*

1. **L. Munzii** Keck. Alkaline clay soil from Cholame Valley to Carrizo Plain,
where locally abundant.

2. **L. Jonesii** Gray. Clay soils, often in serpentine areas, from 2 miles northwest
of Cayucos (where now apparently exterminated) to vicinity of San Luis Obispo.
Limited to this area.

3. **L. platyglossa** (F. & M.) Gray var. **platyglossa.** Tɪᴅʏ Tɪᴘs. Ocean bluffs and
sandy coastal fields from Cambria northward. The nomenclatural type of the species
is a comparatively uncommon maritime race.

Var. **breviseta** Gray. *L. platyglossa* subsp. *campestris* Keck. Common throughout
interior, mostly in sandy soil but also in disintegrated shale or friable clay, reaching
the coast on west side of San Luis Range. At the last named locality the ray flowers
on most of the plants are yellow throughout, without "tidy tips." Similar plants
were seen along Cammatti Canyon.

4. **L. gaillardioides** (H. & A.) DC. North-facing hillsides, Old Creek near Cayucos
(*8462, 8849*). The form found here has yellow rays with white tips, although farther
north plants with entirely yellow rays are common. Also, our plants have the
peduncles recurved when the heads are in bud, but it is not yet known whether
this is a distinguishing feature of the species throughout its range.

5. **L. glandulosa** (Hook.) H. & A. var. **glandulosa.** Wʜɪᴛᴇ Dᴀɪsʏ. In sand near
coast, rapidly disappearing because of land clearance and erection of houses: south
side of Morro Bay; Nipomo Mesa; extending inland to upper Arroyo Grande
(*6858*), and also localized in the granite country east of Santa Margarita. Plants
with pure-white rays apparently also occur in a separate area in the northeastern
part of the county (Cholame to San Juan River), although dried specimens do not
always show the color the rays were when fresh. In the central part of the county,
most of the plants are transitional to var. *lutea,* having cream-colored rays.

Var. **lutea** (Keck) Hoover, n. comb. L. *glandulosa* subsp. *lutea* Keck, Madroño 3:
18. 1935. The extreme form with deep yellow rays is present in an unmixed state
on Caliente Mt. (*8230*). Far more plentiful are plants with cream-colored or pale
yellow rays, intermediate toward var. *glandulosa* but here assigned to var. *lutea* for
geographic reasons. Such plants are common in sands of the interior from Atascadero eastward to La Panza district. Plants with deep yellow rays are of scattered
occurrence, mixed in colonies of the cream-colored form: Huerhuero Creek; Cammatti Canyon; Yaro Creek north of Pozo. In view of the preponderance of intermediate plants and the fact that color of the rays is the only difference, "variety"
is a more appropriate taxonomic category than "subspecies" in this instance. Whatever the color of the rays, the species can be easily recognized by the dark purple
stems.

6. **L. pentachaeta** Gray var. **Hansenii** Jepson. Canyon in Temblor Range near
north end of Elkhorn Plain (*10,403*), in sandy soil, mostly on steep north-facing

slopes. The white-rayed variety was completely absent. In the Sierra Nevada foot-hills, I have noted that this species has a mint-like odor. It has not been observed whether that feature holds true in San Luis Obispo Co.

Var. **albida** (Keck) Hoover, n. comb. *L. pentachaeta* subsp. *albida* Keck, Aliso 4: 107. 1958. Loose soils, usually among trees or shrubs, along our eastern border from Cottonwood Pass to Temblor Range and Carrizo Plain. Polonio Pass is the type locality of the variety.

7. **L. heterotricha** (DC.) H. & A. Frequent on hillsides or sometimes on plains in eastern part, in gypseous clay or decomposed shale. The rays are white in the plants I have seen, but in other localities they may sometimes vary to "pale yellow" as described in the literature.

8. **L. hieracioides** (DC.) H. & A. Common in sandy soils near coast, in brushy and wooded areas, especially following fire; at scattered localities inland to La Panza Range (*7858*). An obvious hybrid between this species and *L. glandulosa* was found growing with the two parent species near Baywood Park (*8369*). For geographic reasons, plants of this area may be assumed to belong to a tetraploid chromosomal race named *L. paniculata* Keck, but there is reason to doubt that the diploid and tetraploid races are entirely separate geographically. A collection made between Mt. Bishop and Cerro Romauldo (*7666*) has exactly the outward appear-ance of *L. hieracioides* of the San Francisco Bay region (which is presumably always diploid), as does also a plant found on the Carpenter Canyon road from San Luis Obispo to Arroyo Grande. On the other hand, a northern collection from Pine Canyon, Mt. Diablo, Contra Costa Co. (*J. T. Howell 15,405*) corresponds in every respect to *L. paniculata*.

16. Hemizonia DC.

Leaves without silky hairs; corollas yellow.
  Upper leaves and bracts not spine-tipped.
    Rays and involucral bracts 5 or fewer.
      Rays 3 or rarely 4; disk-flowers 3 or 4 . . . . .1. *H. Lobbii.*
      Rays 5; disk-flowers usually 6.
        Anthers purple-black.
          Heads mostly in dense clusters, short-peduncled or subsessile; leaves adjacent to heads nearly or quite as long as involucres and more or less overlapping them . . . . .2. *H. fasciculata.*
          Heads solitary at the ends of the numerous branchlets; leaves adjacent to heads much shorter than involucres, slightly if at all overlapping them . . . . .3. *H. pentactis.*
        Anthers yellow . . . . .4. *H. Kelloggii.*
    Rays and involucral bracts 8 or more.
      Anthers yellow.
        Herbage pilose or hirsute; lower leaves coarsely toothed or pin-nately lobed . . . . .5. *H. pallida.*
        Stem glabrous except near heads; leaves glabrous except for the minutely ciliate margin, entire or the largest sparingly toothed.
                           6. *H. Halliana.*
      Anthers black.
        Rays 8 to 13; some of disk-achenes fertile, becoming black and turgid . . . . .7. *H. paniculata.*
        Rays 8, or in subsp. *cruzensis* sometimes 9.

Lower leaves not forming a rosette or conspicuously crowded, mostly disappearing in summer.
    Stem mostly 3 to 5 dm. tall, with ascending branches; corollas deep yellow; rays always 8   subsp. *paniculata.*
    Stem mostly 2 to 3 dm. tall, with widely spreading branches, so that vigorous plants are as broad as tall; corollas pale yellow; rays 8 or 9 .... subsp. *cruzensis.*
Lower leaves forming a conspicuous basal rosette and crowded on lower part of stem, persistent even in age.
                    subsp. *foliosa.*
    Rays 8 to 13, averaging about 10 .... subsp. *increscens.*
    Rays about 15 to 35; disk-achenes all sterile ....8. *H. corymbosa.*
    Heads mostly solitary at tips of branches; rays about 15 to 25.
                    subsp. *corymbosa.*
    Heads densely crowded; rays about 20 to 35.
                    subsp. *macrocephala.*
Upper leaves and bracts spine-tipped.
    Bracts of receptacle rigidly sharp-pointed; pappus none ....9. *H. pungens.*
    Bracts of receptacle not sharp-pointed; pappus present in disk-flowers.
    Plant without dark glands; anthers yellow; pappus-paleae 3 to 5.
                    10. *H. Parryi.*
    Plant with numerous dark stipitate glands; anthers more or less suffused with purple pigment, varying from yellow or brown to black; pappus-paleae 8 to 12 .... 11. *H. Fitchii.*
Leaves, at least the lower, with silky hairs; corollas white ......12. *H. luzulaefolia.*

1. **H. Lobbii** Greene. Hills west of Paso Robles, locally abundant on road to Nacimiento Dam. A plant found along Highway 101 at the turn-off to Avila doubtless represented a casual introduction from farther north.

2. **H. fasciculata** (DC.) T. & G. Mostly in clay soils, coastal area from Cayucos (*7255*) southward. In early summer, hillsides around San Luis Obispo are turned solid yellow by this species in flower.

3. **H. pentactis** (Keck) Keck. Open fields, hillsides, and plains, common and often exceedingly abundant from east base of Santa Lucia Range eastward to Carrizo Plain, southward at least to summit between Pozo and Arroyo Grande. A species difficult to define satisfactorily. The only evident difference from *H. Kelloggii* is the presense of purple pigment in the anthers, and the two, when growing together, are too easily confused. On the other hand, large plants of *H. pentactis* are very similar to *H. ramosissima*, found from coastal Santa Barbara Co. southward. Specimens of unknown geographic origin could hardly be identified with confidence. Nevertheless, the chromosome number of *H. pentactis* is reported as different from that of either *H. Kelloggii* or *H. ramosissima*. The situation calls for much further study, including garden cultures from various sources of all these species.

4. **H. Kelloggii** Greene. Occasional in interior, usually in sandy soil, sometimes in crumbling shale: Cottonwood Pass; Palo Prieto Canyon; Cammatti Canyon; Calf Canyon.

5. **H. pallida** Keck. Dry plains and hills, southern part of Temblor Range south of Crocker Grade (reported by Twisselmann as *H. paniculata*); foothills west of Fellows; east edge of Cuyama Valley near Kern Co. line (*7145*).

6. **H. Halliana** Keck. Cholame Valley, in alkaline clay soil near junction of Cottonwood Pass and Polonio Pass roads. The plants are in good years very plentiful

within a small area. Being found otherwise at a very few localities in San Benito and Monterey Cos., this is one of the species remarkable for occurring in dense but few and widely scattered patches.

7. **H. paniculata** Gray subsp. **paniculata.** Rare locally, in clay soils derived from serpentine: San Bernardo Creek (*9208*); valley west of Mt. Bishop (*9497*). This species furnishes an excellent example of the proper use of the term "subspecies," because the elements composing it, although differing only in unimportant details, are quite distinct and geographically separated. Although there is some broad geographic overlapping between subsp. *paniculata* and subsp. *increscens,* the two in San Luis Obispo Co. are ecologically separated and do not grow adjacent to each other.

Subsp. **cruzensis** Hoover, subsp. nov. Planta 20–30 cm. alta; caule late ramosa; floribus liguliferis 8, interdum 9; corollis pallide luteis.

Plant 20–30 cm. tall; stem widely branched (the spread often as great as the height); ray-flowers 8, sometimes 9; corollas pale yellow.

Between San Carpoforo Creek and Arroyo de la Cruz, August 9, 1947, *Hoover 7361,* type. From the same area are *Robbins & Weiler 4139* in 1959 and *Hoover 9997* in 1966. The plants were abundant on the flat tops of ocean bluffs and were quite uniform in appearance. The main axis of the stem early ceases to grow, to be replaced by widely divergent branches from below. This manner of growth represents an adaptation to the windy maritime habitat.

Subsp. **foliosa** Hoover, subsp. nov. Floribus liguliferis semper 8; foliis valde persistentibus, rosulatis et in parte inferiore caulis congestis.

Ray-flowers always 8; leaves strongly persistent, forming a rosette and crowded on lower part of stem. Pozo, roadside and pasture to west (*6404,* type); sand-hills north of Pozo (*9715*). A unique local race of the upper Salinas Valley, distinct from the coastal forms of *H. paniculata* in the very leafy stems of even old plants. A few plants which were obviously hybrids between this and *H. pentactis,* with the number of ray-flowers varying between 5 and 8, were found where the two parents grew together (*9716*).

Subsp. **increscens** Hall ex Keck. Very common from Morro Bay southward in coastal area, in clay loams or mostly in sandy soils.

8. **H. corymbosa** (DC.) T. & G. subsp. **corymbosa.** Coast Tarweed. Ocean bluffs between San Carpoforo Creek and Arroyo de la Cruz (*7362*), apparently the southern limit for the typical subspecies.

Subsp. **macrocephala** (Nutt.) Keck. Coastal fields and bluffs: Los Osos Valley; common from near Cambria northward. The plants in Los Osos Valley, and some of those near Arroyo de la Cruz, are intermediate, having the heads densely clustered as in subsp. *macrocephala* but within the size range of typical *H. corymbosa.*

9. **H. pungens** (H. & A.) T. & G. Spikeweed. Common in interior, sometimes where water has stood, usually in more or less alkaline soil, often forming dense colonies, from Salinas Valley to Cholame Valley and Cuyama Valley.

10. **H. Parryi** Greene var. **Congdonii** (Rob & Greenm.) Hoover, n. comb. *H. Congdonii* Rob. & Greenm., Bot. Gaz. 22: 169. 1896. Locally plentiful in low valleys just west of San Luis Obispo, in hard clay soil.

11. **H. Fitchii** Gray. Dry fields in interior; San Marcos Creek near San Miguel; between Paso Robles and Creston; Carrizo Plain. Plants found south and east of Santa Margarita apparently also belong to this species but are not entirely identical.

12. **H. luzulaefolia** DC. HAYFIELD TARWEED. Common near coast in clay soils, extending inland to upper Salinas Valley. Most of our plants have the lowest leaves forming a rosette, in that respect conforming to subsp. *rudis* (Benth.) Keck. That subspecies is not sharply defined or separated geographically from the remainder of the species, and it is doubtful to what extent its features are genetically determined.

17. Holocarpha (DC.) Greene

1. **H. Heermannii** (Greene) Keck. Open hills, in firm soils: Cottonwood Pass (*6477*); frequent in Santa Lucia Range from Cambria-Templeton road northward, mostly toward the interior but extending down Santa Rosa Creek nearly to Cambria.

18. Calycadenia DC.

Stems glabrous; corollas yellow; rays 3 to 8; pappus-paleae of disk-flowers very short, truncate ....1. *C. truncata.*
Stems pilose, at least above; corollas white; rays 1 to 4; pappus-paleae of disk-flowers nearly as long as corolla, tapering into awns ....2. *C. villosa.*

1. **C. truncata** DC. Open hillsides and dry meadows, Santa Lucia Range from summit between Cambria and Adelaida northward.

2. **C. villosa** DC. Hard-packed gravelly or sandy soil, rare: 1½ miles northeast of Creston, *H. C. Lee 1005* in 1937; La Panza district. In Monterey Co. found in lower valley of San Antonio River, and probably entering the adjacent portion of San Luis Obispo Co. The La Panza form is not identical with Monterey Co. plants and perhaps should be classified as a distinct species or subspecies.

19. Blepharizonia Greene.

1. **B. laxa** Greene. *B. plumosa* (Kell.) Greene subsp. *viscida* Keck. Occasional in clay or sandy clay in interior: hills west of Paso Robles, and at widely scattered localities eastward to Cottonwood Pass and Caliente Mt. According to the writings of Dr. David D. Keck, *B. laxa* is the same as *B. plumosa,* which does not occur so far south, but Greene's description of *B. laxa* is in several respects inconsistent with true *B. plumosa* and clearly applies to the plants later named *B. plumosa* subsp. *viscida.* In the fall of 1952, this species was notably abundant on the western part of the Caliente Mt. ridge, growing in silty clay with *Antirrhinum ovatum.*

20. Venegasia DC.

1. **V. carpesioides** DC. Along streams or on rocky canyon walls: upper Las Tablas Creek (*Chester Dudley* in 1927); from Lopez Canyon southward.

21. Jaumea Pers.

1. **J. carnosa** (Less.) Gray. Moist places near the seashore, particularly abundant in salt or brackish marshes, as around Morro Bay, at Avila, and Oso Flaco Lake.

22. Helenium L. SNEEZEWEED

Rays 2 to 6 mm. long ....*H. puberulum* var. *puberulum.*
Rays 7 (in intermediates) to 20 mm. long ....var. *Bigelovii.*

1. **H. puberulum** DC. var. **puberulum.** Common in moist places near coast, inland to Santa Lucia Mts.

Var. **Bigelovii** (Gray) Hoover, n. comb. *H. Bigelovii* Gray in Torr., Pac. R. Rep. 4: 107. 1857. The "extreme" form of the variety, with long and showy rays, was collected between San Carpoforo Creek and Arroyo de la Cruz (*7360*). Plants which are intermediate toward var. *puberulum* are much commoner, occurring along coastal streams and in moist hollows among the Nipomo Dunes.

23. Hulsea T. & G.

1. **H. heterochroma** Gray. Local in brushy or wooded areas, sand-hills north of Pozo; east side of La Panza Range.

24. Monolopia DC.

Rays entire or slightly notched at apex, 4 to 7 mm. long.
    Rays obovate, broadly rounded and notched at apex; achenes glabrous or nearly so . . . . .1. *M. gracilens.*
    Rays elliptic or oblong, entire and somewhat pointed at apex; achenes densely strigose . . . . .2. *M. stricta.*
Rays 3-toothed at apex, 8 to 17 mm. long . . . . .3. *M. lanceolata.*

1. **M. gracilens** Gray. San Simeon Creek (*7670*), on shaded slope in loose rocky soil.

2. **M. stricta** Crum. Gypseous or alkaline clay soil, Temblor Range, Carrizo Plain, and hills to the north. Often forms dense, nearly pure patches, but these generally smaller than in *M. lanceolata.*

3. **M. lanceolata** Nutt. Clays, crumbling shales, or calcareous sandy loams, very common from Salinas Valley (vicinity of San Miguel) eastward. Particularly abundant in Temblor Range and in hills around Cholame and Shandon, forming probably the greatest element in the vernal color displays in years of good rainfall.

(*Monolopia major* DC., a species identical with *M. lanceolata* in general aspect, reaches southern Monterey Co., and is therefore to be expected in northern San Luis Obispo Co. It differs from *M. lanceolata* only in having a bowl-shaped involucre with triangular lobes instead of one composed of distinct bracts. If it were mixed in an extensive stand of *M. lanceolata,* it would almost certainly escape detection. Therefore, its presence or absence can be determined only by carefully examining the plants in this area one by one.)

25. Eriophyllum Lag.

Perennials, shrubby or at least slightly woody at base.
    Heads solitary on long peduncles . . . . .1. *E. lanatum.*
    Heads in cyme-like flat-topped inflorescences.
        Involucral bracts 4 to 6 . . . . .2. *E. confertiflorum.*
            Pedicels of heads mostly less than 1 cm. long, rarely up to 2 cm. long; involucre 3 to 4 mm. long, 2 to 4 mm. wide . . . . . .var. *confertiflorum.*
            Some of heads on pedicels 2 to 5 cm. long; involucre 4 to 5 mm. long, 3 to 7 mm. wide . . . . .var. *laxiflorum.*
        Involucral bracts 8 to 12 . . . . .3. *E. staechadifolium.*
  Small annuals.
    Rays present; stems slender, flexible, several from the base in vigorous plants.
                                           4. *E. multicaule.*
    Rays absent; stems short and rigid, or the plant reduced to a dense tuft.
                                            5. *E. Pringlei.*

1. **E. lanatum** (Pursh.) Forbes var. **achillaeoides** (DC.) Jepson. South slope of Cypress Mt. (*8388*), on serpentine in chaparral.

2. **E. confertiflorum** (DC.) Gray var. **confertiflorum.** GOLDEN YARROW. Very common in the hills, sand-dunes and sandy plains near coast (except close to shore where *E. staechadifolium* replaces it) and eastward in wooded or chaparral areas to margin of Carrizo Plain.

Var. **laxiflorum** Gray. Sandstone hills between Shandon and Creston; to Temblor Range and Cuyama Valley.

3. **E. staechadifolium** Lag. var. **artemisiaefolium** (Less.) Macbr. Dunes, margins of beaches and salt-marshes, and bluffs throughout the length of our coast, extending inland only as far as volcanic hills on east side of Morro Bay.

4. **E. multicaule** (DC.) Gray. Common in sandy areas, both near coast and in interior to western edge of Carrizo Plain.

5. **E. Pringlei** Gray. Barren sandy spots from hills between Creston and Shandon to east side of La Panza Range and western margin of Carrizo Plain; sparingly in southern Temblor Range. Almost constantly an associate of *Linanthus Parryae,* which is adapted to the same environment.

26. **Blennosperma** Less

1. **B. nanum** (Hook.) Blake. Low moist places, infrequent: San Luis Obispo (*Condit*), where not recently found and probably exterminated; Carrizo Plain south of Simmler.

27. **Lasthenia** Cass.

Involucre of distinct bracts.
    Leaves all entire or (in *L. debilis*) some of them few-toothed; pappus (except in *L. Fremontii*) uniform, of 1 to 5 paleae tapering into bristles, or none.
        Perennials with a cluster of fibrous roots ....1. *L. macrantha.*
            Stems often decumbent at base; leaves oblanceolate to linear-oblong, mostly obtuse ....var. *macrantha.*
            Stems erect; leaves linear, acute ....var. *Bakeri.*
        Annuals with slender tap-root.
            Involucral bracts 8 to 12 (fewer in dwarfed plants); receptacle conical to dome-shaped.
                Leaves oblong to oblanceolate, or some of them varying to linear, especially on dwarfed plants; involucral bracts obovate to elliptic, becoming succulent in fruit ....2. *L. hirsutula.*
                Leaves narrowly linear; involucral bracts narrowly elliptic, not becoming succulent.
                    Plant strigose throughout; pappus none or of uniform paleae tapering into bristles ....3. *L. chrysostoma.*
                    Lower part of plant essentially glabrous or with a few soft spreading hairs; pappus when present of short scales and slender bristles or of unequal awns ....8. *L. Fremontii.*
        Involucral bracts 3 to 6; receptacle subulate.
            Involucre turbinate or campanulate; rays present.
                Leaves firm, narrowly linear-lanceolate or linear, less than 15 mm. long, shorter than internodes.
                    Plant apparently glabrous or only the peduncles pubescent ....4. *L. leptalea.*
                    Leaves and involucres strigose ....
                          dwarfed form of 3. *L. chrysostoma.*

Leaves flaccid, lanceolate or oblong to linear, sometimes few-
toothed, 1 to 5 cm. long, often as long as internodes or longer;
herbage pubescent . . . . .5. *L. debilis.*
Involucre cylindric-turbinate; rays absent or so much reduced as
to seem absent . . . . .6. *L. microglossa.*
Some of leaves usually pinnately lobed or divided, except in dwarfed plants;
pappus when present not uniform, usually of short scales interspersed with
slender bristles, sometimes of unequal awns, or rarely absent.
Leaves with broad rachis, or when entire mostly over 2 mm. wide; herbage
thinly villous or somewhat woolly, at least from middle upward   7. *L. minor.*
Leaves linear, usually narrowly so, or with linear rachis rarely as much as
2 mm. wide; herbage obscurely pubescent or apparently glabrous, except
on involucres and peduncles . . . . .8. *L. Fremontii.*
Involucre not of distinct bracts, lobed above the middle.
Rays apparently absent ("shorter than the involucre"); pappus present.
9. *L. glaberrima.*
Rays conspicuous; pappus none .10. *L. Ferrisiae.*

1. **L. macrantha** (Gray) Greene. var. **macrantha.** Local on coastal bluffs near
Arroyo de la Cruz.

Var. **Bakeri** (J. T. Howell) Hoover, n. comb. *Baeria Bakeri* J. T. Howell, Leafl.
West. Bot. 1: 7. 1932. Near Arroyo del Oso, south of Arroyo de la Cruz, in moist
meadow near seashore *(9426).*

2. **L. hirsutula** Greene. Sandy soils near beach from vicinity of Cambria northward (the commoner form without pappus); similar plants but with pappus (as in
Greene's original description) were found on the ocean bluff south of Islay Creek
(7526). Living plants differ markedly from *L. chrysostoma,* which may occur in the
same general area, in their very fleshy bracts. Plants intermediate between the two
species, which are reported in the literature and cited by recent authors as a reason
for merging *L. hirsutula* into *L. chrysostoma,* have not been observed locally.

3. **L. chrysostoma** (F. & M.) Greene. GOLDFIELDS. Open or sparsely wooded hills
and plains: sometimes on the coastline (San Simeon; between Cayucos and Cambria) but far more plentiful throughout the interior, often in extensive colorful
stands. Plants of the Temblor Range are often dwarfed, with short internodes and,
when well nourished, with spreading branches from the base.

4. **L. leptalea** (Gray) Ornduff. Forming scattered patches in sterile white sand,
usually around scattered trees of *Quercus agrifolia,* restricted to Salinas River basin:
Cantinas Creek and Bee Rock near Nacimiento Reservoir; Atascadero; sandstone
hills 5 miles east of Creston to northern end of La Panza Range (upper Cammatti
Canyon, *Twisselmann 2608*).

5. **L. debilis** (Greene ex Gray) Ornduff. Locally plentiful in canyons of the Temblor Range near north end of Elkhorn Plain, on steep north-facing sandy slopes.
The species has been collected on Maricopa Grade, Kern Co. (*J. T. Howell 5908*)
and may therefore be more widespread in the southern Temblor Range and Caliente Range.

6. **L. microglossa** (DC.) Greene. Hills bordering Carrizo Plain, rare or overlooked,
mostly on treeless north-facing slopes.

7. **L. minor** (DC.) Ornduff. Mostly in sandy soils: occasional near coast; common
from Salinas Valley eastward, notably abundant in La Panza district. Plants found
on a coastal bluff between Morro Bay and Cayucos tend to have shorter and more
spreading stems, as in maritime variants of many other species.

8. **L. Fremontii** (Torr. ex Gray) Greene. Moist depressions, local south of Simmler on Carrizo Plain. The plants at this locality tend to vary toward the form of the southern San Joaquin Valley which was named *Baeria Fremontii* var. *heterochaeta* Hoover.

9. **L. glaberrima** DC. In winter-wet depressions, near northwest end of Laguna, San Luis Obispo (*7916*); 4 miles southeast of Santa Margarita (*8755*).

10. **L. Ferrisiae** Ornduff. Moist alkaline clay soils, Cholame Valley to Carrizo Plain, where abundant around Soda Lake.

28. **Chaenactis** DC.

Flowers usually white, sometimes varying to pink.
    Plant mostly glabrous, or base of involucres and adjacent peduncles slightly woolly, or involucral bracts minutely woolly-ciliate, not at all glandular; herbage sometimes with scattered wool on lower part; marginal corollas mostly larger than inner ones; pappus of 4 paleae almost as long as corolla, with or without 4 very short outer scales.
        Pappus of 4 rather long paleae only ....1. *C. Fremontii.*
        Pappus with short outer scales as well as longer paleae ....2. *C. Xantiana.*
    Involucres, peduncles, and often the uppermost leaves pubescent and minutely glandular; herbage sometimes slightly woolly below; marginal corollas little if at all larger than the inner; pappus (in the variety found here) much shorter than corolla or even absent ....3. *C. stevioides.*
Flowers bright yellow.
    Leaves rarely crowded at base; reduced leaves present on upper part of plant, extending well up toward heads ....4. *C. glabriuscula.*
        Leaves with usually elongate divisions which are mostly not further divided; involucres 7 to 10 mm. long; pappus about ¾ as long as corolla.
                var. *glabriuscula.*
        Leaves with short obtuse divisions, the larger bipinnatifid; involucres 5 to 7 mm. long; pappus about half as long as corolla ....var. *curta.*
    Leaves usually tufted or crowded at base, or at least confined to lower half of plant; heads on long leafless peduncles above the uppermost leaves 5. *C. lanosa.*
        Lower leaves crowded at base, mostly persistently woolly; flowering heads up to 25 mm. in diameter, usually smaller ....var. *lanosa.*
        Lower leaves less crowded, distributed up to middle of stem or nearly, tending to be glabrate; flowering heads up to 30 mm. in diameter
                var. *denudata.*

1. **C. Fremontii** Gray. On sandy slopes, rare locally: canyon in Temblor Range near north end of Elkhorn Plain (*10,314*); Chalk Mt., (*10,293*). *Chaenactis Fremontii* seems to be identical with *C. Xantiana* except for its pappus and should probably be included in the same species. On a specimen from the north side of the lower Cuyama Valley (*8120*), I have noted "outer pappus evident to obsolete in a single head."

2. **C. Xantiana** Gray. Sandy places in eastern part, widely scattered but not common: Palo Prieto Canyon (*Crum 2102*); region of upper Cammatti Canyon; Caliente Mt., just west of the peak (*8225*); Cuyama Valley (see note above under *C. Fremontii*). On a hill west of Shell Creek, where this species and *C. lanosa* grew together, a few plants were found which were obviously hybrids between the two (*9344*).

3. **C. stevioides** H. & A. var. **brachypappa** (Gray) Hall. Common in eastern part, in crumbling shale, sand, or barren clay: Cottonwood Pass; Palo Prieto Pass; Tem-

blor Range; between San Juan River and Carrizo Plain; south side of Soda Lake; Caliente Mt.; Cuyama Valley. An unusual and remarkable occurrence is on the south shore of Soda Lake, where this is one of the very few plants which thrive in the strongly alkaline soil. All plants of this species within our range have very short pappus, and in some of those on the north side of Cuyama Valley the pappus is absent, at least in the marginal flowers.

4. **C. glabriuscula** DC. var. **glabriuscula.** PINCUSHION. Rocky or sandy soils: frequent in interior from Cottonwood Pass to Paso Robles, Temblor Range, and Cuyama Canyon. In favorable years, the species is notably abundant on gravelly-clay hillsides in La Panza district.

Var. **curta** (Gray) Jepson. Middle Branch of Huerhuero Creek (*6814*); upper Navajo Creek, La Panza Range (*6905*).

5. **C. lanosa** DC. var. **lanosa.** WOOLLY PINCUSHION. Very common in sandy soils of interior, from Salinas Valley eastward to Red Hills and western margin of Carrizo Plain; also upper Arroyo Grande and Nipomo Mesa, where intergrading with var. *denudata.* Because recent authors have consistently included these plants in the same species as *C. glabriuscula,* it is worth noting here that the two in this area are completely distinct. Sometimes *C. glabriuscula* is found on crumbling shale or soil containing some clay, while *C. lanosa* grows in nearly pure sand a short distance away. On the other hand, stands of the two have been seen close together in apparently the same kind of soil. In no case has any evidence of interbreeding been observed in this county.

Var. **denudata** (Nutt.) Hoover, n. comb. *C. denudata* Nutt., Proc. Acad. Philad. II, 4: 21. 1848. In sand near coast: Los Osos (*8960*); north edge of Santa Maria Valley near Nipomo (*7838*); upper Arroyo Grande (*6864*), vegetatively like var. *denudata* but heads smaller, thus intermediate toward var. *lanosa.*

29. **Eatonella** Gray

1. **E. Congdonii** Gray. Sandy soils, southern part of Carrizo Plain and bordering hills.

30. **Rigiopappus** Gray

1. **R. leptocladus** Gray. Rocky or hard-packed soil where not suppressed by competition from other plants: hilly areas in both coastal and interior regions. Common but rarely collected.

31. **Amblyopappus** H. & A.

1. **A. pusillus** H. & A. "Near lighthouse, Port Harford," *Condit* in 1908.

32. **Gutierrezia** Lag.

1. **G. bracteata** Abrams. Sandy or rocky hills in eastern part, usually in calcareous soils, extending down Santa Maria River.

33. **Grindelia** Willd. GUM-PLANT

Tips of involucral bracts erect to curved outward (only the outermost sometimes reflexed or looped).
    Leaves not glandular-dotted, sparsely to conspicuously hairy ...1. *G. hirsutula.*
    Leaves glandular-dotted, glabrous ....2. *G. procera.*
Tips of involucral bracts sharply curved back or curled into a ring.

Involucres less than 25 mm. in diameter; Salinas Valley and eastward.

3. *G. camporum.*

Involucres mostly over 25 mm. in diameter; coastal.

Stems spreading to prostrate; involucre usually subtended by broad leaves

4. *G. latifolia.*

Stems erect or ascending; leaves subtending involucre reduced and narrow, or none ....5. *G. robusta.*

1. **G. hirsutula** H. & A. Gum-Plant. Summit of Cuesta Pass, rocky hills around San Luis Obispo and doubtless elsewhere in western part. Most of the plants here represent a nearly glabrous form called var. *calva* Steyermark. One specimen was identified by J. A. Steyermark as var. *brevisquama* Steyermark. In such a variable complex, these varieties seem too trivial to have practical value. The different forms of *Grindelia* merge gradually into one another. An attempt has been made to refer all our specimens to names which are used in recent publications by David D. Keck, but many plants still seem to fit two or more descriptions equally well. A more useful classification would probably call for including *G. robusta* and *G. latifolia* in *G. hirsutula.*

2. **G. procera** Greene. Cottonwood Pass (*Twisselmann 1566*); Cholame Valley (*Twisselmann 3257*). Not clearly enough differentiated from *G. camporum,* although the two in the San Joaquin Valley seem distinct enough.

3. **G. camporum** Greene. Occasional in interior: 6 miles east of Paso Robles on Shandon road (*6393*); Santa Margarita (*6394*). Possibly introduced from San Joaquin Valley. See note under *G. procera.*

4. **G. latifolia** Kell. Common along the coast, at least from Morro Bay northward, usually in hard-packed soils. A collection from Suey Creek road (*Eastwood 396*), which is nearly glabrous, may perhaps conform most closely to the description of the species. Similar plants have been found near Pismo Beach. Most of the plants along the coastline, although varying greatly, have the distinguishing marks of *G. rubricaulis* var. *platyphylla* f. *villosa* Steyermark.

5. **G. robusta** Nutt. Avila (*Eastwood 13,763*) and doubtless elsewhere in southern coastal region.

34. Chrysopsis (Nutt.) Ell.

Rays present ....1. *C. villosa.*
Rays absent ....2. *C. oregona.*

1. **C. villosa** (Pursh) Nutt. var. **echioides** (Benth.) Gray. Common in sandy soils in interior, mainly in Salinas River basin; upper Arroyo Grande. Plants of this species occasionally seen in rocky places in the Santa Lucia Mts. may also belong to this variety, but specimens for identification are not now available.

2. **C. oregona** (Nutt.) Gray. Occasional in rocky and sandy stream-beds; Las Tablas Creek; upper Salinas River; Huasna River. A collection from the bed of the Salinas River at the mouth of Rocky Canyon is somewhat hairy and perhaps comes nearest var. *scaberrima* Gray. The remainder of our plants are virtually glabrous and so represent typical *C. oregona.*

35. Stenotopsis Rydb.

1. **S. linearifolia** (DC.) Rydb. Common in hills throughout eastern part, in decomposed shale or sandstone, extending to vicinity of Creston and probably farther westward.

36. Hazardia Greene

Flowers about 10 to 30 in a head; involucre turbinate in anthesis, its bracts gradually widened upward and abruptly pointed, the tips green, thickened and glandular
1. H. squarrosa.

Flowers 4 to 8 in a head; involucre subcylindric in anthesis, becoming narrowly turbinate as the bracts spread in fruit; bracts linear-acuminate, with a longitudinal green stripe near apex but the tips not thickened or obviously glandular
2. H. stenolepis.

1. **H. squarrosa** (H. & A.) Greene. Rocky places, common near coast, occasionally inland to east slope of Santa Lucia Range (Atascadero Creek) and lower part of Cuyama Canyon.

2. **H. stenolepis** (Hall) Hoover, n. comb. *Haplopappus squarrosus* H. & A. subsp. *stenolepis* Hall, Carnegie Inst. Wash. Publ. 389: 253. 1928. Hard-packed or rocky soils in interior: Cottonwood Pass; Temblor Range; frequent in upper Salinas River basin from Parkhill district east of Santa Margarita to summit between Pozo and Arroyo Grande. Probably also on Caliente Mt., as it has been found in the mountains of Santa Barbara Co. south of Cuyama Valley. Occupying a different area from that of *H. squarrosa* and probably entirely distinct. No intermediates have been found.

37. Isocoma Nutt.

Green tips of involucral bracts flat, not thickened . . . . .1. *I. veneta.*
   Stems erect or ascending; leaves oblanceolate or oblong, firm but not succulent
var. vernonioides.

   Stems decumbent or prostrate; leaves broadly oblanceolate to obovate, brittle-succulent . . . . .var. *sedoides.*
Green tips of involucral bracts thickened, containing a resin-pocket   2. *I. acradenia.*

1. **I. veneta** (H.B.K.) Greene var. **vernonioides** (Nutt.) Jepson. Common on coastal bluffs and dunes, borders of salt-marshes, and occasionally as far inland as edge of San Luis Valley.

Var. **sedoides** (Greene) Jepson. Mouth of Coon Creek (*9313*); coast north of Cambria (Point Sierra Nevada, *10,000*).

2. **I. acradenia** (Greene) Greene. More or less alkaline soils in interior: San Miguel; Creston; very common eastward from Cholame Valley south to Cuyama Valley. Most of our plants have the entire leaves attributed to typical *I. acradenia* of the Mohave Desert, but a few of them have toothed leaves as in *Haplopappus acradenius* subsp. *bracteosus* (Greene) Hall. Comparison of specimens does not provide support for that subspecies as a distinguishable entity.

38. Ericameria Nutt.

Leaves narrowly linear or filiform.
   Leaves on main branches 15 mm. long or less, bearing in the axils short branchlets with fascicled leaves . . . . .1. *E. ericoides.*
      Axillary fascicles usually nearly or quite as long as subtending leaves; rays mostly 2 to 5 . . . . .var. *ericoides.*
      Axillary fascicles usually shorter than subtending leaves; rays mostly 5 to 8
var. pachylepis.

Leaves 3 to 6 cm. long, without fascicled leaves in the axils . .2. *E. arborescens.*
Leaves obovate or broadly oblanceolate to spatulate . . . . .2. *E. cuneata.*

1. **E. ericoides** (Less.) Jepson var. **ericoides.** MOCK HEATHER. Common in coastal sands, especially south of Morro Bay and on Nipomo Dunes. Very showy in the fall. The strictly typical form with glabrous achenes is rare here: Morro Bay, *Eastwood* in 1928; south of Hazard Canyon, *9695.* Most of the shrubs represent a trivial variant which is identical except for its hairy achenes; it may be designated f. **Blakei** (Wolf) Hoover, n. comb. *Haplopappus ericoides* subsp. *Blakei* Wolf, Occ. Papers Rancho Santa Ana Bot. Gard. 1: 87. 1938.

Var. **pachylepis** (Hall) Hoover, n. comb. *Haplopappus Palmeri* Gray subsp. *pachylepis* Hall, Carnegie Inst. Publ. 389: 267. 1928. *H. Palmeri* var. *pachylepis* Munz. Sandy soils back from the coast, intergrading with var. *ericoides*: upper Arroyo Grande; Huasna River to Santa Maria River; extending up Cuyama Canyon to Chalk Mt. on north side of Cuyama Valley. The achenes are usually strigose, but some plants in the same localities have nearly glabrous achenes as in the nomenclatural type of *E. ericoides.* Rays sometimes as many as 8. These shrubs resemble *E. pinifolia* (Gray) Hall in appearance, but differ from undoubted *E. pinifolia* in having the outer involucral bracts somewhat broader and not attenuate-caudate but merely acute. Because of variation in the plants and the consequent lack of reliable distinguishing characters, it is probable that *E. ericoides, E. Palmeri,* and *E. pinifolia* will eventually be classified as a single species.

2. **E. arborescens** (Gray) Greene. Mostly on rocky slopes, especially after clearing: occasional in Santa Lucia Range; upper Arroyo Grande.

3. **E. cuneata** (Gray) McCl. var. **spathulata** (Gray) Hall. Among rocks in interior mountains, rare: Caldwell Mesa; Big Rocks near head of American Canyon, La Panza Range (*8597*); 2 miles northeast of Freeborn Mt. (*R. T. Orr* in 1949); sandstone rocks at western edge of Carrizo Plain; Temblor Range in Kern Co.

39. **Chrysothamnus** Nutt.

Stems white-tomentose; leaves persistent; outer bracts of involucre pubescent or tomentose . . . . *C. nauseosus* var. *hololeucus.*
Stems yellowish green, thinly tomentose; leaves mostly deciduous at anthesis; involucres glabrous . . . . var. *mohavensis.*

1. **C. nauseosus** (Pall.) Britt. var. **hololeucus** (Gray) Hall. RABBIT BRUSH. Sandy flood-bed of Cuyama River and adjacent slopes, formerly extending down the canyon to Pioneer Grove (where now flooded by Twitchell Reservoir). Twisselmann reports it from Recruit Canyon (Crocker Grade) in the Temblor Range and as a recently introduced plant in the La Panza Range.

Var. **mohavensis** (Greene) Hall. Caliente Mt., plentiful in sandy soil along summit ridge.

40. **Solidago** L. GOLDENROD

Stem widely branched above, each branch ending in a rounded or flat-topped cymose cluster of heads . . . . .1. *S. occidentalis.*
Stem not branched below the cylindric to ovoid or pyramidal paniculate inflorescence, or in very vigorous plants sometimes with branches bearing paniculately arranged heads.
    Herbage not glandular and glutinous; involucres 3 to 5 mm. long.
        Middle leaves on flowering stems elliptic to broadly oblanceolate, the upper often smaller but mostly the same shape.

> Herbage pubescent or puberulent, often grayish; leaves entire or the
> largest finely serrate; involucre 4 to 5 mm. long ......2. *S. californica.*
> Herbage glabrous or sparsely and obscurely puberulent; larger leaves
> mostly coarsely serrate; involucre 3 to 4 mm. long ....3. *S. canadensis.*
> Middle leaves on flowering stems linear-elliptic or linear-oblanceolate, the
> upper reduced and linear ....4. *S. Guiradonis.*
> Inflorescence cylindric or subcylindric, 2 to 4 cm. in diameter, bearing
> relatively few heads ....var. *Guiradonis.*
> Inflorescence rounded to pyramidal, 4 to 15 cm. in diameter, mostly
> with long branches bearing many heads ....var. *luxurians.*
> Involucres (at least) and often other parts glandular, somewhat glutinous; in-
> volucres 5 to 7 mm. long ....5. *S. spathulata.*

1. **S. occidentalis** (Nutt.) T. & G. Frequent in permanently moist places, mostly near coast, occasionally inland (Creston, *6384*).

2. **S. californica** Nutt. CALIFORNIA GOLDENROD. Occasional in dry or moist, usually sandy, soil in western part, extending inland at least to Salinas Valley; reported from Temblor Range by Twisselmann.

3. **S. canadensis** L. San Carpoforo Creek (*7358*), to Arroyo de la Cruz, by roadside near ocean (*7408*), evidently the form named *S. lepida* var. *fallax* Fernald; Santa Rita Creek west of Templeton (*9541*). Although our plant is doubtless not typical *S. canadensis,* a new combination is not to be made without an extensive study of all variants of the species.

4. **S. Guiradonis** Gray var. **Guiradonis.** Wet soil, upper Arroyo Grande (*10,097*), grading into the following variety.

Var. **luxurians** (Hall) Hoover, n. comb. *S. confinis* f. *luxurians* Hall, U. C. Publ. Bot. 3: 46. 1907. *S. confinis* Gray. *S. confinis* var. *luxurians* Jepson. Common in wet ground near coast from Morro Bay southward. I would include under this name all plants which have been called *S. confinis,* as they differ from *S. Guiradonis* in no respect except their more luxuriant growth.

5. **S. spathulata** DC. Occasional in dry sandy soil: Hills south of Morro Bay, extending to Coon Creek, apparently the southern limit for the species; hill south of Arroyo de la Cruz.

41. **Heterotheca** Cass.

1. **H. grandiflora** Nutt. Common in sandy areas south of Morro Bay and on Nipomo Mesa; also along Salinas River and rarely eastward. Probably native in the localities indicated, but also spreading as a weed in disturbed soil, especially by roadsides.

42. **Conyza** L.

Herbage bright green, hirsute to glabrous, not strigose; involucres glabrous or only the outer bracts pubescent, 3 to 4 mm. long ....1. *C. canadensis.*
Herbage grayish green, strigose and also more or less hirsute; involucres densely hairy, 4 to 5 mm. long ....2. *C. bonariensis.*

1. **C. canadensis** (L.) Cronquist. HORSEWEED. Common weed in low valleys, along streams, and in cultivated fields, mostly in western part.

2. **C. bonariensis** (L.) Cronquist. Weed in cultivated ground, in towns, or along streams: Cambria; San Luis Obispo; Cuyama River; reported by Twisselmann (as *Erigeron crispus*) as abundant around Shandon.

43. **Chaetopappa** DC.

Rays bright yellow, showy, many . . . . 1. *C. fragilis.*
Rays usually none; when present pink, inconspicuous, 3 or fewer.
    Stem simple or with few branches from base or lower part; involucre 4 to 5
    mm. long; disk-flowers dull red . . . . 2. *C. exilis.*
    Stem usually diffusely branched; involucre 2 to 3 mm. long; disk-flowers white
    or "reddish-tinged" . . . . 3. *C. alsinoides.*

    1. **C. fragilis** (Bdg.) Keck. Caliente Mt. (*8216, 8228*), local in sandy spots along
ridge west of peak. The rays expand in the morning sunshine and in the afternoon
curl backward into a ring.

    2. **C. exilis** (Gray) Keck. Forming patches on open hills in interior. Cottonwood
Pass; Palo Prieto Canyon; Fernandez Creek; Highland district.

    3. **C. alsinoides** (Greene) Keck. Moist places: Paso Robles; Arroyo Grande. Appar-
ently rare but more probably overlooked.

44. **Lessingia** Cham.

Corollas yellow, with or without a purple throat, sometimes (in *L. tenuis*) turning
purple or white in age.
    Involucral bracts in about 5 or 6 series, green, rarely purplish-tinged and then
    only in coastal plants . . . . 1. *L. germanorum.*
        Cauline leaves, except the reduced uppermost ones, pinnately lobed or
        coarsely toothed.
            Stems mostly slender, rarely as much as 3 dm. tall; branchlets with
            rather few widely spaced reduced leaves (coastal).
                Involucres and upper leaves with microscopic glands but with very
                few large spheroidal glands . . . . var. *germanorum.*
                Involucres and upper leaves conspicuously glandular, the large
                spheroidal glands numerous . . . . var. *pectinata.*
            Stems stout, usually at least 3 dm. tall; branchlets under flower-heads
            with many overlapping reduced leaves (Salinas Valley eastward).
                                  var. *vallicola.*
        Cauline leaves mostly entire, a few of the lower sometimes sparingly
        toothed.
            Style-branches with appendages less than 0.6 mm. long (mainly Salinas
            River basin).
                Plant glabrate and markedly glandular above the base.
                                   var. *tenuipes.*
                Plant persistently tomentose throughout . . . . var. *tomentosa.*
            Style-branches with slender appendages over 0.6 mm. long (Caliente
            Mt. and Cuyama Valley) . . . . var. *Lemmonii.*
    Involucral bracts in about 3 or 4 series, mostly purple . . . . 2. *L. tenuis.*
Corollas white to lavender . . . . 3. *L. nemaclada.*

    1. **L. germanorum** Cham. Various forms of this species are found in sandy soils
in nearly all parts of the county. The best marked form of each variety is distinctive, but intermediate plants, or plants which fit the descriptions of two or more
varieties equally well, are more plentiful than the extremes. In addition, the geographic separation of plants showing the features of the different varieties is less
clearly marked than is suggested by previous references. Furthermore, a single plant
may change so much during its development that different stages of growth may be

given different names. The approximate distribution of each variety is indicated as follows.

Var. **germanorum.** Price Canyon (*6343, 9170* in part); Santa Maria River bed (*Eastwood 304*). These specimens, from within the area occupied chiefly by var. *pectinata*, resemble plants from San Francisco, the type locality of the species, in their almost complete lack of large glands.

Var. **pectinata** (Greene) J. T. Howell. Coastal area from Morro Bay southward.

Var. **vallicola** J. T. Howell. From Salinas Valley eastward to Temblor Range.

Var. **tenuipes** J. T. Howell. East side of Santa Lucia Range, eastward to Temblor Range (according to Twisselmann); also Cuyama Canyon.

Var. **tomentosa** (Greene) J. T. Howell. Plants found at Creston (*10,065*) fit the description of this variety, which otherwise is reported only from the "southwestern Colorado Desert."

Var. **Lemmonii** (Gray) J. T. Howell. Caliente Mt. (*8329*); Chalk Mt., Cuyama Valley (*10,130*). These plants could be characterized as intermediate toward var. *Peirsonii* J. T. Howell, in that they are persistently tomentose up to the heads, though not actually on the involucres. Both varieties are distinguished from var. *tenuipes*, which they resemble in appearance, by the long appendage on the style-branches. In available specimens, it is difficult to find the style-branches, not to mention measuring the appendages. The differentiation between var. *Lemmonii* and var. *tenuipes* is therefore most conveniently made on a geographic basis.

2. **L. tenuis** (Gray) Cov. var. **Jaredii** Jepson. Gravelly or sandy places, east side of La Panza Range to west edge of Carrizo Plain. The leaves are mostly entire as in typical *L. tenuis*, but on some of the plants are sparingly toothed as in *L. parvula* Greene (attributed to "the interior of Monterey and San Luis Obispo Counties"), which is not here regarded as distinct. The variety *Jaredii*, although it is based only upon a peculiarity in the pappus, is geographically significant, as it apparently replaces typical *L. tenuis* entirely in this area. The type locality of the variety is given as "Estrella," but more probably Jared obtained the plant on an excursion to Carrizo Plain.

3. **L. nemaclada** Greene var. **albiflora** (Eastw.) J. T. Howell. Messa Ridge and Drake Ridge, Temblor Range in Kern Co. (*Twisselmann*); therefore undoubtedly in the adjacent portion of San Luis Obispo Co.

45. **Aster L.**

Perennials with rhizomes; rays much exceeding the pappus.
    Leaves mostly sharply serrate; heads in a flat-topped cyme which is broader than long . . . . 1. *A. radulinus.*
    Leaves entire or crenulate, or sometimes the largest ones serrate; inflorescence paniculate, racemose, or cymose, often widely branched but always longer than broad.
        Herbage, except near heads, usually glabrous or nearly so, sometimes pubescent; inflorescence loosely paniculate or cymose, on vigorous stems widely branched . . . . 2. *A. chilensis.*
        Herbage densely pubescent throughout with soft spreading hairs; heads in racemes, or in panicles with very short branches . . . . 3. *A. bernardinus.*
Annuals with slender tap-root; rays inconspicuous, only slightly exceeding the pappus.
                                                          4. *A. exilis.*

1. **A. radulinus** Gray. Occasional in Santa Lucia Range: Klau district (*7377*); Poly Canyon, San Luis Obispo.

2. **A. chilensis** Nees. Along streams, around springs, and in dry soil in low valleys: common near coast and inland at least to Paso Robles.

3. **A. bernardinus** Hall. Edges of road, just north of Creston, *Hardham 2600.*

4. **A. exilis** Ell. Moist places, between Mt. San Luis and Mt. Bishop (*7556*). Probably more widely distributed in the county.

46. Corethrogyne DC.

Stems prostrate or nearly so, the peduncles curved upward; involucre usually white-woolly at maturity, rarely glabrate .... 1. *C. leucophylla.*
Stems erect to spreading; involucre not white-woolly, usually glandular.
2. C. filaginifolia.

    Leaves mostly white-woolly, the upper often glabrate and glandular.
        Inflorescence slightly glandular .... var. *filaginifolia.*
        Inflorescence conspicuously glandular.
            Stems erect or at maturity bending over from weight of heads; mainly in interior .... var. *virgata.*
            Stems spreading or ascending; on coastal dunes .... var. *robusta.*
    Plant bright green and glandular except at base .... var. *viscidula.*

1. **C. leucophylla** (Lindl.) Jepson. Shifting sand-dunes near Piedras Blancas Point (*6453*) and northward. Some forms of *C. filaginifolia* resemble these plants rather closely in one respect or another, but *C. leucophylla* shows a combination of features which is not exactly duplicated in any variety of *C. filaginifolia.*

2. **C. filaginifolia** (H. & A.) Nutt. var. **filaginifolia.** Plants which perhaps represent the typical variety are relatively uncommon near the coast from Cambria to the vicinity of San Luis Obispo. The species is remarkably variable.

Var. **virgata** (Benth.) Gray. By far the commonest variety, if only because the name is used for such a wide range of variants. It extends from the coast eastward in hilly areas to Temblor Range, Caliente Mt., and east end of Cuyama Valley. Common in Santa Lucia Range, and the only variety occurring to the east.

Var. **robusta** Greene. Common near coast from Morro Bay southward, mainly on sand-dunes, sometimes on rocky slopes.

Var. **viscidula** (Greene) Keck. Coast road just south of Hazard Canyon (*9696*), growing with var. *robusta* and intergrading with it.

47. Erigeron L.

Rays present.
    Rhizomes not developed, or sometimes short woody ones in *E. glaucus.*
        Plants without a caudex, the erect stem arising from a basal leaf-rosette; heads in a usually compound cyme, 2.5 cm. or less in diameter.
1. E. philadelphicus.
        Base of stem forming an erect to prostrate elongated caudex; heads solitary at ends of simple stems or a few long branches, 3 cm. or more in diameter.
2. E. glaucus.
    Plants with slender creeping rhizomes.
        Heads solitary on simple erect stems or branches from the base; leaves mostly oblanceolate; leaves below heads reduced or absent 3. *E. sanctarum.*
        Heads cymosely arranged or rarely solitary; leaves narrow; stems leafy up to the heads.
            Stems glabrous or puberulent above; leaves short-pubescent or nearly glabrous; heads 12 to 20 mm. in diameter; achenes hairy 4. *E. foliosus.*

Leaves linear-oblong or linear-oblanceolate . . . . var. *foliosus*.
Leaves narrowly linear or filiform . . . . var. *stenophyllus*.
Stems and leaves densely puberulent with spreading white hairs;
heads 20 to 30 mm. in diameter; achenes glabrous 5. *E. Blochmaniae*.
Rays absent . . . . 6. *E. petrophilus*.

1. **E. philadelphicus** L. Moist places near coast, extending inland at least to upper Arroyo Grande.

2. **E. glaucus** Ker. SEASIDE DAISY. Coastal bluffs and rocks, common from Cambria northward; south of Point Buchon.

3. **E. sanctarum** Wats. Occasional in firm sandy soil near coast: hill south of Arroyo de la Cruz; openings in pine woods at Cambria; head of Carpenter Canyon, Arroyo Grande.

4. **E. foliosus** Nutt. var. **foliosus**. Near coast, in rocky places, infrequent: San Luis Mt.; Reservoir Canyon; Suey Creek (*Eastwood 430*).

Var. **stenophyllus** (Nutt.) Gray. Common in sandy soils of interior, from Salinas Valley to Palo Prieto Canyon and La Panza Range.

5. **E. Blochmaniae** Greene. Coastal dunes, usually in shifting sand, near Morro Bay and from Pismo Beach southward.

6. **E. petrophilus** Greene. Rooted in rock crevices, between Rocky Butte and Pine Mt., Santa Lucia Range (*8000, 8402*).

48. **Baccharis** L.

Leaves obtuse or abruptly pointed, most or all of them sharply toothed, the largest 4 cm. long or less.
Leaves oblanceolate or oblong to linear; pistillate involucre 6 to 8 mm. long, its bracts loosely imbricated . . . . 1. *B. Plummerae*.
Plant visibly pubescent; leaves oblanceolate or oblong . . subsp. *Plummerae*.
Plant glabrate, only the youngest parts puberulent; leaves linear.
subsp. *glabrata*.
Leaves obovate or broadly oblanceolate; pistillate involucre 3 to 5 mm. long, its bracts closely imbricated . . . . 2. *B. pilularis*.
Low, mat-forming shrub . . . . var. *pilularis*.
Erect, widely branched shrub . . . . var. *consanguinea*.
Leaves "willow-like," acute or acuminate, entire or shortly toothed, over 4 cm. long except the reduced upper ones or those on short lateral flowering branches.
Herbaceous perennial . . . . 3. *B. Douglasii*.
Shrubs.
Heads in a terminal cyme . . . . 4. *B. glutinosa*.
Heads in small cymes on short leafy lateral branches . . . . 5. *B. viminea*.

1. **B. Plummerae** Gray subsp. **Plummerae**. Shaded rocky slopes, lower Cuyama Canyon (*6535*), where now submerged in Twitchell Reservoir. This collection is identical with Santa Barbara Co. plants, representing the typical form of the species.

Subsp. **glabrata** Hoover, subsp. nov. Planta glabrata, foliis linearibus. Forks of San Simeon Creek on dry shaded rocky slope, *Hoover 7756*, type (pistillate), *7757* (staminate). This distinct-looking plant of the Santa Lucia Range does not seem to differ enough to rank as a species. It may prove to be more widespread, although known so far only at the type locality.

2. **B. pilularis** DC. var. **pilularis**. On the coast from Arroyo de la Cruz northward are low spreading shrubs which are referable to this variety, and others intermediate between it and var. *consanguinea*.

Var. **consanguinea** (DC.) Ktze. Coyote Brush. Very common on mostly north-facing hills and sand-dunes near coast, occasional inland in canyon-bottoms and along dry stream-courses as far east as the hills between Salinas River and La Panza Range.

3. **B. Douglasii** DC. In moist places, in dry stream-beds, or sometimes on sand-dunes, from coast inland to Salinas River near Santa Margarita; Cuyama Valley.

4. **B. glutinosa** Pers. Moist places and dry stream-beds in eastern part, extending down Cuyama River to coastal area; Huasna River.

5. **B. viminea** DC. Mule Fat. Occasional along moist or dry stream-courses in all parts of county. The nature of its occurrence suggests that it may be merely a vernal phase of *B. glutinosa*. Many species of *Compositae* show changes in appearance from one season to another which are considerably greater than the differences separating *B. viminea* from *B. glutinosa*.

49. Anthemis L.

1. **A. Cotula** L. Mayweed. Dog-Fennel. Common near coast around towns and in overgrazed pastures and cultivated fields, mostly in fertile clay soils; also in Salinas Valley and occasionally eastward to Carrizo Plain.

50. Chrysanthemum L.

1. **C. segetum** L. Fairly well established in sandy soils around Morro Bay.

51. Achillea L. Yarrow

Stems less than 1 m. tall.
    Leaf-segments thin, not fleshy, linear; widespread away from coastline.
                                                      A. Millefolium var. *californica*.
    Leaf-segments thick, slightly fleshy, the terminal ones ovate to oblanceolate; coastal . . . . var. *arenicola*.
Stems 1 to 2 m. tall . . . . var. *gigantea*.

1. **A. Millefolium** L. var. **californica** (Pollard) Jepson. Yarrow. Common on wooded hills, mostly on coastal hills and in Santa Lucia Range but also scattered eastward even as far as Palo Prieto Canyon and Temblor Range.

Var. **arenicola** (Heller) Nobs. The coastal plants which are common on shifting dunes, or sometimes in marshy places, seem to belong to this variety, or at least approach it.

Var. **gigantea** (Pollard) Nobs. Cholame (*Eastwood 13,902*). Since this variety is found mainly in the San Joaquin Valley, its occurrence along streams in the Cholame region is not surprising.

52. Matricaria L.

Heads at maturity 6 to 11 mm. wide; achenes bearing 2 very narrow oil-pockets which extend nearly to base of achene . . . . 1. *M. matricarioides*.
Heads at maturity 10 to 15 mm. wide; achenes bearing above the middle 2 broad oil-pockets which extend upward into the very short lobes of the pappus.
                                                      2. *M. occidentalis*.

1. **M. matricarioides** (Less.) Porter. Pineapple Weed. Common in nearly all parts of the county, most frequently in disturbed ground and by roadsides.

2. **M. occidentalis** Greene. Alkaline clay soil near Soda Lake on Carrizo Plain (*9761, 9796*).

53. **Soliva** R. & P.

1. **S. sessilis** R. & P. Moist sandy soils near coast: Cambria (*Eastwood 13,633*); Price Canyon (*7491*); upper Arroyo Grande. Probably overlooked rather than rare.

54. **Cotula** L.

Leaves varying from entire to pinnately lobed with broad rachis; stems fleshy; disk bright yellow ....1. *C. coronopifolia*.
Leaves finely divided; stems slender; disk green ....2. *C. australis*.

1. **C. coronopifolia** L. BRASS BUTTONS. In wet places, common especially near coast; "common in permanently wet places in the Temblor Mountains," according to Twisselmann.

2. **C. australis** (Sieber) Hook. f. Garden weed at Cambria, San Luis Obispo, and probably elsewhere.

55. **Artemisia** L.

Shrubs.
    Leaves cuneate, mostly 3-toothed at apex ....1. *A. tridentata*.
    Leaves with narrowly linear or filiform divisions, or filiform and entire.
                                          2. *A. californica*.
Herbs (woody only at base in *A. Dracunculus*).
    Perennials with rhizomes or a woody root-crown; leaves entire to few-lobed.
        Leaves dark green on upper surface, closely white-tomentose on lower surface ....3. *A. Douglasiana*.
        Leaves bright green on both surfaces ....4. *A. Dracunculus*.
    Biennial (or annual?) with tap-root; leaves finely divided ....5. *A. biennis*.

1. **A. tridentata** Nutt. SAGEBRUSH. Sandy flood-plains in Cuyama Valley, extending along Cuyama River far down the canyon.

2. **A. californica** Less. CALIFORNIA SAGEBRUSH. Common on coastal hills, sand-flats, and dunes; at scattered localities inland to Temblor Range.

3. **A. Douglasiana** Besser. MUGWORT. In canyon bottoms and along dry or moist stream-courses in western part, eastward at least to Creston. This plant is reported to be an effective remedy for poison oak.

4. **A. Dracunculus** L. TARRAGON. Canyon bottoms and dry sandy hillsides, well distributed through the county but seldom abundant. Most plentiful just back of coastal dunes.

5. **A. biennis** Willd. Low moist places in western part: Paso Robles; Osos Creek estuary; Oceano; Corbett Canyon, Arroyo Grande.

56. **Arnica** L.

1. **A. discoidea** Benth. Rare in Santa Lucia Mts.: Pine Mt., in dry rocky woods (*8018*).

57. **Senecio** L.

Plants not vines; leaves not palmately lobed.
    Shrubs.
        Leaves simple or more often pinnately divided, grayish, with some thin tomentum persistent; rays about 10 to 13 ....1. *S. Douglasii*.

Leaves simple and entire, bright green, glabrous or with a little wool in some of the axils; rays about 5 to 8 . . . . .2. *S. Blochmaniae.*
Herbaceous plants.
Perennials with fibrous roots.
Leaves simple, coarsely or finely toothed; rays usually none.
Stem usually solitary; herbage glabrate; cauline leaves reduced.
3. *S. aronicoides.*
Stems forming a clump; herbage thinly woolly at maturity; large leaves distributed well up the stem . . . . .4. *S. astephanus.*
Leaves pinnate; rays showy . . . . .5. *S. Breweri.*
Annuals with a tap-root.
Rays present, showy . . . . .6. *S. californicus.*
Rays none, or so short as to seem absent.
Main involucral bracts usually with a very small black tip, with several small conspicuously black-tipped ones at base 7. *S. vulgaris.*
Involucral bracts not black-tipped, with no reduced ones at base or a few minute ones . . . . .8. *S. aphanactis.*
Herbaceous vine, climbing over trees and shrubs; leaves palmately lobed.
9. *S. mikanioides.*

1. **S. Douglasii** DC. Dry stream-beds and gravelly or sandy hills from Santa Lucia Range eastward; particularly common along Salinas River and many of its tributaries; upper Arroyo Grande; Cuyama River. Replaced near coast by *S. Blochmaniae.*

2. **S. Blochmaniae** Greene. Coastal sands north of Morro Bay (where perhaps now exterminated) southward nearly to Islay Creek, and from Pismo Beach into western Santa Barbara County. Locally abundant in some places and showy when in bloom.

3. **S. aronicoides** DC. In oak woods, Santa Rita Creek west of Templeton (*8469*).

4. **S. astephanus** Greene. Chaparral or wooded places in Santa Lucia Range; apparently localized in the area between Atascadero and Morro Bay.

5. **S. Breweri** Davy. Wooded hills or open north-facing slopes, frequent from hills west of Paso Robles eastward to Temblor Range. Here growing almost always in soils derived from decomposing shale. The type locality is "Atascadero Ranch, Santa Margarita Valley."

6. **S. californicus** DC. In sand near coast from San Simeon Creek southward; not generally common and largely destroyed in places where it formerly grew, but plentiful in a few places. Rare in the interior: 5 miles east of Creston; Cottonwood Pass. A local maritime race near Cambria has short stout stems and unusually large flower-heads.

7. **S. vulgaris** L. GROUNDSEL. Introduced weed mainly in cultivated ground, notably in gardens at San Luis Obispo and other towns; sometimes along streams.

8. **S. aphanactis** Greene. Clay soils which are mostly free of more vigorous competing vegetation, in areas of serpentine rock around San Luis Obispo; barren gravelly or sandy slopes east of Creston. This species is one of several having a rather wide distribution but known in relatively few widely scattered localities.

9. **S. mikanioides** Otto. GERMAN IVY. Climbing on trees, mostly willows, along coastal streams, especially common northward.

58. **Crocidium** Hook.

1. **C. multicaule** Hook. "Headwaters of San Juan River, foot of Castle Mountain, damp sand, slopes above stream, north exposure," *Hardham 11,007* in 1964.

59. **Lepidospartum** Gray

1. **L. squamatum** Gray. Frequent in sandy stream-beds, rarely spreading to higher ground: Santa Maria River; Lopez Canyon; and especially common eastward, as along Salinas River and its tributaries and in Cuyama Valley.

60. **Pluchea** Cass.

1. **P. sericea** (Nutt.) Cov. Arrow-Weed. Sandy soil along Cuyama River.

61. **Gnaphalium** L. Everlasting Flower

Inflorescence cymose or paniculate, varying from open to densely congested.
 Heads more than 3 mm. long, without long subtending leaves.
  Plants persistently woolly, at least on the backs of the leaves, from moderately glandular to not visibly so.
   Leaves white-woolly on upper surface, though sometimes less than on the lower.
    Mostly annual or biennial, with rather slender roots, sometimes perennial; heads containing 150 or more flowers.
     Heads 3 to 3.5 mm. long; corollas reddish-tipped.
                1. *G. luteo-album.*
     Heads 4 to 6 mm. long; corollas yellow, not reddish-tipped.
                 2. *G. chilense.*
    Short-lived perennials, when old with many withered leaves near base, with somewhat woody tap-root; heads containing 35 to 55 flowers.
     Plant white-woolly or grayish, not noticeably glandular or fragrant; leaves mostly oblanceolate . . . 3. *G. microcephalum.*
     Plant at maturity somewhat yellowish, glandular and fragrant; leaves linear or slightly widened upward, the lowest often broader . . . . 4. *G. beneolens.*
   Leaves dark green and glabrate on upper surface, contrasting with the densely white-woolly lower surface . . . . 5. *G. bicolor.*
  Plants conspicuously glandular and bright green at maturity, the wool then more or less deciduous.
   Inflorescence usually compact and rounded; heads at anthesis subglobose; leaves (except the reduced upper ones) oblanceolate to linear-oblong . . . . 6. *G. californicum.*
   Inflorescence usually open and elongate; heads at anthesis cylindric or narrowly ovoid; leaves linear-acuminate or the lowest ones broader.
                7. *G. ramosissimum.*
 Heads hardly 3 mm. long, in dense terminal clusters, with subtending leaves which equal or exceed them . . . . 8. *G. palustre.*
Inflorescence spike-like, the heads (except the terminal) in dense axillary clusters, the clusters either closely or rather widely spaced on the stem . . . . 9. *G. purpureum.*

 1. **G. luteo-album** L. Common weed in moist places in western part.
 2. **G. chilense** Sprengel. Frequent near coast, mostly on stabilized dunes. Rarely inland to east slope of Santa Lucia Range (Rinconada Mine); not known farther eastward locally, although beyond our limits it extends to the Rocky Mountains.
 3. **G. microcephalum** Nutt. Mostly on rocky slopes, frequent from coast eastward at least to La Panza Range.

4. **G. beneolens** Dav. Found locally in about the same areas as *G. microcephalum,* though perhaps more often in loose sandy soils. Both have been too little collected to warrant definite statements about their relative abundance and detailed distribution.

5. **G. bicolor** Bioletti. Stabilized dunes, canyons, and north-facing slopes in western **part.**

6. **G. californicum** DC. CALIFORNIA EVERLASTING. Coastal dunes, dry rocky hills, and chaparral areas, inland at least to La Panza Range; the commonest species of the genus locally. Occasional plants near the coast have the involucres a beautiful pink instead of the usual white; this form may prove to be of horticultural value. Certain coastal specimens provisionally included in this species have larger leaves less narrowed toward the apex. They are glandular as in *G. californicum* but more noticeably woolly. These plants, which may represent an unnamed species, call for further field study.

7. **G. ramosissimum** Nutt. Common in sandy soil near coast, often in live-oak woods, extending to Santa Lucia Range.

8. **G. palustre** Nutt. Common in winter-moist places in western part, extending into interior at scattered localities.

9. **G. purpureum** L. Grassland areas, pastures, and openings in woods near coast, at least from San Luis Range northward.

62. Anaphalis DC.

1. **A. margaritacea** (L.) Gray. PEARLY EVERLASTING. Canyon bottoms and north slopes near coast: Coon Creek in San Luis Range; common from Cambria northward.

63. Stylocline Nutt.

Heads with scarious bracts ....1. *S. gnaphalioides.*
Heads without scarious bracts but with 5 stellately spreading bracts ending in hooked spines ....2. *S. filaginea.*

1. **S. gnaphalioides** Nutt. Sandy places, common in both coastal and interior regions; mostly following fires.

2. **S. filaginea** Gray. Barren clay or sandy-clay soils; widely scattered in interior, the plants here never numerous: Paso Robles; 14 miles east of Creston; Syncline Hill; Temblor Mts. and Choice Valley Hills, according to Twisselmann.

64. Micropus L.

Heads covered with loose wool ....*M. californicus* var. *californicus.*
Heads covered with closely appressed wool ....var. *subvestitus.*

1. **M. californicus** F. & M. var. **californicus.** Common on hillsides or in open woods in nearly all parts of county, but scarce or absent in coastal localities.

Var. **subvestitus** Gray. Firm soil on open coastal hillsides: south of San Carpoforo Creek; Mt. Bishop; Arroyo Grande (the type locality). This plant and var. *californicus* differ as much in appearance as do many pairs of *Compositae* which are classified as distinct species, and in this county they seem to be geographically separated. The question as to whether var. *subvestitus* should be called a separate species calls for further study.

65. **Filago** L.

Subtending leaves conspicuously surpassing the heads . . . . .1. *F. gallica.*
Subtending leaves scarcely if at all surpassing the heads . . . . .2. *F. californica.*

1. **F. gallica** L. Firm sandy soils in western part: Cambria; Price Canyon; Huasna district.

2. **F. californica** Nutt. Sandy or gravelly soils, frequent in all parts of county.

66. **Psilocarphus** Nutt.

Heads 6 to 10 mm. in diameter . . . . .1. *P. brevissimus.*
Heads 3 to 6 mm. in diameter.
    Leaves mostly 10 to 22 mm. long, linear or narrowly oblanceolate 2. *P. oregonus.*
    Leaves mostly 4 to 10 mm. long, spatulate or oblong to ovate . . . . .3. *P. tenellus.*
        Leaves spatulate or oblanceolate . . . . .var. *tenellus.*
        Leaves elliptic-ovate to ovate or broadly oblong . . . . .var. *tenuis.*

1. **P. brevissimus** Nutt. Low moist places, often in beds of vernal pools, in eastern part, Cholame Valley to Carrizo Plain.

2. **P. oregonus** Nutt. San Luis Obispo Co., according to Cronquist (Res. Stud. St. Coll. Wash. 18: 85).

3. **P. tenellus** Nutt. var. **tenellus.** Bare soil, often but not always in winter-flooded places, near coast and inland to Stoney Creek. Uncommon or overlooked.

Var. **tenuis** (Eastw.) Cronquist. "6 miles west of Painted Rock Ranch" (Res. Stud. St. Coll. Wash. 18: 89).

67. **Evax** Gaertn.

Heads mostly solitary or few together in the axils.
    Stems erect or ascending; leaf-blades 5 to 15 mm. long, with petioles about as long . . . . .1. *E. sparsiflora*
    Stems with prostrate branches, or simple and less than 1 cm. tall, or the plant reduced to a small tuft; leaf-blades not over 7 mm. long, sessile or with short petioles . . . . .2. *E. acaulis.*
Heads many in dense clusters at ends of branches or in the center of a reduced tuft
                                              3. *E. caulescens.*

1. **E. sparsiflora** (Gray) Jepson. Occasional in fertile clay soils near coast; not definitely known in interior, though perhaps present there.

2. **E. acaulis** (Kell.) Greene. Bare, usually sandy, soils in upper Salinas basin, particularly in or near La Panza Range. Very easily overlooked.

3. **E. caulescens** Gray var. **humilis** (Greene) Jepson. Clay soil in Cholame Valley (*7648*).

68. **Eupatorium** L.

1. **E. adenophorum** Spreng. Garden plant from Mexico, established in See Canyon, where it has persisted for a number of years.

69. **Brickellia** Ell.

Heads mostly in small lateral dense cymes or racemosely arranged on the branches; involucral bracts with green veins extending to the tips, scarious between the veins, not curved outward . . . . .1. *B. californica.*
Heads mostly solitary (sometimes 2 or 3) at ends of short divergent branches; involucral bracts with green veins not reaching the bright green tips, the outer ones with outwardly curved tips . . . . .2. *B. Nevinii.*

1. **B. californica** Gray. Frequent in rocky stream-beds and on lower slopes of rocky hills, back from the coast and eastward to La Panza Range and Cuyama Canyon. The plants are pleasantly fragrant, especially on warm windless evenings, and are worth planting in gardens on that account.

2. **B. Nevinii** Gray. Caliente Mt., on sandstone (*8267, 8331*). Fresh plants have a mint-like fragrance.

70. **Silybum** Adans.

1. **S. marianum** (L.) Gaertn. MILK THISTLE. Common in fertile soils, usually clay, in coastal region; less common in Salinas Valley.

71. **Cynara** L.

1. **C. Scolymus** L. ARTICHOKE. Occasionally escaping from cultivation near coast; persistent for many years in vacant lots at San Luis Obispo.

72. **Cirsium** Adans.

Involucre closely subtended by several leaf-like bracts with broad rachis (see also *C. loncholepis,* in which the crowded heads are not widely separated from the upper leaves).
    Herbage white-tomentose; involucral bracts usually spine-margined, sometimes the marginal spines apparently absent or concealed by wool 1. *C. rhothophilum.*
    Herbage green, not tomentose on upper leaf-surfaces; involucral bracts entire
        2. *C. brevistylum.*
Involucre not closely subtended by leaf-like bracts, or with 1 or 2 reduced to a narrow spine-bordered rachis.
    Leaves not decurrent.
        Upper leaf-surfaces not glandular; heads erect; involucral bracts narrowly linear-lanceolate or linear-acuminate.
            Herbage woolly, often partly glabrate at maturity but the backs of the leaves persistently woolly.
                Involucral bracts frequently glandular along midrib; corollas white to pink or pale lavender-pink ....3. *C. californicum.*
                Involucral bracts not glandular on midrib; corollas bright lavender-pink to deep purple or bright red.
                    Involucres densely and loosely arachnoid-woolly; corolla most commonly bright rose-purple, varying to dull purple in var. *compacta* ....4. *C. occidentale.*
                    Stem over 3 dm. tall, openly branched above
                        var. *occidentale.*
                    Stem 1 to 3 dm. tall, branched near base or reduced and densely leafy ....var. *compacta.*
                    Involucres thinly arachnoid-woolly or, if densely woolly, the wool more appressed (tomentose rather than arachnoid); corolla usually bright red or bright red-purple 5. *C. proteanum.*
            Herbage not woolly ....6. *C. loncholepis.*
        Upper leaf-surfaces glandular; heads nodding in bud; involucral bracts lanceolate or the outer ovate ....7. *C. fontinale.*
    Leaves decurrent, the stems thus with spine-margined wings .....8. *C. vulgare.*

1. **C. rhothophilum** Blake. Rare on low shifting dunes bordering beaches: Pismo Beach; south of Oso Flaco Lake. Otherwise known only in western Santa Barbara Co.

2. **C. brevistylum** Cronquist. Mostly in moist places, common near coast, extending inland through Santa Lucia Mts. In the pine woods near Cambria, plants have been seen at least 10 feet (3 meters) tall.

3. **C. californicum** Gray. Frequent in interior, from Red Hills near Shandon to La Panza Range and summit between Pozo and Arroyo Grande. The plants within this limited area closely resemble specimens from the region of the type locality of *C. californicum* (Knight's Ferry, Stanislaus Co.) and can readily be separated from *C. proteanum*.

4. **C. occidentale** (Nutt.) Jepson var. **occidentale**. Common in coastal region, mostly in sandy soils. Typical *C. occidentale* is here interpreted as being relatively tall, like plants from the vicinity of Santa Barbara, the type locality.

Var. **compacta** Hoover, var. nov. Planta 1–3 dm. alta; caule saepissime e basi paucis ramis. (Plant 1–3 dm. tall; stem usually branched from base). Type: Point Sierra Nevada, north of Arroyo de la Cruz, June 13, 1964, *Hoover 9087*. This low-growing variety is apparently restricted to the coast of San Luis Obispo Co., from the vicinity of Cambria northward.

5. **C. proteanum** J. T. Howell. Mostly in gravelly or stony soil, common in Santa Lucia Range and eastward to Salinas River; also along eastern border from Cottonwood Pass to Temblor Range. Uncommon, if not entirely absent, in the intermediate area, where *C. californicum* replaces it. The plants vary among themselves. Some resemble closely the coastal *C. occidentale,* while others hardly differ from some plants of *C. californicum*. A large part of this variation may well be the result of hybridization.

6. **C. loncholepis** Petrak. Moist hollows among coastal dunes, at least from Black Lake southward.

7. **C. fontinale** (Greene) Jepson var. **obispoense** J. T. Howell. Rare in wet places in areas of serpentine rock: San Simeon Creek; upper Pennington Creek; Chorro Creek; Perfumo Canyon.

8. **C. vulgare** (Savi) Tenore. Bull Thistle. Common in low valleys and in moist places along streams, especially near coast, becoming widely scattered inland. Isolated plants often appear in disturbed ground in dry places.

73. **Carduus** L.

1. **C. pycnocephalus** L. Weed in heavily grazed areas and along roadsides, becoming common in hills around San Luis Obispo.

74. **Carthamus** L.

1. **C. baeticus** (Boiss. & Reut.) Nyman. Scarce weed by roadsides and in fields: Toro Creek Road (*Frey 482* in 1959).

75. **Cnicus** L.

1. **C. benedictus** L. Blessed Thistle. Sparingly established in sandy soil: Calf Canyon; upper Arroyo Grande.

76. **Centaurea** L.

Involucral bracts without spines.
    Perennial with creeping rhizomes; marginal corollas not enlarged 1. *C. repens*.
    Annual; marginal corollas enlarged, with long spreading lobes ...2. *C. Cyanus*.

Involucral bracts tipped with a spine.
    Biennial with thick tap-root; flowers purple ....3. *C. Calcitrapa.*
    Annuals with slender root; flowers yellow.
        Heads about 15 to 20 mm. long; spines dark purple, 5 to 10 mm. long
            4. *C. melitensis.*
        Heads about 20 to 25 mm. long; spines yellow, 10 to 20 mm. long
            5. *C. solstitialis.*

1. **C. repens** L. Russian Knapweed. Weed of fertile soils in valleys: Atascadero; Laguna near San Luis Obispo (*6330* in 1946); the latter infestation has perhaps since been eradicated.

2. **C. Cyanus** L. Cornflower. Garden flower, spontaneous in sandy fields near Atascadero but perhaps not permanently established.

3. **C. Calcitrapa** L. Purple Star-Thistle. Camp San Luis Obispo, in hard ground by roadside (*7281* in 1947); perhaps not permanently established.

4. **C. melitensis** L. Common weed in overgrazed pastures, grain-fields, city lots, and sandy chaparral areas in all parts of the county.

5. **C. solstitialis** L. Yellow Star-Thistle. Established at widely scattered places: San Luis Obispo; Santa Margarita; San Juan River near La Panza Ranch. This weed has not yet become abundant here as it is in the San Francisco Bay region.

77. **Perezia** Lag.

1. **P. microcephala** (DC.) Gray. Dry gravelly and sandy soils: west of Parkfield in Monterey Co. (the northern limit for the species); hills west of Paso Robles to La Panza Range and southward; San Luis Obispo to Nipomo Mesa. In midsummer the purple flowers make a remarkable display, as on the road from Paso Robles to Nacimiento Dam.

78. **Cichorium** L.

1. **C. Intybus** L. Chicory. Low valleys, roadsides, and along streams: sparingly introduced in Salinas Valley and near San Luis Obispo.

79. **Stephanomeria** Nutt.

Perennials with rhizomes or a densely leafy caudex; stems branched at base as well as above.
    Stems at base forming a stout caudex densely clothed with old leaves; heads containing about 10 to 12 flowers; involucre over 1 cm. long ..1. *S. cichoriacea.*
    Stems from rhizomes, not forming a caudex; heads containing 5 or 6 flowers; involucre less than 1 cm. long ....2. *S. pauciflora.*
Plants with a tap-root, annual (or possibly *S. carotifera* biennial or perennial); stem branched above (when cut off or grazed, branched at base).
    Herbage entirely glabrous or apparently so.
        Root fleshy, stouter than base of stem; root-crown bearing small woolly buds ....3. *S. carotifera.*
        Root more slender than base of stem; root-crown not bearing buds.
            Herbage bright green or purplish-tinged; pappus-bristles completely deciduous ....4. *S. virgata.*
            Herbage silvery-green; pappus-bristles deciduous above the base, leaving a crown of short bristles ....5. *S. coronaria.*
    Herbage finely appressed-tomentose throughout ....6. *S. tomentosa.*

1. **S. cichoriacea** Gray. Rocky slopes, frequent in Santa Lucia Range, often on serpentine; Perfumo Canyon in San Luis Range.

2. **S. pauciflora** (Torr.) A. Nelson. Dry places in eastern part, uncommon or overlooked: between San Juan River and Carrizo Plain; Cuyama Valley; Temblor Range (reported by Twisselmann as *S. runcinata* Nutt.).

3. **S. carotifera** Hoover. Rocky slopes on serpentine, shale, or sandstone: San Bernardo Creek near Morro Bay; south of San Luis Obispo and eastward to upper Arroyo Grande. The nature of the relationship between this and *S. virgata* needs investigation. The two occur in the same area and often in the same kinds of habitats, but have not yet been found growing together. *S. carotifera* is unknown outside San Luis Obispo Co.

4. **S. virgata** Benth. Common in sandy, gravelly, or stony soils near coast, extending inland to Santa Lucia Mts. Reported by Twisselmann from Temblor Range, but on geographic grounds the occurrence of *S. coronaria* is more probable.

5. **S. coronaria** Greene. Sandy soils, or sometimes in calcareous clays, from Salinas Valley eastward and southward. The plants here called *S. coronaria* seem to replace *S. virgata* completely in the central and eastern parts of San Luis Obispo Co., but plants apparently typical of *S. virgata* are not entirely coastal elsewhere, being found in at least the northern San Joaquin Valley and the Sierra Nevada.

6. **S. tomentosa** Greene. Edge of beach north of Morro Bay (*6278, 6463* in 1946). Since the State Division of Beaches and Parks has been improving this area, the species has recently been sought unsuccessfully.

80. **Malacothrix DC.**

Involucral bracts broad, scarious with dark midrib, broadly rounded at apex

1. *M. Coulteri.*

Involucral bracts narrow, with narrow scarious margins, obtuse or acute to acuminate.
  Annuals with slender tap-root; largest leaves in a basal rosette, the cauline reduced or absent (except in *M. Clevelandii*).
    Flowering heads 3 to 5 cm. across; involucre 10 to 18 mm. long; stems scapose and bearing a single head, or few-branched.
      Leaves persistently woolly or cobwebby, at least at base (or in Morro Bay form on the lobes); heads all solitary on scapes ..2. *M. californica.*
      Leaves glabrous (wool often present on buds and sometimes on mature involucres); earliest stem usually scapose, the later ones branching and bearing reduced leaves and 2 to 5 heads, or in starved plants unbranched but still bearing reduced leaves ....3. *M. glabrata.*
    Flowering heads 0.5 to 3 cm. across; involucre 4 to 9 mm. long; stems (except in starved plants) branching and bearing several or many heads.
      Stems narrowly branched; leaves glabrous; flowers yellow; corollas exserted from involucre by not more than 2 mm. .....4. *M. Clevelandii.*
      Stems usually widely branched; leaves with tufts of wool on margin; flowers white, or corollas purplish on back (sometimes pale yellow outside our area); corollas exserted from involucre by 2 to 10 mm. (rarely less than 5 mm.) ....5. *M. floccifera.*
  Perennials with fleshy tap-root or rhizomes; basal rosette absent at maturity; stems leafy.
    Plant with fleshy tap-root; flowers yellow ....6. *M. incana.*
    Plant with rhizomes; flowers white, or corollas purplish on back

7. *M. saxatilis.*

1. **M. Coulteri** Harvey & Gray. Snake's Head. Common from vicinity of San Miguel eastward in drier areas, mostly in calcareous or gypseous soils, sometimes in sand. The species is variable in size of heads, color of flowers, and degree to which the leaves are lobed.

2. **M. californica** DC. In loose sand. Two different plants are known by this name. The commoner form, and the only one known outside San Luis Obispo Co., is widely distributed in the interior. It has creamy-yellow flowers and very narrow, almost thread-like leaves or leaf-divisions covered with copious cobwebby wool. On the south side of Morro Bay, the plants have deep yellow flowers, and the leaves, which bear scattered tufts of wool along the margins, have a narrow rachis with comparatively short and broad lobes. The plants of Nipomo Mesa (now very largely destroyed) are partly intermediate but mostly of the widely distributed form. It is not known which form most closely resembles the type specimen of the species, or exactly where Douglas collected the type.

3. **M. glabrata** Gray. Occasional in sandy soil in eastern part: Shell Creek hills; San Juan River; southern part of Temblor Range; Cuyama Valley.

4. **M. Clevelandii** Gray. Gravelly or sandy chaparral areas, usually appearing after fires, from coast to La Panza Range.

5. **M. floccifera** (DC.) Blake. Gravelly soils away from coast, not common in the county: Cottonwood Pass; east side of La Panza Range; ridge southeast of Cuesta Pass.

6. **M. incana** (Nutt.) T. & G. Sand-dunes west of Morro Bay and from Pismo Beach southward. Glabrous plants have been called *M. succulenta* Elmer or *M. incana* var. *succulenta* (Elmer) E. Williams. There is no geographic separation, and intermediate plants are probably more numerous than the extremes.

7. **M. saxatilis** (Nutt.) T. & G. var. **commutata** (T. & G.) Ferris. Generally on crumbling shale: eastern foothills of Santa Lucia Range southward to near Atascadero; sporadically found elsewhere, as on Cuesta Grade and along Arroyo Seco in the Huasna district. The plants vary, but all those in the county are regarded as forming a coherent entity. Var. *commutata* as generally understood has dentate leaves. The name *M. altissima* Greene refers to plants of this relationship having markedly lobed leaves; such plants often grow mixed with the typical form of var. *commutata*.

81. Sonchus L. Sow-Thistle

Stem not conspicuously hollow, usually branched below inflorescence, not (or seldom?) with stalked glands; marginal spines of leaves short and weak; achenes moderately flattened, transversely wrinkled .... 1. *S. oleraceus.*
Stem stout and hollow, erect and simple, bearing stalked glands just below inflorescence (always?); marginal spines of leaves rather long and rigid; achenes strongly flattened, thin-margined, not transversely wrinkled .... 2. *S. asper.*

1. **S. oleraceus** L. Sow-Thistle. Common weed in gardens and cultivated fields, most abundant in western part.

2. **S. asper** (L.) Hill. Generally less common than *S. oleraceus* and more often found as scattered solitary plants, but extending into uncultivated areas where the other species is scarce or absent. No specimens have been seen from the area east of the Santa Lucia Range.

82. **Hieracium** L. Hawkweed

1. **H. argutum** Nutt. Rocky places or under trees: San Luis Range (Coon Creek and Price Canyon); Santa Lucia Mts.; Pine Mountain of La Panza Range. The plants of the San Luis Range, at least, are typical *H. argutum*. Those from farther inland may prove to belong, wholly or in part, to var. *Parishii*, which is said to occur in Monterey Co. They are not yet represented by adequate collections.

83. **Hedypnois** Mill.

1. **H. cretica** (L.) Willd. Well established and plentiful in coastal area from Cambria to Cayucos.

84. **Scorzonella** Nutt.

1. **S. paludosa** Greene. Open grassy hills and moist places near coast from Arroyo de la Cruz northward.

85. **Rafinesquia** Nutt.

1. **R. californica** Nutt. Mostly on dry wooded hillsides, frequent from coast eastward to La Panza Range and Red Hills.

86. **Tragopogon** L.

Flowers yellow . . . . .1. *T. pratensis.*
Flowers purple . . . . .2. *T. porrifolius.*

1. **T. pratensis** L. Meadow Salsify. Near San Luis Obispo: "one-fourth mile north of the west entrance to California State Polytechnic College," *Frey 169* in 1957.
2. **T. porrifolius** L. Salsify. Well established in fertile clay soils around San Luis Obispo and probably elsewhere; reported by Twisselmann from Eben McMillan ranch (Shandon Hills).

87. **Hypochoeris** L.

Annual; leaves glabrous . . . . .1. *H. glabra.*
Perennial; leaves hirsute . . . . .2. *H. radicata.*

1. **H. glabra** L. Smooth Cat's-Ear. Common weed, widely distributed in mostly sandy soils.
2. **H. radicata** L. Gosmore. Hairy Cat's-Ear. Rather moist ground near coast, often in cultivated fields and overgrazed pastures, especially plentiful from Cambria northward.

88. **Picris** L.

1. **P. echioides** L. Bristly Ox-Tongue. Common weed in low valleys and moist places near coast, rarely extending inland.

89. **Lactuca** L. Lettuce

Stems and midribs of backs of leaves more or less hispid or prickly; leaves denticulate and usually with broad lobes; inflorescence widely branching . . . . .1. *L. Serriola.*
Stems and leaves glabrous; leaves entire or with few entire lobes; inflorescence racemose or narrowly paniculate, with short-peduncled heads and small lateral clusters of heads . . . . .2. *L. saligna.*

1. **L. Serriola** L. Prickly Lettuce. Common weed in all parts of the county, mostly in cultivated ground. Although seeming at first glance nothing but a pest, this species is a significant forage plant for both cattle and deer. The form with leaves merely toothed rather than lobed (f. *integrifolia* Bogenhard) is rare here but has been seen at San Luis Obispo and is said by Twisselmann to occur in the Temblor Range.

2. **L. saligna** L. Abundant in clay soils in coastal area; occasional inland to summit south of Pozo and reported also from Twisselmann Ranch.

90. **Microseris** D. Don

Leaves closely spaced along lower part of stem, not forming a tight rosette; heads erect in bud; achenes black; pappus shining white .... 1. *M. linearifolia.*
Leaves all in a rosette, the heads on scapes (in *M. heterocarpa* the leaves sometimes more widely spaced); heads nodding in bud; achenes not black, usually brown or gray; pappus not pure white.
    Pappus-paleae longer than the awns which terminate them.
        Paleae bifid at base of awn .... 2. *M. heterocarpa.*
        Paleae tapering into awn, not bifid .... 3. *M. acuminata.*
    Pappus-paleae shorter than the awns.
        Paleae normally 2 to 6 mm. long.
            Achenes constricted just below the summit .... 4. *M. Douglasii.*
            Achenes not constricted just below the summit .... 5. *M. Bigelovii.*
        Paleae rarely over 2 mm. long, sometimes almost obsolete.
            Pappus-paleae curved, the margins incurved at maturity.[7]
                Paleae villous or scabrous .... 4. *M. Douglasii* var. *tenella.*
                Paleae smooth or minutely scabrous .... 6. *M. campestris.*
            Pappus-paleae straight and flat at maturity at least in upper half
                                                  7. *M. elegans.*

1. **M. linearifolia** (Nutt.) Sch. Bip. *Uropappus linearifolius* Nutt. Common in sandy soils, or sometimes on rocky hills, in all parts of county.

2. **M. heterocarpa** (Nutt.) Chambers. *Uropappus Lindleyi* Nutt., as interpreted by most California authors. Fertile loams and clays, apparently not generally common but found at scattered localities in both interior and coastal regions.

3. **M. acuminata** Greene. In red stony soil, 18 miles east of Creston on La Panza road (*7777*).

4. **M. Douglasii** (DC.) Sch. Bip. var. **Douglasii.** Widely distributed in clay or loam soils, occasional near coast but mainly inland and notably common in the region of Carrizo Plain. The species is variable. Coastal plants have, on the average, smaller heads than inland ones, and their leaves are entire or merely dentate rather than lobed. A collection from an ocean bluff between Morro Bay and Cayucos (*7844*) conforms rather closely to Chambers's description of subsp. *platycarpha* (Gray) Chambers, and possibly should be included under that name.

Var. **tenella** (Gray) Hoover, n. comb. *Calais tenella* Gray, Pac. R. Rep. 4: 114. 1857. Coast between Morro Bay and Cayucos (*7845*), to vicinity of San Luis Obispo. See note under *M. campestris.*

---

7. This section of the key is adapted from diagnostic characters given by K. L. Chambers. The differentiation of San Luis Obispo Co. specimens on the basis of these characters is very difficult, if not quite impossible, because the characters are not clearly contrasting, and no other differences are readily apparent.

5. **M. Bigelovii** (Gray) Sch. Bip. Mainly in more or less sandy soils on ocean bluffs, from Cambria northward. Replaced south of that point by *M. Douglasii* or its variants; and interspecific hybrids, although not definitely identified, may well occur in this area. Referring to *M. Bigelovii* and *M. Douglasii*, Chambers stated that "in the vicinity of San Luis Obispo and Morro Bay, the differences between these taxa are less clear-cut, and there the blurring of specific distinctions is very likely the result of introgression."

6. **M. campestris** Greene. Cholame and Carrizo Plain, according to Chambers. A collection from the west base of La Panza Range (*7496*) seems to belong here. An attempt has been made to name all specimens according to the classification of Chambers (Contr. Dudley Herb. 4: 207–312), but the lack of clear-cut distinguishing characters makes it uncertain whether this attempt has been successful. Chambers reports *M. campestris* as tetraploid, but qualifies this generalization by the following statement. "It must be admitted that diploid biotypes have been found that are impossible to distinguish morphologically from some forms of *Microseris campestris,* but as yet these biotypes are known only from a very limited area in coastal San Luis Obispo County quite removed ecologically and geographically from the tetraploid species."

7. **M. elegans** Greene. 1 mile south of Atascadero and Carrizo Plain, according to Chambers; Temblor Range, according to Twisselmann; upper Arroyo Grande (*10,424*), around salt springs. Apparently not distinguishable except by microscopic details from *M. campestris,* which reportedly occurs in many of the same localities, but *M. elegans* is said by Chambers to be diploid, whereas *M. campestris* is tetraploid.

<h3 style="text-align:center">91. Anisocoma T. & G.</h3>

1. **A. acaulis** T. & G. In loose sand along ridge of Caliente Mt. (*8212*).

<h3 style="text-align:center">92. Agoseris Raf.</h3>

Perennials with somewhat fleshy tap-root.
    Lobes of leaves retrorse; achenes abruptly beaked from a truncate summit
                                                                  1. *A. retrorsa.*
    Lobes of leaves (if any) ascending or diverging at no more than a right angle; achenes tapering into the beak.
        Beak of achene at least twice as long as body.
            Flowering heads about 3 to 4 cm. across; corollas conspicuously surpassing involucre ....2. *A. grandiflora.*
            Flowering heads about 2 cm. across; corollas scarcely spreading beyond involucre ....3. *A. plebeia.*
        Beak of achene less than twice as long as body ....4. *A. apargioides.*
Annuals with slender tap-root.
        Flowering heads about 3 cm. across; ligules of corollas about as long as the longest involucral bracts ....5. *A. californica.*
        Flowering heads about 1 to 1.5 cm. across; ligules of corollas less than half as long as longest involucral bracts ....6. *A. heterophylla.*

1. **A. retrorsa** (Benth.) Greene. Rare in woods along highest ridges of Santa Lucia, La Panza, and Caliente Ranges.

2. **A. grandiflora** (Nutt.) Greene. The only occurrence known in the county is at Gillis Canyon summit, Eben McMillan ranch (*Twisselmann 1288*).

3. **A. plebeia** (Greene) Greene. Frequent in open woods from coastal hills eastward at least to La Panza Range.

4. **A. apargioides** (Less.) Greene. Sandy or rocky places near coast from Cambria northward. Specimens corresponding to descriptions of *A. hirsuta* (Hook.) Greene (with beak longer than body of achene) are, in agreement with recent authors, included in this species. The available specimens from this region are too few for a proper evaluation of the subspecies and variety distinguished by Quentin Jones and by P. A. Munz (Abrams, Ill. Fl. Pac. St. 4: 566). Variants of the species, in any event, do not appear to be segregated according to a recognizable geographic pattern.

5. **A. californica** (Nutt.) Hoover, n. comb. *Cryptopleura californica* Nutt., Trans. Am. Phil. Soc. II. 7: 431. 1841. *A. major* Jepson ex Greene. Very common on open plains and hills of the interior, especially in upper Salinas Valley; absent from the driest areas, such as the southern part of Carrizo Plain. This presumably is the plant which Twisselmann reported under the name *A. heterophylla* var. *crenulata* as "common throughout the Temblor Range."

6. **A. heterophylla** (Nutt.) Greene. Sharply distinct from *A. californica* locally and far less common: summit between Middle and East Branches of Huerhuero Creek (*6816*). Although *A. californica* was collected close by, there was no indication of interbreeding between them.

93. **Taraxacum** Wiggers

1. **T. vulgare** (Lam.) Schrank. Dandelion. Common and widespread weed in moist places, especially in lawns and gardens.

Excluded Species

Tetradymia canescens DC. var. *inermis* (Nutt.) Gray. Reported by McMinn (Ill. Man. Cal. Shrubs 619) as collected by Palmer on Clark Creek, 10 miles from San Luis Obispo. Clark Creek is near Los Osos. I have seen the Palmer collection and identify it as an immature specimen of *Ericameria ericoides*.

Addenda

Species not previously reported from the county are marked with an asterisk.

Isoetaceae

* *Isoetes Howellii* Engelm. O'Brien Lake, San Andreas Fault at summit of Palo Prieto Canyon, *Twisselmann 16,322*.

Pinaceae

Abies bracteata (Don) Nutt. Neil Havlik reports a grove of about seventy trees on a tributary of Estrada Creek on a north-facing slope in a narrow canyon.

Gramineae

* *Heleochloa schoenoides* (L.) Host. Twisselmann Lake at head of Palo Prieto Canyon.

Potamogetonaceae

Potamogeton pectinatus L. O'Brien Lake, San Andreas Fault, *Twisselmann 16,320*.

* *Potamogeton foliosus* Raf. Laguna at San Luis Obispo, *Hoover 11,552* in 1969.

Amaryllidaceae

Allium fimbriatum Wats. var. *diabloense* Ownbey & Aase. Serpentine outcrop in Palo Prieto Canyon, according to Twisselmann.

Berberidaceae

* *Mahonia dictyota* (Jepson) Fedde. According to Neil Havlik, there are five small clumps in a large rock outcrop about ½ mile north-northwest of Pine Spring Campground in the La Panza Range.

Mahonia pinnata (Lag.) Fedde. Neil Havlik has observed a number of additional occurrences which show that this species is not rare locally but rather well distributed within our coastal area.

Callitrichaceae

* *Callitriche trochlearis* Fasset. Peter Rubtzoff in a recent paper has cited, under this name, specimens from the northwestern part of the county. Whether or not all plants of the genus in this area belong to this species cannot at present be determined.

Hippocastanaceae

Aesculus californica (Spach) Nutt. Neil Havlik states that there is a small grove about 1 mile east of Marquart Park on old Highway 46; also two large trees and several smaller ones on Estrada Creek.

Elatinaceae

* *Elatine heterandra* Mason. Vernal pool near Highway 46, Estrella Plains, 1 mile east of Paso Robles, *Twisselmann 15,218*. Identified by Peter Rubtzoff.

Lythraceae

* *Ammannia coccinea* Rottb. Twisselmann Lake, San Andreas Fault at head of Palo Prieto Canyon, *Twisselmann 16,323*. Single plant in a dense stand of marsh growth.

Styracaceae

Styrax officinalis L. var. *fulvescens* (Eastw.) M. & J. "Common at north end of Pine Ridge, growing on sandstone soil," Neil Havlik. This confirms the collection made by H. C. Lee many years before at approximately the same locality.

Labiatae

Trichostema ovatum Curran. East end of Cuyama Valley near Kern Co. line, *Hoover 11,646* in 1969. In alluvial soil the plants were vigorous and had many branches. On adjacent gravelly hills they were dwarfed. It is of special interest that the species was sought without success at this locality several times between 1946 and 1969, but the widespread species *T. lanceolatum* appeared regularly.

Salvia sonomensis Greene. North end of Pine Ridge northwest of Caldwell Mesa, according to Neil Havlik.

Martyniaceae

Martynia louisianica Mill. Locally plentiful 1 mile west of Paso Robles, *Hoover 11,542* in 1969.

Orobanchaceae

Orobanche fasciculata Nutt. Serpentine outcrop in Palo Prieto Canyon, *Twisselmann 15,226.*

Compositae

* *Holocarpha virgata* (Gray) Keck. 12 miles northwest of Paso Robles on road to Nacimiento Dam, *Hoover 11,541* and *11,581* in 1969. Also collected near San Antonio Dam in Monterey Co.

Glossary

Excluded are many terms which it is assumed would be learned in an elementary botany course. Included are some common words which have a botanical meaning different from their meaning in the vernacular.

Aberrant. **Differing from the normal or usual form of a species or variety.**

Achene. A one-seeded dry fruit, similar to a nut but smaller.

Acicular. Shaped like a needle, as the leaves of pines.

Acuminate. Tapering to a narrow point.

Acute. With apex forming an angle less than 90 degrees.

Alternate. Attached singly at each node of a stem.

Annual. Completing its life cycle in a year or less, then dying.

Anthesis. The period during which a flower is expanded.

Appressed. Lying close to the surface rather than spreading (used mostly with reference to hairs).

Ascending. Growing upward but not erect.

Attenuate. Prolonged into a narrow point.

Auricle. One of a pair of lobes at the base of a leaf.

Auriculate. Having auricles.

Awn. A stiff, frequently flattened, bristle.

Axil. The angle between a leaf and the stem.

Axillary. In the axils.

Axis. The central branch of a stem, or the stem to which the flowers are attached in an elongate inflorescence.

Banner. The uppermost petal in the flower of the *Leguminosae.*

Biennial. Living for 2 years, flowering and fruiting in the second year.

Bifid. Split into two lobes.

Bilabiate. Divided into 2 "lips," or distinct upper and lower parts.

Bipinnate. Pinnate, with the divisions again pinnate.

Bipinnatifid. Pinnately lobed, with the lobes pinnately lobulate.

Blade. The flat expanded portion of a leaf.

Bract. A leaf, or one of a group of leaves, which subtends a flower or a group of flowers (often highly specialized, reduced, or shaped or colored differently from the other leaves).

Bractlet. In the *Umbelliferae,* the bracts subtending one of the simple umbels within a compound umbel; in general, a smaller bract closer to the flower than the leaves which are called bracts.

Bur. A dry fruit with spines.

Callous. Hardened; firmer than the adjacent parts.

Campanulate. Shaped like a conventional bell.

Canescent. White-hairy.

Capsule. A dry dehiscent fruit formed from a compound pistil.

Caudate. With the end resembling a tail; similar to attenuate.

Caudex. A thickened above-ground stem, bearing crowded leaves, usually spreading along the surface of the ground.

Cauline. Borne on the stem above the base, as contrasted with basal leaves.

Ciliate. Hairy on the margin.

Clavate. Club-shaped, thickened upward.

Claw. The narrowed base of a petal, corresponding to the petiole of a leaf.

Cleft. Lobed with sharp sinuses between the lobes, as if cut with scissors.

Compressed. Flattened.

Cone-scale. One of the parts which make up the cone of a pine or other conifer.

Cordate. Shaped like a valentine or conventionalized heart, with rounded lobes at base.

Corm. A short, thick, vertical underground stem bearing a cluster of fibrous roots and giving rise to an above-ground flowering stem; in vernacular usage generally incorrectly called a bulb.

Corymb. Like a raceme except that the pedicels of the lower flowers are elongate, thus forming an inflorescence with a flat or rounded top. Some botanists call a flat-topped cyme a corymb.

Corymbose. Having an outline like a corymb, as the arrangement of heads in some *Compositae.*

Crenate. With blunt-pointed or rounded teeth on the margin.

Cuneate. Wedge-shaped; widened upward from a pointed base.

Cuspidate. Abruptly narrowed (not gradually tapering) into a point.

Cyme. An inflorescence in which the earliest flower to open terminates the main axis, the later flowers borne on branches from below.

Deciduous. Falling off. Deciduous tree: one which is without leaves for a certain period each year.

Decumbent. With base lying on the ground but the tips ascending.

Decurrent. Extending down the stem; used to describe leaf-bases which form wings on the stem below the attachment of the leaf.

Deltoid. Triangular with broad base.

Dentate. With sharp teeth pointing outward.

Denticulate. With fine teeth pointing outward.

Diffuse. With many widely spreading branches.

Dioecious. Producing staminate and pistillate flowers (or male and female gametes) on separate plants.

Disk. In *Compositae,* the central part of the head which bears flowers with regular corollas.

Distinct. Of taxonomic groups, distinguishable from one another and apparently not connected genetically (intersterile). Of parts of plants, not united, even at the base.

Elliptic. Broadest at the middle, narrowed toward both ends.

Emarginate. With a notch at the apex.

Endemic. Limited to a particular geographic area. When this area is very small, the plant is called a narrow endemic.

Exserted. Extending beyond the surrounding parts; for example, stamens from a corolla.

Filiform. Thread-like; very narrow and almost as thick as wide.

Flaccid. Limp, as contrasted with firm.

Foliage leaves. Ordinary leaves as distinguished from bracts or other specialized types of leaves.

Free. Not united with an adjacent part.

Fused. Grown together into a single structure; united.

Galea. A narrow upper corolla-lip, extending beyond the lower lip, in certain *Scrophulariaceae.*

Glabrate. Becoming glabrous.

Glabrous. Without hairs.

Gland. A structure which secretes a fluid, which is often sticky or odorous.

Glandular. Bearing glands.

Glaucous. Having the surface covered with a white wax; as, for example, the berries of elder.

Glumes. In *Gramineae,* the two (sometimes only one) bracts which subtend a spikelet.

Glutinous. Sticky, like glue or syrup.

Hastate. With a pair of outwardly pointing lobes at the base.

Head. An inflorescence in which the flowers are sessile and crowded on a flat or slightly elongate common receptacle.

Helicoid. Like a snail's shell; coiled. *Helicoid cyme.* A sort of inflorescence which is coiled when in bud.

Herb. A plant in which the stems develop little woody tissue and do not persist for more than a year.

Herbaceous. With the characteristics of an herb.

Herbage. Leaves and stems.

Hirsute. With firm spreading hairs.

Hispid. Similar to hirsute, but the hairs stiffer.

Imbricated. Overlapping like the shingles of a roof; usually applied to the involucral bracts of some *Compositae,* when the inner bracts are longer than the outer.

Included. Not extending beyond the surrounding parts, as contrasted with exserted.

Inflated. Hollow and enlarged, as if blown up.

Inflorescence. A group of flowers on a plant, or the portion of a plant where the flowers are borne. See accompanying diagram for some common kinds of inflorescences.

Introduced. Not native in the region but brought from somewhere else.

Involucre. A group of bracts, or a structure formed of fused bracts, surrounding a head or umbel, or rarely a single flower.

Involute. Rolled inward from both edges.

Keel. In general, a longitudinal ridge suggesting the keel of a boat. In *Leguminosae,* a structure formed by the 2 lower petals in the flower.

Lanceolate. Rather narrow and broadest near the base, like the blade of a lance.

Leaflet. A division of a compound leaf.

Lemma. The outer of the 2 bracts which enclose an individual flower in the *Gramineae.*

Ligule. A tongue-shaped structure. In *Gramineae* and some other families (mostly monocotyledons), an outgrowth from the surface of a leaf where the leaf diverges from the stem, usually appressed to the stem or partly folded around it. In *Compositae,* the one-sided limb of an irregular corolla, which extends outward from the head.

Limb. The expanded portion of a corolla, as distinguished from the tube.

Linear. Narrow with approximately parallel sides.

Lip. The upper or the lower part of a corolla or calyx which has definite (and nearly always unlike) upper and lower parts.

Lobe. A portion of any part of a plant which extends out from the main body of the structure. Lobed leaves are distinguished from compound leaves by having the parts (lobes) not completely separated.

Lobulate. Describing lobes which are parted into secondary lobes.

Lobule. A lobe of a lobe.

Maritime. On or near the sea.

Markedly. To such a degree that the feature described is easily visible; contrasted with *obscurely.*

Membranous. Thin and soft.

Nomenclatural type. The specimen to which a particular name was first applied. At present, the author of a species or variety designates a certain specimen as the "type," but this has been the general practice only since about 1900. By extension, the expression "nomenclatural type" is sometimes used in this work to include plants which closely resemble the actual type collection.

Native. Present, or presumably present, in the region prior to recorded history; contrasted with introduced.

Nutlet. A small hard portion of a fruit containing one seed.

Oblanceolate. Narrow, but gradually widened toward the apex.

Oblique. Not straight across, but slanting from one side to the other.

Oblong. Of intermediate width and with approximately parallel sides.

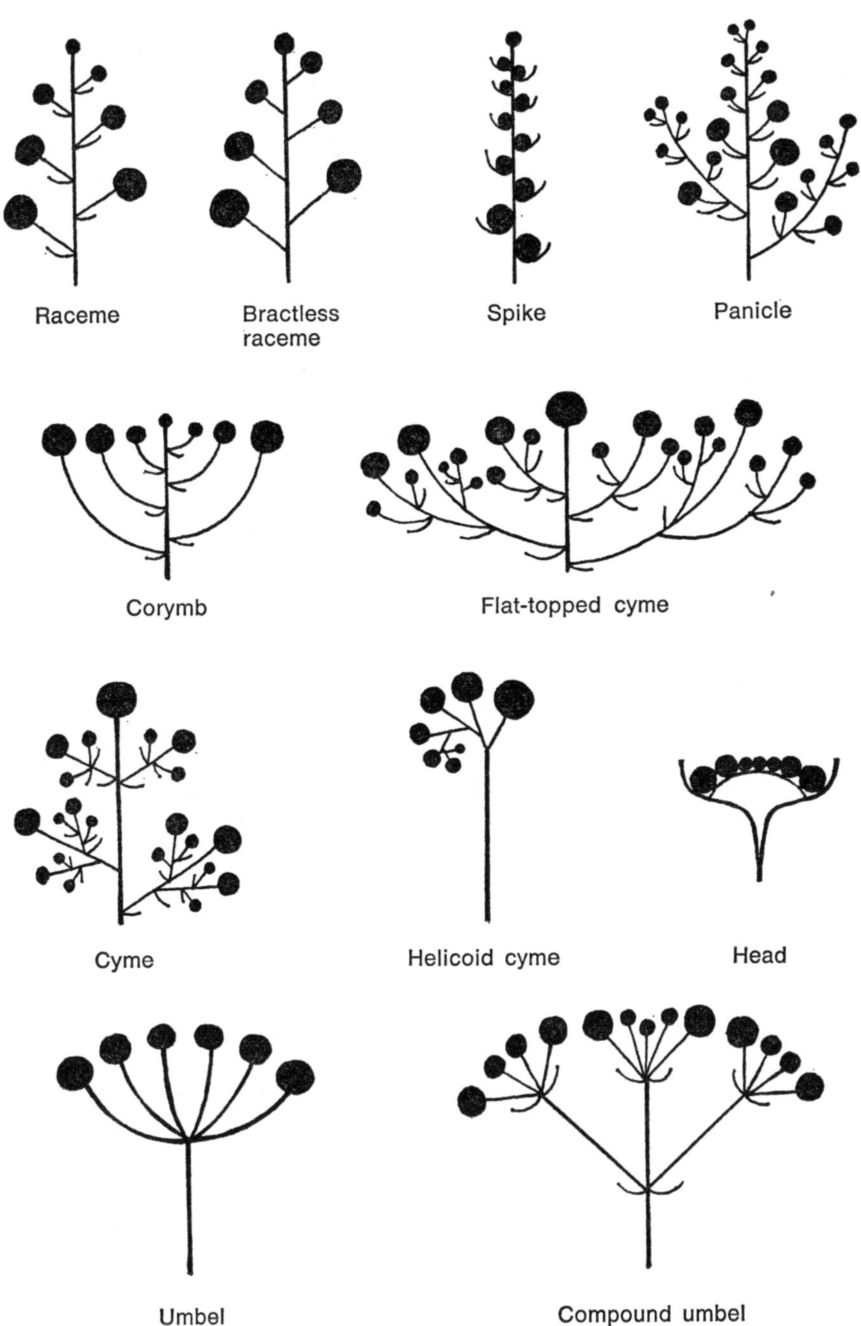

DIAGRAM OF SOME COMMON KINDS OF INFLORESCENCES

The largest circles represent the earliest flowers to open; smaller circles represent progressively later flowers.

Obovate. In outline like a hen's egg, with the broader end toward the apex.

Obscurely. To a degree or in a manner such that the described feature is not readily apparent. For example, *obscurely pubescent,* with hairs which are not easy to find. Contrasted with *markedly* or *conspicuously*.

Obsolete. Missing; used when other plants of the species or related species have the part in question.

Orbicular. Approximately circular, about as broad as long.

Our, ours. Belonging to the geographical region covered by a book; in this case, San Luis Obispo County.

Oval. Broad with rounded ends.

Ovate. In outline like a hen's egg, with the broader end at the base.

Ovoid. Shaped like a hen's egg. In general usage, "ovoid" refers to three-dimensional objects, and "ovate" to flat surfaces such as the outline of a leaf.

Palea. In *Gramineae,* the inner of the 2 bracts which enclose an individual flower. In *Compositae,* one of the parts of the pappus when these are flat instead of bristle-like or hair-like.

Panicle. An inflorescence which is branched and in which the branches resemble racemes.

Papillate. Bearing minute nipple-shaped protuberances.

Pappus. The part of an individual flower in the *Compositae* which corresponds (at least in position) with the calyx.

Pedicel. The stalk of an individual flower in an inflorescence, or of a head in the *Compositae* when the inflorescence is branched.

Peduncle. The stalk of an inflorescence or of a solitary flower.

Perennial. Any plant which lives more than 2 years. This includes all woody plants. Herbaceous perennials can usually be distinguished from annuals by the presence of the next year's buds or the dead stubs of the last year's stems at the base, by a thicker root-crown, or by the presence of spe-

cialized structures such as rhizomes, corms, or bulbs.

Perfect. Descriptive of a flower in which stamens and pistil are both present.

Persistent. Remaining attached to the part from which it grows; for example, leaves on a stem, a style on an ovary, etc. *Deciduous* is the opposite.

Pilose. With soft spreading hairs.

Pinna. One of the primary divisions of a pinnate leaf; used mostly in describing ferns.

Pinnule. One of the divisions of a pinna, when it is compound.

Pinnatifid. Pinnately lobed.

Plumose. Bearing fine side-branches or hairs, like a feather.

Prostrate. Spreading on the surface of the ground.

Puberulent. Minutely pubescent.

Pubescence. The hairy covering of any part of a plant.

Pubescent. Often used to mean hairy in general; more specifically, downy, like the surface of a peach.

Pustulate. Bearing surface swellings like blisters.

Raceme. An elongate inflorescence in which the flowers are on pedicels, the lowest flowers opening first.

Racemose. Arranged as in a raceme; for example, the heads of some *Compositae.*

Rachis. The main axis of a structure; used mostly with reference to pinnately lobed or compound leaves.

Ray. The expanded corolla-limb of some *Compositae* which extends outward from the margin of the head; also called a ligule.

Ray-flower. In *Compositae,* a flower bearing a ray.

Receptacle. The short portion of a stem to which the flower-parts are attached. Also, in *Compositae,* the enlarged end of a stem which bears the flowers of a head (this properly called a common receptacle, because the flowers of the head possess it "in common").

Recurved. Curved so that the tip points toward the base.

Reflexed. Directed backward or downward.

Reniform. Kidney-shaped.

Reticulate. Showing a network pattern; for example, the veins in many sorts of leaves.

Retrorse. Pointing backward, as the hairs on the stems of some species.

Revolute. With margins rolled back.

Rib. A longitudinal ridge. *Ribbed.* Having ribs.

Rosette. A group of leaves growing from a stem in which the nodes are so close together that internodes are not evident (like the petals of a garden rose).

Rotate. Spreading regularly from the center like the spokes of a wheel.

Sac. A more or less inflated structure.

Saccate. Inflated; enlarged and hollow.

Sagittate. Shaped like an arrowhead, with 2 acute lobes at the base.

Scabrous. Bearing short stiff projections or hairs, and hence rough to the touch.

Scape. A leafless flowering stem arising from the base of a plant.

Scarious. Thin, dry, and translucent.

Series. In *Compositae*, the involucral bracts which form a whorl or which are of equal length are called a series.

Serrate. With toothed margin, the teeth pointed and directed toward the apex.

Sessile. In general, without a stalk. Of leaves, without a petiole.

Sheath. The lower part of a leaf which is rolled around the stem; used mostly with reference to the *Gramineae*.

Simple. Unbranched. Of leaves, not divided into leaflets.

Sinus. The space or indentation between lobes or teeth.

Spatulate. Spoon-shaped; with short, broad blade tapering downward into a narrow petiole.

Spicate. Arranged as in a spike.

Spike. A kind of elongate inflorescence in which the flowers are sessile.

Spikelet. The unit of the inflorescence in grasses and sedges, usually sub-tended by 2 glumes. A spikelet may contain one to many flowers.

Spine. A sharp-pointed thorn-like structure which is a specialized leaf or part of a leaf. Compare *thorn*.

Spinose-dentate. With sharp-pointed rigid teeth.

Spur. A projection of a petal, sepal, corolla, or calyx which points backward.

Stellate. With parts spreading like the rays of a conventionalized star.

Sterile. Not producing seeds.

Stipe. A stalk; nearly always used for the narrowed base of an ovary above its attachment to the receptacle; sometimes applied to the petioles of ferns.

Stipitate. Having a stipe.

Striate. Marked with longitudinal lines.

Strigose. With appressed hairs.

Succulent. Thick, juicy, and soft, as distinguished from thin or from woody.

Symmetrical. With parts spreading equally from a common center or from a lengthened axis.

Tap-root. A solitary root descending vertically; if branched, the branches are smaller than the main root.

Terete. Narrowly cylindric; round in cross-section as distinguished from flattened.

Thorn. A sharp-pointed branch of a stem. Compare *spine*.

Throat. The upper enlarged portion of a corolla-tube. When a distinction is made between throat and tube, the word "tube" is used for the slender lower portion.

Tomentose. Felty; covered with tomentum.

Tomentum. A dense covering of soft hairs, like felt.

Topotype. Plants of a species or subordinate group occurring at the type locality.

Truncate. With the end (apex or base) forming a straight transverse line.

Tube. The portion of a corolla or calyx below the point where the lobes separate or the limb expands; not necessarily tubular in form.

Tuberous. Thick and fleshy; applied to roots.

Turbinate. Top-shaped; enlarged upward from a pointed base.

Type. A standard of comparison. (1) The prevailing or most familiar form of a species or variety, as distinguished from variants. (2) The specimen or taxonomic group to which a particular name was first applied. See *nomenclatural type.*

Type collection. The collection designated by the author of a name, or by a later author, as the type of a species or subordinate group.

Type locality. The place where the type collection was made.

Typical. Having the characteristics of a type, in either of the senses given above.

Umbel. An inflorescence in which the pedicels of the flowers arise from the same point.

Undulate. With margins wavy rather than flat.

Villous. Shaggy; with dense long hairs.

Viscid. Sticky.

Wing. A thin edge or projection on a stem, fruit, seed, etc.; also, in *Leguminosae,* one of the 2 petals under the banner on either side of the keel.

Index